# Agent-Based Social Systems
# Volume 6

Series Editor: Hiroshi Deguchi, Yokohama, Japan

# ABSS—Agent-Based Social Systems

This series is intended to further the creation of the science of agent-based social systems, a field that is establishing itself as a transdisciplinary and cross-cultural science. The series will cover a broad spectrum of sciences, such as social systems theory, sociology, business administration, management information science, organization science, computational mathematical organization theory, economics, evolutionary economics, international political science, jurisprudence, policy science, socioinformation studies, cognitive science, artificial intelligence, complex adaptive systems theory, philosophy of science, and other related disciplines.

The series will provide a systematic study of the various new cross-cultural arenas of the human sciences. Such an approach has been successfully tried several times in the history of the modern science of humanities and systems and has helped to create such important conceptual frameworks and theories as cybernetics, synergetics, general systems theory, cognitive science, and complex adaptive systems.

We want to create a conceptual framework and design theory for socioeconomic systems of the twenty-first century in a cross-cultural and transdisciplinary context. For this purpose we plan to take an agent-based approach. Developed over the last decade, agent-based modeling is a new trend within the social sciences and is a child of the modern sciences of humanities and systems. In this series the term "agent-based" is used across a broad spectrum that includes not only the classical usage of the normative and rational agent but also an interpretive and subjective agent. We seek the antinomy of the macro and micro, subjective and rational, functional and structural, bottom-up and top-down, global and local, and structure and agency within the social sciences. Agent-based modeling includes both sides of these opposites. "Agent" is our grounding for modeling; simulation, theory, and real-world grounding are also required.

As an approach, agent-based simulation is an important tool for the new experimental fields of the social sciences; it can be used to provide explanations and decision support for real-world problems, and its theories include both conceptual and mathematical ones. A conceptual approach is vital for creating new frameworks of the worldview, and the mathematical approach is essential to clarify the logical structure of any new framework or model. Exploration of several different ways of real-world grounding is required for this approach. Other issues to be considered in the series include the systems design of this century's global and local socioeconomic systems.

T. Terano, H. Kita, S. Takahashi,
H. Deguchi (Eds.)

# Agent-Based Approaches in Economic and Social Complex Systems V

## Post-Proceedings of The AESCS International Workshop 2007

With 123 Figures

Springer

Takao Terano, Ph.D.
Professor, Interdisciplinary Graduate School of Science and Engineering
Tokyo Institute of Technology
4259 Nagatsuta-cho, Midori-ku, Yokohama 226-8502, Japan

Hajime Kita, Dr. Eng.
Professor, Academic Center for Computing and Media Studies
Kyoto University
Yoshida-Nihonmatsu-cho, Sakyo-ku, Kyoto 606-8501, Japan

Shingo Takahashi, Ph.D.
Professor, Department of Industrial and Management Systems Engineering
Waseda University
3-4-1 Okubo, Shinjuku-ku, Tokyo 169-8555, Japan

Hiroshi Deguchi, Ph.D.
Professor, Interdisciplinary Graduate School of Science and Engineering
Tokyo Institute of Technology
4259 Nagatsuta-cho, Midori-ku, Yokohama 226-8502, Japan

ISSN 1861-0803
ISBN 978-4-431-99857-0
e-ISBN 978-4-431-87435-5



Springer is a part of Springer Science+Business Media
springer.com

Printed in Japan

Printed on acid-free paper

# Preface

This volume contains papers selected from presentations at the AESCS'07, which was held at Waseda University, Tokyo, Japan, August 29–30, 2007. The workshop was the fifth in a series of Pacific Rim activities in emerging interdisciplinary areas of social and computational sciences. Since the fourth AESCS, in 2004, the workshop has been held as a regular meeting of the Pacific Asian Association for Agent-Based Social Sciences (PAAA). In 2006, it was extended as the First World Congress of Social Simulation (WCSS06) in collaboration with PAAA, NAACSOS, and ESSA, the three regional societies in the field in the Asia-Pacific, North American, and European areas, respectively. Following the success of AESCS'05 and WCSS06, AESCS'07 received 29 submissions of original papers. Each paper was reviewed by at least two program committee members of AESCS'07, and in the workshop we had 25 presentations. At AESCS'07, we also had one plenary talk by Prof. Shu-Heng Chen, National Chengchi University, Taiwan, and two invited talks, by Dr. Gaku Yamamoto of IBM Research and by Dr. Richard Warren, Air Force Research Laboratory, USA.

For this volume, we selected 21 papers from among those presented in the workshop, along with two invited papers by Prof. Shu-Heng Chen and Dr. Richard Warren.

Contributions cover various areas of the social sciences such as the market, finance, and other topics in economics, organization and management, marketing, and sociology. Contributions also deal with subjects more closely related to engineering, such as production and traffic. The progress of study in this field shows that researchers have started to construct more complex and realistic models with implications for policy making and engineering design, as well as simplified models to elucidate principles of social systems. This trend shows the growing importance of the field both in the social sciences and in engineering. Further contributions from social sciences and computer science through interdisciplinary study are anticipated in such a promising field.

| | |
|---|---|
| AESCS'07 Workshop Chair | Takao Terano |
| AESCS'07 Organizing Committee Chair | Shingo Takahashi |
| AESCS'07 Program Committee Chair | Hajime Kita |
| AESCS'07 Publication Chair | Hiroshi Deguchi |

# Committees and Chairs of AESCS'07

**Workshop Chair**

**Takao Terano**, Tokyo Institute of Technology, Japan

**Program Committee**

**Chair** **Hajime Kita**, Kyoto University, Japan
**Akira Namatame**, National Defence Academy, Japan
**David Batten**, Commonwealth Scientific and Industrial Research Organisation, Australia
**David Yeung**, St. Petersburg State University and Hong Kong Baptist University, Hong Kong
**Hao Lee**, The Kyoto College of Graduate Studies for Informatics, Japan
**Hiroshi Deguchi**, Tokyo Institute of Technology, Japan
**Hiroyuki Matsui**, Kyoto University, Japan
**Isamu Okada**, Soka University, Japan
**Isao Ono**, Tokyo Institute of Technology, Japan
**Keiki Takadama**, The University of Electro-Communications, Japan
**Keiko Zaima**, Kyoto Sangyo University, Japan
**Kyoichi Kijima**, Tokyo Institute of Technology, Japan
**Masayuki Ishinishi**, Ministry of Defense, Japan
**Naoki Shiba**, Nihon University, Japan
**Norman Y. Foo**, University of New South Wales, Australia
**Philippa Pattison**, University of Melbourne, Australia
**Ryo Sato**, University of Tsukuba, Japan
**Shingo Takahashi**, Waseda University, Japan
**Sung-Bae Cho**, Yonsei University, Korea
**Takao Terano**, Tokyo Institute of Technology, Japan
**Toru Ishida**, Kyoto University, Japan
**Toshiji Kawagoe**, Future University-Hakodate, Japan
**Toshiyuki Kaneda**, Nagoya Institute of Technology, Japan
**Yosihiro Nakajima**, Osaka City University, Japan
**Yusuke Koyama**, Tokyo Institute of Technology, Japan
**Yutaka Nakai**, Shibaura Institute of Technology, Japan
**Hiroshi Sato**, National Defense Academy, Japan
**Kazuhisa Taniguchi**, Kinki University, Japan
**Yoshinori Shiozawa**, Osaka City University, Japan
**Yusuke Arai**, Tokyo Institute of Technology, Japan
**Reiko Hishiyama**, Waseda University, Japan
**Naoki Mori**, Osaka Prefecture University, Japan
**Toshiya Kaihara**, Kobe University, Japan

**Thomas Lux**, University of Kiel, Germany
**Hideyuki Mizuta**, IBM JAPAN, Japan
**Keiji Suzuki**, Future University-Hakodate, Japan
**Shu-Heng Chen**, National Chengchi University, Taiwan

**Organizing Committee**

**Chair** **Shingo Takahashi**, Waseda University, Japan
**Reiko Hishiyama**, Waseda University, Japan
**Takashi Yamada**, Tokyo Institute of Technology, Japan
**Yusuke Arai**, Tokyo Institute of Technology, Japan
**Hiroyuki Matsui**, Kyoto University, Japan
**Naoki Shiba**, Nihon University, Japan
**Hiroshi Deguchi**, Tokyo Institute of Technology, Japan
**Yusuke Koyama**, Tokyo Institute of Technology, Japan
**Yusuke Goto**, Waseda University, Japan
**Kotaro Ohori**, Waseda University, Japan

**Publication Chair**

**Hiroshi Deguchi**, Tokyo Institute of Technology, Japan

# Acknowledgement

Publication of this volume is partly supported by the 21st-Century COE Program "Creation of Agent-Based Social Systems Sciences (ABSSS)" of the Tokyo Institute of Technology. We also wish to thank the Air Force Office of Scientific Research, Asian Office of Aerospace Research and Development (AFOSR/AOARD) for their contribution to the success of this workshop. AFOSR/AOARD support is not intended to express or imply endorsement by the U.S. Federal Government.

# Contents

## Fundamentals of Agent-Based and Evolutionary Approaches



## Production, Services and Urban Systems



## Agent-Based Approaches to Social Systems

**Market and Economy I**

# Plenary Talk

# Genetic Programming and Agent-Based Computational Economics: From Autonomous Agents to Product Innovation

Shu-Heng Chen

**Abstract** Despite their great development over the last decade, most ACE (agent-based computational economics) models have been generally weak in demonstrating discovery or novelty-generation processes. In this sense, they are not very distinct from their counterparts in neo-classical economics. One way to make progress is to enable autonomous agents to discover the modular structure of their surroundings, and hence they can adapt by using modules. This is almost equivalent to causing their "brain" or "mind" to be designed in a modular way. By this standard, simple genetic programming is not an adequate design for autonomous agents; however, augmenting it with automatic defined terminals (ADTs) may do the job. This paper provides initial research with evidence showing the results of using ADTs to design autonomous agents.

## 1 Introduction

GP maintains a unique position when compared with other computational intelligence tools in modeling autonomous agents. Basically, there are two distinguishing features of using GP in modeling autonomous agents. First, in a sense, GP provides agents with a larger degree of autonomy. Second, it provides us with a concrete picture to visualize the learning process or the discovery process as a growing process, i.e., that of growing the evolving hierarchies of building blocks (subroutines) from an immense space of subroutines.

---

Shu-Heng Chen
AI-ECON Research Center, Department of Economics, National Chengchi University, Taipei, Taiwan, e-mail: chchen@nccu.edu.tw

## 1.1 Autonomy

The first feature, a larger degree of autonomy, has two implications. First, it lessens the burden of model-builders in their intervention or supervisory efforts over these agents. Second, it implies a larger degree of freedom left for agents to explore the environment around them, and a better chance for us to watch how they adapt and what they learn.

The first implication is important when model-builders themselves know very little about the structure of the environment in which their agents are placed, and hence they do not even know how to supervise these agents in a well-defined manner; in particular, they do not want to misinform these agents with biased information. The second implication is even more important because what they learn or discover may be non-trivial for us. In this case, we are taking lessons from them. Alternatively, it makes us able to have the novelties- or surprises-generating processes, an essential element of any complex adaptive system. By observing and making sense of what agents learned, we as outsiders are also able to learn.

## 1.2 Learning

The second feature is also appealing because it enables us to give an alternative interpretation of what we mean by *learning*. Learning is a highly interdisciplinary concept, which concerns many disciplines, ranging from psychology, education, neural sciences, cognitive sciences, mathematics and statistics, to information sciences.

Its meaning in economics also varies. In some situations, it is very trivial and means nothing more than *making a choice* repeatedly under the same or a very similar environment with the same options. There are a number of learning algorithms corresponding to this simple case. The most famous one is *reinforcement learning*, and the other equally familiar and related one is the *discrete choice model* associated with the Boltzmann-Gibbs distribution. These learning algorithms only involve a very simple stimulus-reaction mechanism, and the development of sophisticated reasoning is not required, at least, not explicitly.

In some other situations, learning means the attempt to find out the law between the causes and the effects, the mapping between the inputs and outputs, and the underlying mechanism by which observations are generated. It is more like a scientific learning. The feedforward neural networks (FNNs) represent such a kind of learning. Numerous mathematical analyses of neural networks show that FNNs are universal function approximators, even though to build such an approximation process is another issue.

However, these two kinds of learning, the stimulus-reaction learning and the scientific learning, may cover only a very limited part of what we generally experience about learning. What has been missing is the idea of the *building block*, which connects what we have learned before to what we are learning now or what we will learn in the near future. In considering the learning of mathematics as an example, we cannot study differential equations without having calculus as the prerequisite. If we perceive learning as a walk along a *ladder* which makes us move higher and become more experienced at each step, then the kind of learning which we are interested in is developmental learning, and genetic programming is one of the learning algorithms which are able to demonstrate this feature.

## 2 Genetic Programming and Economics

Genetic programming is a methodological innovation in economic. It is so because it captures three essential elements in the making of economics. The three elements are *constant changes* from inner nature to outer forms, *evolving populations* of decision rules, and *modularity*. These three elements have been initiated by three prominent economists at different times. Two of them, Herbert Simon and Robert Lucas, are Nobel Laureates, and the one, who is not, died in 1924 when the Nobel Prize had not yet existed, but who is generally regarded as the father of the neo-classical economics. In what follows, we shall go through them in chronological order.

### *2.1 Alfred Marshall*

The first connection between GP and economics is the idea of *constant change*. Its origin can be traced back to the late 19th century. Alfred Marshall [20] wrote:

> Economics, like biology, deals with a matter, of which the inner nature and constitution, as well as outer form, are constantly changing. (Ibid, p. 772)

He also wrote

> The Mecca of the economists lies in economic biology rather than in economic dynamics. (Ibid, p. xiv)

Alfred Marshall is regarded as a pioneer in starting the dialogue between economics and biology, whose legacy has been further pursued in a branch of economics, referred to as *Evolutionary Economics*. To have an idea of the *constant change* of the inner nature, the constitution, and the outer form of a matter, one can think of the evolution of technology, from its primitive form to its state of the art.[1] Nevertheless, this picture of constant change has not been demonstrated in any model known to economists before the advent of GP. Even the leading economists in Evolutionary Economics did not provide us with a tool to simulate this developmental-biology-like process.

[1] For example, see [3], in particular, Figures 1.3 and 1.4.

## 2.2 *Robert Lucas*

The second connection between GP and economics is the idea of *evolving populations.* [19] provided a notion of an economic agent.

> In general terms, we view or model an individual as *a collection of decision rules* (rules that dictate the action to be taken in given situations) and a set of preferences used to evaluate the outcomes arising from particular situation-action combinations. (Ibid, p.217, Italics Added.)

Immediately after the *static description* of the economic agent, Lucas continued to add an *adaptive* (*evolutionary*) version of it.

> These decision rules are continuously under review and revision: new decision rules are tried and tested against experience, and rules that produce desirable outcomes supplant those that do not. (Ibid, p.217).

So, according to Lucas, the essence of an economic agent is *a collection of decision rules which are adapting (evolving) based on a set of preferences.* In brief, it is an idea of an *evolving population.*

If we suppose that an evolving population is the essence of the economic agent, then it seems important to know whether we economists know any operational procedure to substantiate this essence. Back in 1986, the answer was absolutely *no.* That certainly does not mean that we did not know anything about evolving one *decision rule.* On the contrary, since the late 1970s, the literature related to bounded rationality in macroeconomics has introduced a number of techniques to evolve a single decision rule (a single equation or a single system of equations): recursive regression, Kalman filtering, and Bayesian updating, to name a few. [25] made an extensive survey of this subject. However, these techniques shed little light on how to build a Lucasian agent, especially since what we wanted to evolve was not a single decision rule but a population of decision rules.

In fact, it may sound a little surprising that economists in those days rarely considered an individual as a population of decision rules, not to mention attending to the details of its evolution. Therefore, all the basic issues pertaining to models of the evolving population received little, if any, attention. For example, how does the agent *initialize* a population of decision rules? Once the agent has a population of decision rules, which one should they follow? Furthermore, in what ways should this population of decision rules "be continuously under review and revision"? Should we review and revise them one by one because they are independent, or modify them together because they may be correlated with each other? Moreover, if there are some "new decision rules to be tried," how do we generate (or find) these new rules? What are the relationships between these new rules and the old ones? Finally, it is also not clear how "rules that produce desirable outcomes should supplant those that do not."

### 2.2.1 John Holland

There is one way to explain why economists are not interested in, and hence not good at, dealing with a population of decision rules: economists used to derive the decision rule for the agent *deductively*, and the deductive approach usually led to only one solution (decision rule), which is the *optimal* one. There was simply no need for a population of decision rules.

We do not know exactly when or how the idea of the *evolving population of decision rules* began to attract economists, but John Holland's contribution to *genetic algorithms* definitely exerted a great influence. In 1991, John Holland and John Miller published a sketch of the *artificial adaptive agent* [16], where they stated

> ...an agent may be represented by a single string, or it may consist of *a set of strings* corresponding to *a range of potential behaviors*. For example, a string that determines an oligopolist's production decision could either represent a single firm operating in a population of other firms, or it could represent *one of many possible decision rules* for a given firm. (Ibid, p. 367; Italics added.)

Now, formally, each decision rule is represented by a *string*, and, at each point in time, agents may have a set of strings characterizing a range of potential behaviors. In this sense, the agents' behavior is no longer deterministic; instead there are many decision rules competing before the final one is chosen.[2]

### 2.2.2 John Koza

It is interesting to note that the (binary) strings initiated by Holland were originally motivated by an analogy to machine codes. After decoding, they can be computer programs written in a specific language, say, LISP or FORTRAN. Therefore, when a GA is used to evolve a population of binary strings, it behaves as if it were used to evolve *a population of computer programs*. If a decision rule is explicit enough not to cause any confusion in implementation, then one should be able to write it in a computer program. It is the *population of computer programs* (or their machine codes) which provides the most general representation of the *population of decision rules*. However, the equivalence between computer programs and machine codes *breaks down* when what is coded consists of the parameters of decision rules rather than the decision rules (programs) themselves, as we often see in economic applications with GAs. The original meaning of evolving binary strings as evolving computer programs is lost.

The loss of the original function of GAs has finally been noticed by John Koza. He chose the language LISP as the medium for the programs created by genetic programming (GP) because the syntax of LISP allows computer programs to be manipulated easily like the bitstrings in GAs, so that the same genetic operations

---

[2] Whether or not the mind of an agent can simultaneously have many different competing ideas or solutions is certainly an issue not in the realm of conventional economics, but a subject long studied in psychology, neuroscience, and the philosophy of the mind. See also [21].

used on bitstrings in GAs can also be applied to GP. Genetic programming simulates the biological evolution of *a society of computer programs*. Each of these computer programs can be matched to a solution to a problem. This structure provides us with an operational procedure of the Lucasian agent. First, a collection of decision rules is now represented by a society of computer programs. Second, the review and revision process is implemented as a process of natural selection when the genetic operators are applied to evolve the society of computer programs.

### *2.3 Herbert Simon*

The third connection of GP to economics is the idea of complexity, in particular, the Simonian notion of complexity [26], i.e., *hierarchy*. Herbert Simon viewed hierarchy as a general principle of complex structures. Hierarchy, he argued, emerges almost inevitably through a wide variety of evolutionary processes, for the simple reason that hierarchical structures are *stable*. To demonstrate the importance of a *hierarchical structure* or *modular structure* in production, Simon offered his well-known story about a competition between Hora and Tempus, two imaginary watchmakers. In this story, Hora prospered because he used the modular structure in his design of watches, whereas Tempus failed to prosper because his design was not modular. Therefore, the story is mainly about a lesson: the advantage of using a modular design in production.

Modularity is becoming more important today because of the increased complexity of modern technology. Using the computer industry as an example, [2] shows that the industry has experienced previously unimaginable levels of innovation and growth because it embraced the concept of modularity. [17] also asserts that embracing the principle of modular design can enable organizations to respond rapidly to market needs and allow the changes to take place in a cost-effective manner.

## 3 What is Missing in ACE?

The three ideas individually already had an impact on the later development of economics. For example, after Marshall, through the additional efforts made by Thorstein Veblen, Armen Alchian, Richard Nelson, Sidney Winter, and many others, the ideas of evolution have been brought into the modeling of economics. Recently, much of this progress has been further made in *agent-based computational economics* (ACE), where we can see how the Lucasian agent has been brought into evolutionary economics via genetic programming [9, 10, 11, 8, 7, 12].

However, the central element on *constant change in the inner nature and outer form* has largely been missing in this literature. As we have seen above, Simon's work on modularity also concerns evolution. How Simon's view of evolution in terms of *modularity* can be related to Marshall's view of evolution in terms of *con-*

*stant change* is also missing in the literature, even though a reflection of human history does indicate that our economy evolves toward higher and higher degree of complexity and novelty. The idea of *hierarchical modularity* should then play a central role as the economy evolves with these features. Nevertheless, not many ACE models are able to deliver this feature, including those so-called *agent-based economic models of innovation*.[3] To fill the void, there are a number of research questions that need to be addressed. One of these is an in-depth investigation of the relationship between complexity and diversity. The other issue, as a continuation of what has been said in Section 1.2, concerns a learning algorithm enabling our autonomous agents to learn in a developmental or accumulation process through which unsupervised discovering can be expected.

## *3.1 Complexity and Diversity*

The diversity which we discuss in this section is restricted to the production side, in particular, the *product diversity*. It could more broadly include other related kinds of diversity, such as process diversity, organizational diversity, and job diversity, but it is still restricted to the production aspect. This restriction may drive our attention away from other important diversity issues which may appear in the context of, for example, biodiversity, cultural diversity, anthropological diversity, etc.[23, 27]. The reason for making such a restriction is to have a sharp focus on modularity.

Like complexity, diversity is involved because it is an important feature observed in the evolutionary process. Studies have shown that the development of our economy is accompanied by constant increases in product diversity.[4] However, in addition to that, what concerns us more is that the two ideas, diversity and complexity, may not be disentangled. Intuitively speaking, the more diversified an economy is, the more complex it becomes.[5] Assume that without being able to manage the level of complexity required to match a certain level of diversity, the further pursuit of diversity is technologically infeasible; in other words, the incompetence to cope with increasing complexity can be a potential barrier to the realization of a greater diversity. Then the following issue becomes important: if complexity is an inevitable consequence of diversity, and diversity is welfare-enhancing, how can the economy manage its complexity while enjoying the fruits of diversity? Simon already gave us the key for the solution, i.e., using modular design. However, what is lacking is a demonstration of how this modular design can emerge from the economy.

---

[3] For a survey on this literature, see [14].

[4] According to an EPA (Environmental Protection Agency) study conducted in conjunction with the U.N. Task Force On Global Developmental Impact, consumer-product diversity now exceeds biodiversity. See *Onion*, October 21, 1998, Issue 34-12. http://www.theonion.com/content/node/38901

[5] Of course, this statement can not be made seriously without a clear notion of complexity. What we, therefore, propose here is something similar to algorithmic complexity, while with a modification in order to take cognitive constraints of human agents into account.

### 3.2 Learning of Hierarchical Modularity

One key element to see the emergence of modular design is to have autonomous agents so that they can constantly discover useful modules (building blocks). The next question is how such autonomous agents can be designed. This leads us to some further thinking on learning, given what we have already discussed in Section 1.2.

What do we mean that we *learned*? How do we make sense of what we learn? How do we know or feel confident that we are *learning*? Must sensible learning be *incremental* (i.e., in a developmental process)? If sensible learning is incremental, then how do we compare learning at different stages? What is the role of building blocks or functional modularity in this learning process? How do building blocks or modules help agents to learn and hence to manage the complexity given their severe cognitive constraints?

## 4 Toward a New Design of Autonomous Agents

### 4.1 Gram-Schmidt Orthogonalization Process

The *Gram-Schmidt orthogonalization process*, well taught in linear algebra or functional analysis, provides us with a kind of developmental learning. In fact, mathematicians also use the term "innovation" for the orthogonal elements (residuals) extracted from projections. This is because, along this process, each innovation implies the discovery of a new *basis*, which is equivalent to the discovery of a new space. The basis may be taken as a kind of building block. The developmental learning defined by the Gram-Schmidt orthogonalization process can, therefore, be used to think of how to construct a similar discovery or learning process driven by GP.

### 4.2 Automatically Defined Terminals

Although GP can have a hierarchical modular structure, the simple genetic programming is not good at using the modular structure. The standard crossover and mutation can easily destroy the already established structure, which may cause the whole discovery or learning process to be non-incremental and non-progressive. This problem is well-known in the GP literature, and has been extensively studied with various treatments [1, 15, 18, 24]. Motivated by these earlier studies, [6] proposes *automatically defined terminals* (ADTs) as a way to enhance GP to find structured solutions.

An ADT, as shown in Fig. 1, is very similar to the automatically defined function (ADF) [18]. It itself has a fixed structure, in this case, a tree with a depth of two. The root of an ADT can be any function from the primitives (function set), while its

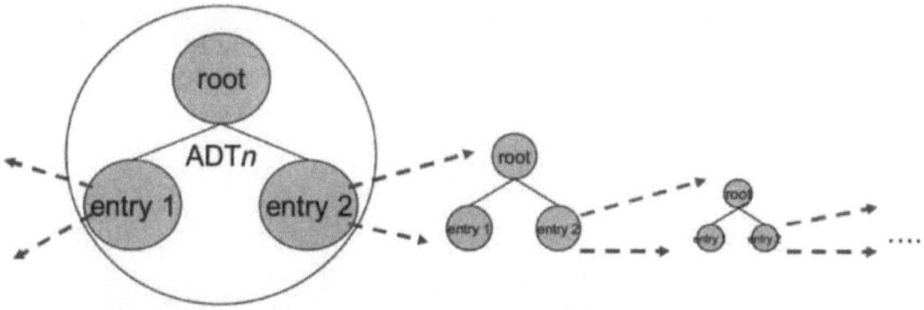


**Fig. 1** Automatically defined terminals.

leaf can be either a terminal from the primitives (terminal set) or can be any existing ADTs. In this way, it shares the same spirit as an ADT, namely, *simplification*, *reuse*, and *encapsulation*. The last item is particularly important because it means that whatever is inside an ADT will not be further interrupted by crossover and mutation.

In this way, ADTs can be considered to be the part of learning in which we have great confidence, and which leaves no room for doubt. Through ADTs we distinguish what is considered to be *knowledge* from what is still in a trial-and-error process. Only the former can then be taken as the building blocks (modules), but not the latter.[6] Without ADTs or equivalents, simple genetic programming is essentially not designed to develop building blocks; therefore, it is not very good at finding the modular structure inherent in the problem.

## 4.3 Modular Economy

[6] tested the idea of augmented GP (augmented with ADTs) in a *modular economy*. The modular economy, which is first proposed in [5], is an economy whose demand side and supply side both have a *decomposable* structure. The decomposability of the supply side, i.e., production, has already received intensive treatments in the literature (See Section 2.3). On the demand side, the modular economy implies a market composed of a set of consumers with *modular preference*. Therefore, it is based on a crucial assumption that *the preference of consumers can be decomposable*. This is, indeed, a big assumption, since its validity has received very little attention in the literature. The closest study which may shed light on this assumption is that the study of *neurocognitive modularity*. The recent progress in neuroscience has allowed us to identify a number of brain modules at various levels of granularity. In addition, various hypotheses regarding *the modularity of mind* also exist, such as the famous *massive modularity hypothesis* [28, 13]. Nevertheless, whether or not one can build preference modules upon the brain/mind modules is still an open issue.

[6] One criterion for modules is their persistence as identifiable units for long enough time spans or generations [22].

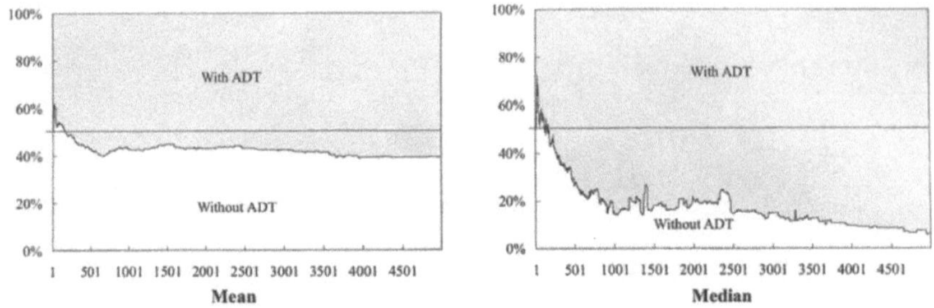


**Fig. 2** Modularity and competitiveness.

In the modular economy, the assumption of modular preference is made as a dual relation to the assumption of modular production. Nevertheless, whether in reality the two can have a nice mapping, e.g., a one-to-one relation, is an issue related to the distinction between *structural modularity* and *functional modularity*. While in the literature, this distinction has been well noticed and discussed, "recent progress in developmental genetics has led to remarkable insights into the molecular mechanisms of morphogenesis, but has at the same time blurred the clear distinction between structure and function." ([4], p. 10)

The modular economy initiated by [5] does not distinguish two kinds of modularity, and they are assumed to be the same. One may argue that the notion of modularity suitable for preference is structural, i.e., *what it is*, whereas the one suitable for production is process, i.e., *what is does*. However, this understanding may be partial. Using the LISP parse-tree representation, [5] actually integrated the two kinds of modularity. Therefore, consider drinking coffee with sugar as an example. Coffee and sugar are modules for both production and consumption. Nevertheless, for the former, producers *add* sugar to coffee to deliver the final product, whereas for the latter, the consumers drink the mixture by knowing the existence of both components or by "seeing" the development of the product.

Within this modular economy, [6] considered an economy with two oligopolistic firms. While both of these firms are autonomous, they are designed differently. One firm is designed with simple GP (SGP), whereas the other firm is designed with augmented GP (AGP). These two different designs match the two watchmakers considered by [26]. The modular preferences of consumers not only define the search space for firms, but a search space with different hierarchies. While it is easier to meet consumers' needs with very low-end products, the resultant profits are negligible. To gain higher profits, firms have to satisfy consumers up to higher hierarchies. However, consumers become more and more heterogeneous when their preference are compared at higher and higher hierarchies, which calls for a greater diversity of products.[7]

[7] If the consumers' preferences are randomly generated, then it is easy to see this property through the combinatoric mathematics. On the other hand, in the parlance of economics, moving along the hierarchical preferences means traveling through different regimes, from a primitive manufacturing

The figures show the simulation results of the competing firms in the modular economy based on 100 runs. The main statistics displayed are the mean and median market shares of two competing firms. It can be seen that the AGP firm (the firm using modular design, ADTs) performs better than the SGP firm (the firm not using modular design), as Simon predicted.

## 5 Concluding Remarks

The design of autonomous agents plays a pivotal role in the further development of agent-based models in economics. The essence of autonomous agents is to own the automatic-discovery capability. This leads us to have a more fundamental thinking of *what to learn* and *how to learn* in light of the evolution of the real economy, in particular, the constant change of the production economy, the product, the technology and the organization. This paper has shown that Simon's notion of *near decomposability* provides an important direction for us to work with, i.e., a *modular economy*. Needless to say, the empirical content and operational details of the proposed modular economy need to be further addressed. Nevertheless, the modular economy guides us to grasping the key to the promising design of autonomous agents. In this paper, we suggest the use of *automatic defined terminals* in GP to design autonomous agents. The agent-based economic models composed of these autonomous agents can, therefore, feature a process of constant change with incessant novelty-findings, which is what the history of our human economy has evidenced.

**Acknowledgements** An earlier draft of this paper was first prepared as an invited talk delivered at the Argentina Academy of Sciences on May 9, 2007, and was then delivered as a keynote speech at *The 5th International Workshop on Agent-based Approaches in Economic and Social Complex Systems* (**AESCS'07**), Waseda University, Tokyo, Japan, on August 29-30, 2007. The author is grateful to Prof. Ana Marostica, Daniel Heymann, Julio Olivera, Hajime Kita and Takao Terano for their superb arrangements regarding the invitation. With great sadness, I learn that Prof. Ana Marostica passed away in Dec. 2007. This paper is, therefore, written in memory of her, in particular her enthusiastic devotion to academia and to friends.

## References

1. Angeline P, Pollack J (1993) Evolutionary module acquisition. Proceedings of the 2nd Annual Conference on Evolutionary Programming. MIT Press, Cambridge, 154–163
2. Baldwin C, Clark K (2000) Design rules: The power of modularity, Vol. 1. MIT Press, Cambridge
3. Basalla G (1988) *The evolution of technology*. Cambridge University Press, Cambridge

---

economy to a quality service economy, from mass production of homogeneous goods to limited production of massive heterogeneous customized products.

4. Callebaut W (2005) The ubiquity of modularity. In: Callebaut W, Rasskin-Gutman D (eds) Understanding the development and evolution of natural complex systems, MIT Press, Cambridage
5. Chen S.-H, Chie B.-T (2004) Agent-based economic modeling of the evolution of technology: The relevance of functional modularity and genetic programming. International Journal of Modern Physics B 18(17-19): 2376-2386
6. Chen S.-H, Chie B.-T (2007) Modularity, product innovation, and consumer satisfaction: An agent-based approach. In: Yin H, Tino P, Corchado E, Byrne W, Yao X (eds.), Intelligent Data Engineering and Automated Learning, Lecture Notes in Computer Science (LNCS 4881), Springer, 1053-1062.
7. Chen S.-H, Liao C.-C (2005) Agent-based computational modeling of the stock price-volume relation. Information Sciences 170:75–100
8. Chen S.-H, Tai C.-C (2003) Trading restrictions, price dynamics, and allocative efficiency in double auction markets: Analysis based on agent-based modeling and simulations. Advances in Complex Systems 6(3):283–302
9. Chen S.-H, Yeh C.-H (1996) Genetic programming learning and the cobweb model. In: Angeline P. (ed.), Advances in Genetic Programming, Vol. 2, MIT Press, Cambridge, Chap. 22, 443–466
10. Chen S.-H, Yeh C.-H (2001) Evolving traders and the business school with genetic programming: A new architecture of the agent-based artificial stock market. Journal of Economic Dynamics and Control 25: 363–393
11. Chen S.-H, Yeh C.-H (2002) On the emergent properties of artificial stock markets: The efficient market hypothesis and the rational expectations hypothesis. Journal of Economic Behavior and Organization 49(2): 217–239.
12. Chen S.-H, Liao C.-C, Chou P.-J (2008) On the plausibility of sunspot equilibria: Simulations based on agent-based artificial stock markets. Journal of Economic Interaction and Coordination 3(1):25-41
13. Dawkins R (1976) The selfish gene. Oxford University Press, Oxford.
14. Dawid H (2006) Agent-based models of innovation and technological change. In: Tesfatsion L, Judd K (eds.), Handbook of computational economics, Vol. 2, North Holland, Amsterdam. 1187-1233
15. Hoang T.-H, Daryl E, McKay R, Nguyen, X.H. (2007) Developmental evaluation in genetic programming: the tAG-based framework. International Journal of Knowledge-based and Intelligent Engineering Systems 12(1):69-82
16. Holland J, Miller J (1991) Artificial adaptive agents in economic theory. American Economic Review 81(2):365-370
17. Kamrani A (2002) Product design for modularity, Springer
18. Koza J (1994) Genetic programming II–Automatic discovery of reusable programs. The MIT Press, Cambridge
19. Lucas R (1986) Adaptive behaviour and economic theory. In: Hogarth R, Reder M (eds) Rational choice: The contrast between economics and psychology. University of Chicago Press, Chicago, 217-242
20. Marshall A (1924) Principles of economics, MacMillan, New York
21. Minsky M (1988) Society of mind, Simon and Schuster, New York
22. Muller G, Newman S eds (2003) Origination of organismal form: Beyond the gene in developmental and evolutionary biology. MIT Press, Cambridge.
23. Page S (2007) The difference: How the power of diversity creates better groups, firms, schools, and societies, Princeton University Press, Princeton
24. Rosca J, Ballard D (1994) Hierarchical self-organizaion in genetic programming. In: Rouveirol C, Sebag M. (eds.), Proceedings of the Eleventh International Conference on Machine Learning. Morgan Kaufmann, San Fransisco
25. Sargent T (1993) Bounded rationality in macroeconomics. Oxford Press, Oxford.
26. Simon H (1965) The architecture of complexity. General systems 10: 63-76
27. Spradlin L, Parsons R (2007) Diversity matters: Understanding diversity in schools, Wadsworth Publishing
28. Williams G (1966) Adaptation and natural selection. Princeton University Press, Princeton

# Invited Talk

# Simulating the Emergence of Complex Cultural Beliefs

M. Afzal Upal and Rik Warren

**Abstract** This paper describes the architecture of a multiagent society designed to model the dynamics of cultural knowledge. It argues that knowledge-rich agent-based social simulations are needed to understand and model the cultural dynamics of natural and artificial societies. The methodology is illustrated with the help of the Multiagent Wumpus World (MWW) testbed in which agents (1) have a causal model of the environment, (2) are goal-directed, and (3) can communicate and share information. We also present results of experiments conducted using a version of MWW. One results is the emergence of the Pareto 80/20 principle in which the 20% most communicative agents account for 80% of all communications.

## 1 Introduction

Arguably, human and animal societies are some of the most complex systems given the complex structure of social connections among people as well as complex patterns of the distribution of shared cultural beliefs. Clearly, these social patterns arise out of the interactions of a number of individuals. It is unclear, however, what the individual cognitive tendencies and interaction rules are that give rise to this complexity.

Agent-based social simulation (ABS) offers a principled way of studying how the micro-changes in individual cognitive tendencies and local interaction patterns affect macro-social patterns. The key idea behind the ABS approach is to encapsulate each member of a population in a software module (called

M. Afzal Upal
Cognitive Science, Occidental College, Los Angeles, CA, e-mail: upal@oxy.edu

Rik Warren
U.S. Air Force Research Laboratory, Wright-Patterson AFB, Ohio, U.S.A. e-mail: richard.warren@wpafb.af.mil

an agent) to build bottom-up models of human or animal societies. The ABS models focus on interactions between agents and, for the most part, abstract away the internal cognitive structure of the agents. This allows the ABS researchers to tease apart the micro-macro causal links by carefully making one local change at a time and by analyzing its impact on the emergent social patterns. Thomas Schelling, one of the early pioneers of the ABS approach, designed 1500 agents that lived on a 500 x 500 board [14]. The agent's cognitive structure consisted of one simple inference rule, namely, "if the proportion of your different-colored neighbors is above a tolerance threshold then move, otherwise stay." He showed that even populations entirely consisting of agents with high tolerance end up living in segregated neighborhoods.

Since Schelling's pioneering work, the ABS systems have been used to discover possible explanations of a number of social patterns. Thus we now know the local interaction patterns that can give rise to complex patterns of social networks. For instance, we know that if individuals prefer to establish connections with well-connected individuals then the society is likely to have scale free networks [5]. Simple agent structure is beneficial for discovering micro-macro links because it results in simulations that can be tractably run but also because having fewer local variables makes it easy to identify the micro phenomena that results in the social patterns.

However, the ABS approach has had less success in being able to discover the reasons for the emergence of complex patterns of shared beliefs that characterize human cultures. This paper outlines how the traditional ABS approach can be enhanced to study the emergence of cultural knowledge.

## 2 Cultural Knowledge

While defining culture is a notoriously difficult exercise, most cultural scientists consider shared knowledge to be a crucial aspect of culture [6]. Thus in order to understand culture, we need to understand: how cultural ideas are created; how they come to be widely shared; and how they evolve as they are transmitted within and across populations. Agent-based social simulations appear to be well suited to answer such questions by discovering the cognitive tendencies and individual behaviors that can result in the emergence of cultural knowledge. However, a few ABS systems that have attempted to model beliefs dynamics have assumed overly simplistic models of individual cognition and knowledge representation. These systems [3, 7, 8] model an agent belief as a single bit and belief change involves flipping the bit from 0 to 1 or vice versa often to match the beliefs of the neighbors. This severely limits these systems as they are unable to model most real world distributed systems applications. Complex patterns of shared beliefs such as those that characterize people's cultural and religious beliefs are also not likely to emerge out of such systems because the ABS agents are not even able to represent them. Thus existing ABS systems cannot be used to explore or model belief dynamics in human societies.

Traditionally, artificial intelligence and cognitive modeling have studied how individuals form and modify complex belief structures [7, 8, 1] but have, for the most part, ignored agent interactions assuming single agents living unperturbed in closed worlds. Artificial intelligence research on the classical planning problem illustrates this approach well [2]. Given knowledge about: (a) the current state of world, (b) the goals that the agent desires to achieve, and (c) the generalized actions that the agent can take in the world, the planning problem is to compute an ordered sequence of action instances that the agent can execute to attain its goals. Classical AI planning research assumes that the planning agent is acting alone in the world so that the world does not change while the agent is figuring out what to do next because if that happens, the agent's plan may not be executable. In the worst case, if the world continues to change the agent may never be able to act as it will always be computing the plan for the changed situation. Abstracting away other actors allows AI researchers to eliminate additional sources of complexity to focus on complex reasoning processes that go on inside the heads of individuals and result in the rich knowledge structures such as plans. This has led to the development of successful game playing programs that work in environments with limited or no interaction with other agents. However, this approach is not useful for modeling cultural dynamics because these dynamics are by their very nature products of the interaction of a large number of agents.

## 3 KBS: Knowledge-rich Agent-based Social Simulation

Clearly, to simulate belief dynamics in human societies, we need to develop knowledge-rich agent-based social simulation systems (KBS) [18]. Agents in these systems must have rich knowledge representation and reasoning capabilities and they must be able to interact with other agents which are present in their environment. Such simulation systems must overcome computational tractability concerns without abstracting away the agent's internal cognitive structure (as done by ABS systems) or ignoring interactions with other agents (as done by much of traditional AI & CM work). Furthermore, to be able to tell us something about belief dynamics in human societies, agents in such systems must model the cognitive tendencies that people are known to possess. We believe that people's ability to communicate, comprehend information, and integrate the newly received information into their existing knowledge structures is crucial to understanding the formation, propagation, and evolution of cultural beliefs. We have designed a knowledge-rich multi-agent society, called CCI (for communicate, comprehend, and integrate), to model these processes.

The challenge for a knowledge-rich agent-based social simulation architecture, such as CCI, is that of overcoming the computational intractability problems to create an implementation that can be run in real time. Drawing inspiration from early artificial intelligence work that progressed by designing synthetic "toy-domains" such as the Blocksworld [11], we argue that synthetic computer-games-like environments that are rich enough to exercise the enhanced knowledge representation and reasoning capabilities of KBS agents, and yet are not so complex as to make the simulation intractable and the results impossible to analyze and understand, are needed to make progress in the study and modeling of cultural dynamics.

## 4 Communicating, Comprehending, & Integrating (CCI) Agents

The CCI agents are goal directed and plan sequences of actions to achieve their goals. Agents attempt to build accurate models of their environment by acquiring information about cause-effect relationships among various environmental stimuli. At each instant, agents sense their environment and decide the best action to take in a given situation. The possible actions an agent can undertake include comprehension actions, speech actions, and movement actions.

The CCI agents are comprehension driven. They attempt to explain their observations using their existing knowledge and their causal reasoning engine. On observing an effect (OE), an agent searches for a cause (C) that could have produced the effect. If multiple causes are available then the agent may have to reason to eliminate some of the possible causes to select the most likely cause for the current observations. The assumed cause (AC) allows the agent to make some further predictions about the unobserved effects of the assumed cause. The assumed effects (AEs) deduced from ACs are added to the agent's world model which helps the agent form expectations about aspects of the world that the agent has not observed yet. Agent may also be able to observe causes. The observed causes (OCs) allow the agent to predict the effects (PEs) of those causes.

Agents also sense actions performed by other agents that are in the vicinity of the observing agent and attempt to comprehend those actions. Other agents are assumed to be intentional agents and, hence, causes of their actions are those agent's intentions. The CCI agents ignore the information received from others if they cannot find any justification for it. Inferring these intentions allows the observing agent to make predictions about the future behavior of the agent.

An agent A may decide to send a message M to an agent B that happens to be within listening distance if it believes that sending B the message M will result in changing B's mental state to cause it to perform an action C which can help A achieve some of its goals.

At every instant, agents consult their knowledge-base to form expectations about the future. If these expectations are violated, they attempt to explain the reasons for these violations and if they can find those explanations, they revise their world model.

## 5 Ecological Psychology

The concepts of agent-based simulation and CCI agents have their origin in computer science but clearly use a somewhat psychological language. This is not an accident or mere metaphor, but rather is intended to provide tools for psychological research. Caution must be taken with psychological terms since within the field of psychology, there are many traditions and each has its own idiosyncratic lexicon and meanings even when the words are the same.

One tradition in modern psychology which might mesh nicely with agent-based simulation and CCI agents is the "ecological approach" to perception and action advocated by J.J. Gibson [10]. We are exploring parallels and dualities between the two approaches. For example, in the ecological approach, organisms are not passive receivers of information but rather active seekers. Organisms do not simply perceive in order to act but also act in order to perceive. Perception and action are not seen as a one-way flow from stimulus to receptor to sensation to perception to cognition and finally to action. There is much attendant baggage, many conundrums, and obstacles with such a unidirectional approach. Rather, perception and action are intimately related in a perception-action cycle.

Another key concept of the ecological approach is that of an "affordance." "The affordances of the environment are what it offers the animal, what it provides or furnishes, either for good or ill [10] (p. 127)." Affordances arise out of a mutual relationship between an animal and its environment. The relationship aspect is underscored by considering that what is nutritious for one animal might be poisonous to another. Affordances let us address what is important for an animal or person. They let us focus on what is needed to survive and to go beyond just survival to actions which result in the exploitation of the physical and social environment. This exploitation can be positive as well as negative and can lead to mastery and thriving as well as to ruin, that is, to life in a real and dynamic world.

In short, affordances capture the animal-specific meanings and values of the environment and capture them in a way that demystifies them and makes them tractable. It is here that the domains of agent-based simulation, CCI agents, and a meaningful, purpose-oriented psychology might intersect and interact. A significant advantage afforded to agent-based modeling and computational cultural dynamics is that the concepts of the ecological approach to psychology promise to permit computational approaches that are rich, tractable, and relevant to real-world psychological events. A small microcosm can make this clearer:

## 6 A CCI Society: Multiagent Wumpus World (MWW)

We have designed the first version of a CCI society by embedding it into an AI domain called the Multiagent Wumpus World (MWW) [13]. Multiagent Wumpus World, shown in Figure 1, is an extension of Russell and Norvig's single agent Wumpus World and is inspired by the well known Minesweeper [12] game where an agent's objective is to navigate a minefield while looking for rewards.

MWW has the same basic configuration as the single agent Wumpus World (WW). MWW is an N x N board game with a number of wumpuses and treasures that are randomly placed in various cells. Wumpuses emit stench and treasures glitter. Stench and glitter can be sensed in the horizontal and vertical neighbors of the cell containing a wumpus or a treasure. Similar to the single agent WW, once the world is created, its configuration remains unchanged, i.e., the wumpuses and treasures remain where they are throughout the duration of the game. Unlike the single agent version, MWW is inhabited by a number of agents randomly placed in various cells at the start of the

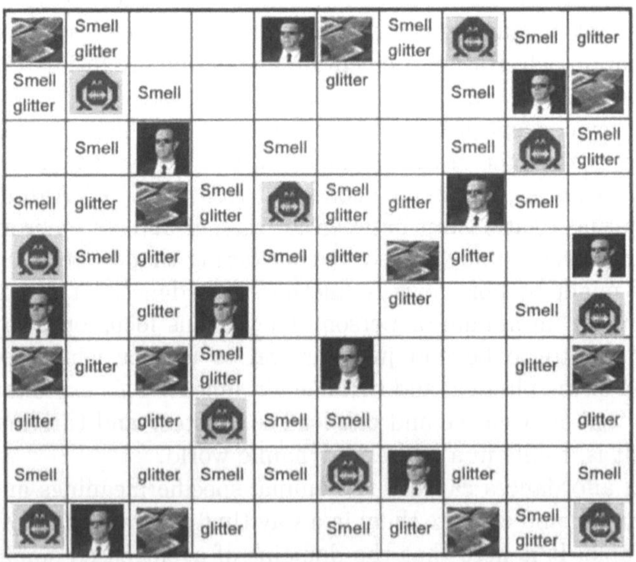


**Fig. 1** A 10 × 10 version of the Multiagent Wumpus World (MWW) domain. This version has 10 agents, 10 Wumpuses, and 10 Treasures.

simulation. An agent dies if it visits a cell containing a wumpus. When that happens, a new agent is created and placed at random on the board.

The MWW has several features that make it especially useful for simulating the emergence of complex social beliefs:

- Agents have a causal model of the environment.
- Agents are goal-directed.
- Agents can communicate and share information.

## *6.1 Agents Have a Causal Model of the Environment*

The MWW agents have a causal model of their environment. They know that stench is caused by the presence of a wumpus in a neighboring cell while glitter is caused by the presence of treasure in a neighboring cell. Agents sense their environment and attempt to explain the contents of each cell they observe. While causes (such as wumpuses and treasures) explain themselves, effects (such as stench and glitter) do not. The occurrence of effects can only be explained by the occurrence of causes that could have produced the observed effects, e.g., glitter can be explained by the presence of a treasure in a neighboring cell while stench can be explained by the presence of a wumpus in a neighboring cell. An observed effect, however, could have been caused by many unobserved causes e.g., the stench in cell (2,2) observed in Figure 2 could be explained by the presence of a wumpus in any of the four cells:

- (1, 2)
- (3, 2)
- (2, 1)
- (2, 3)

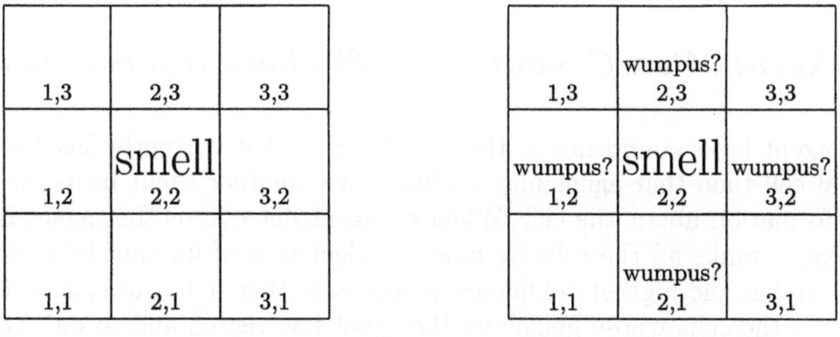


**Fig. 2** Left panel: A part of the MWW. Right panel: Possible cause(s) for smell.

An agent may have reasons to eliminate some of these explanations or to prefer some of them over the others. The MWW agents use their existing knowledge to select the best explanation. An agent's knowledge base contains both the game rules as well as their world model. A world model contains agent's observations and past explanations. The observations record information (stench, glitter, treasure, wumpus, or nothing) the agent observed in each cell visited in the past. The MWW agents use their past observations and game knowledge to eliminate some possible explanations, e.g., if an agent sensing stench in cell (2,2) has visited the cell (1,3) in the past and did not find sense any glitter there, then it can eliminate "wumpus at (2,3)" as a possible explanation because if there were a wumpus at (2,3) there would be stench in cell (1,3). Lack of stench at (1,3) means that there cannot be a wumpus at (2,3). Agents use their knowledge base to form expectations about the cells that they have not visited, e.g., if the agent adopts the explanation that there is a wumpus in cell (2,1) then it can form the expectation that there will be stench in cells (1,1) and (3,1). In each simulation round, an agent has to decide whether to take an action or to stay motionless. Possible actions include the action to:

- move to the vertically or horizontally adjacent neighboring cell
- send a message to another agent present in the same cell as the agent, and
- process a message that the agent has received from another agent.

## *6.2 Agents are Goal-Directed*

The MWW agents are goal directed agents that aim to visit all treasure cells on the board while avoiding wumpuses. Agents create a plan to visit all treasure cells they know about. The plan must not include any cells that contain wumpuses in them. To do this successfully and economically, agents need to acquire information and exchange that information with other agents.

## *6.3 Agents Can Communicate & Share Information*

If an agent lacks confidence in the knowledge that it currently has about a critical cell then that agent may decide to ask another agent in its vicinity for information about the cell. When an agent detects another agent in its vicinity, it ranks all the cells by how confident it is of its knowledge about a cell. It has the highest confidence in the cells that it has already visited. Next are the cells whose neighbors the agent has visited and so on. Agents also rank cells by how critical it is to find out information about that cell. The order in which the cells are to be visited determines the criticality, e.g., if a cell is the next to be visited then finding information about that cell is assigned the highest priority while a cell that is not planned to be visited for another 10 rounds gets low priority. The agents then use an information seeking function that takes the two rankings (confidence and criticality) as inputs and decides what cell (if any) to seek information about.

Once the first agent has sent the request for information, the second agent may also request information about a cell from the first agent in turn. A negotiation between the two agents then ensues and communication takes place only if the both agents find the communication beneficial. This way information about MWW can be transmitted throughout the population and after some time $t$, the agents may come to have some shared beliefs. We believe that studying the emergent patterns of shared beliefs can get us closer to the aim of developing computational predictive models of human cultural dynamics.

## 7 Experiments, Results & Discussion

Our experimental methodology involves designing progressively richer versions of MWW and studying the impact of each local change to see how changes in agent's internal cognitive structure result in changes in the patterns of shared beliefs. Previously, we [18, 16, 17] have reported the results of a number of experiments. Upal [18] reported that the version of a $10 \times 10$ MWW with 10-agents was most challenging for CCI agents when it contained 10 randomly distributed wumpuses and treasures compared with MWWs containing 5 or 20 wumpuses and treasures. This is the version we used in the subsequent experiments. Upal [16] found that even without any communication, false beliefs generated in such a society have a particular structure to them; they are more likely to be about objects and events whose presence is harder to confirm or disconfirm. Upal & Sama [17] reported that that communication does not eliminate or even decrease the prevalence of such false beliefs. There is some evidence to suggest that in human societies, people are also more likely to have false beliefs about unconfirmable entities and events. Bainbridge and Stark [4] made confirmability the core of their theory of religion to argue that religious beliefs are unconfirmable algorithms to achieve rewards that are highly desired by people yet cannot be obtained. Similarly, there is some evidence to suggest that many false ethnic stereotypes people have are about things that are harder to confirm or disconfirm such as the sexual practices of the neighboring tribes [15].

While, our previous work has focused on similarities between agent's beliefs, this paper focuses on differences that automatically emerge among the MWW agents. We used a $10 \times 10$ world with 10 agents, wumpuses, and treasures and ran the world to 300 rounds.

Our results show that the MWW agents do not all equally engage in communication. Figure 3 shows that, as expected, agents that live longer engage in more communication. However, there is no linear relationship between agent age and the number of communication patterns. This is because random spatial patterns (i.e., an agent that happens to be born in or happen to have traveled on routes that happened to be frequented by other agents) also determine the number of communication opportunities that agents get.

Figure 4 shows that the majority of agents communicate only once or never communicate while a very small number of agents communicate 10 times or more. Figure 5 is a cumulative plot of the data in Figure 4 starting with the most communicative agents and ending with the least. It shows that the distribution of agent communications follows Pareto's Principle: around 80% of all communication is carried out by around 20% of the agents. This is yet another instance of the "vital few and trivial many" principle that indicates that a few agents are responsible for a vast majority of the social impact. This principle appears to work in a variety of social domains, e.g., 80% of profits are created by 20% of all employees, 80% of help calls are by 20% of all customers, and 80% of the profits are due to the work of 20% of all employees [5].

The emergence of the 80/20 principle for agent communications in the wumpus world has a number of implications. This means that a small number of agents may get many more opportunities to spread their message than the vast majority of agents and they may come to disproportionately influence the beliefs of the MWW population. True as well as false beliefs of such agents are more likely to be spread than true and false beliefs of other agents. These so-called "influentials" have been well studied in business and information diffusion circles to study how to design strategies to diffuse a message to a population [9].

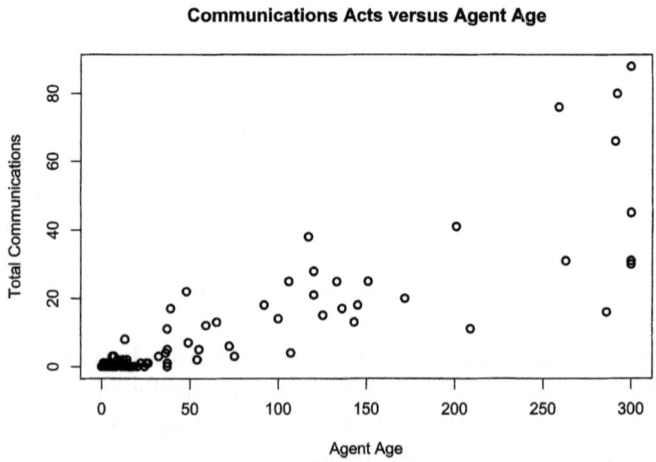


**Fig. 3** The number of communication acts (both requests-for-information and answers to such requests) agents engage in plotted against agent age.

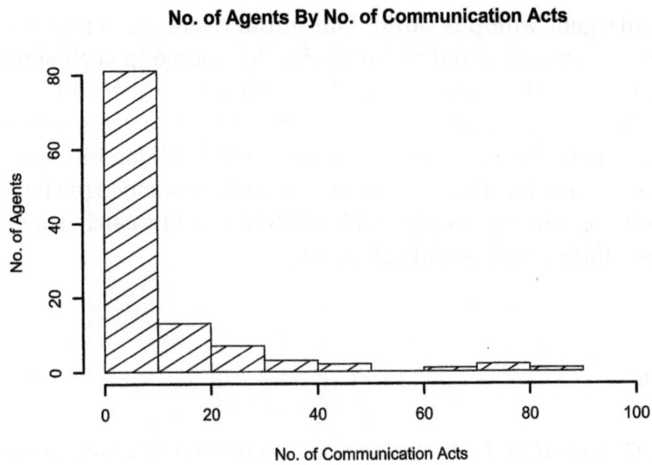


**Fig. 4** A majority of agents communicate only once or never communicate while a very small number of agents communicate 10 times or more.

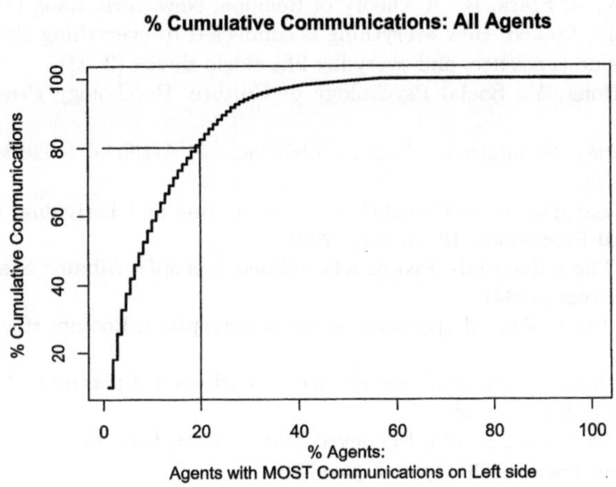


**Fig. 5** Cumulative graph of all communications starting with the most communicative agents and progressing through the least. Auxiliary lines indicate that the 20% most communicative agents account for 80% of all communications.

# 8 Conclusions

There is a need for a knowledge-rich agent-based social simulation methodology to understand the cultural dynamics of natural and artificial societies. Such societies can be simulated in real time by designing synthetic toy worlds such as the multiagent wumpus world. Our initial results are encouraging and show that various aspects of natural societies do emerge in such simulations.

We are enhancing the capabilities of our CCI agents to allow them to model other agents present in their environment and to reason about their beliefs and intentions. We are also enhancing the MWW domain by designing different types of agents. This will allow us to test sociocognitive theories involving differences among people such as differences in wealth, social status, communicating ability, and social influence.

## References

1. Alchourrn, C., Grdenfors, P., & Makinson, D.: On the logic of theory change: Partial meet contraction and revision functions. J. Symbolic Logic, 50, 510-530, 1985.
2. Allen, J.: Natural Language Understanding. Menlo Park, CA: Benjamin Cummings (1987).
3. Bainbridge, W.: Neural Network Models of Religious Belief. Sociological Perspectives, 38, 483-495, 1995.
4. Bainbridge, W., & Stark, R.: A Theory of Religion. New York: Lang (1987).
5. Barabasi, A. L.: Linked: How everything is connected to everything else and what it means for business, science, and everyday life. Basic Books (2003).
6. Chiu, C. & Hong, Y.: Social Psychology of Culture. Psychology Press, New York (2006).
7. Doran, J. (1998). Simulating collective misbeliefs. J. Artificial Societies and Social Simulation, 1(1).
8. Epstein, J.: Learning to be thoughtless: Social norms and individual computation. Computational Economics, 18(1), 9-24, 2001.
9. Gabriel, W.: The influentials: People who influence people. Albany: State University of New York Press (1994).
10. Gibson, J.J.: The ecological approach to visual perception. Boston: Houghton Mifflin (1979).
11. Gupta, N., & Nau, D.S.: On the complexity of blocks-world planning. Artificial Intelligence, 56(2-3), 223-254, 1992.
12. Minesweeper (computer game). Retrieved May 5, 2008, from `http://en.wikipedia.org/wiki/Minesweeper_%28computer_game%29`
13. Russell, S., & Norvig, P.: Artificial Intelligence: A Modern Approach, 2nd ed. Englewood Cliffs, NJ: Prentice Hall (2003).
14. Schelling, T.: Dynamic models of segregation. J. Mathematical Sociology, 1, 143-186 (1977).
15. Smith, L.: Sects and Death in the Middle East. The Weekly Standard, (2006).
16. Upal, M. A.: The structure of false social beliefs, in Proceedings of the First IEEE International Symposium on Artificial Life, 282-286, Piscataway, NJ: IEEE Press (2007).
17. Upal, M.A., & Sama, R.: Effect of Communication on the Distribution of False Social Beliefs, in Proceedings of the International Conference on Cognitive Modeling.
18. Upal, M.A., & Sun, R.: Cognitive Modeling and Agent-based Social Simulation: Papers from the AAAI-06 Workshop (ISBN 978-1-57735-284-6), Menlo Park, CA: AAAI Press (2006).

# Organization and Management

# Synchronization in Mobile Agents and Effects of Network Topology

Masaru Aoyagi and Akira Namatame

## 1 Introduction

Reynolds developed a method that creates realistic simulations of bird flocking [12][13]. Traditionally, in order to simulate a flock of birds, the simulation would consider the path of each bird individually. However, in Raynolds method, there is no central authority for each flock. Instead, local interaction rules between the adjacent birds would be used to determine the flocking behavior. This model is known as "boids model". In boids model, there are three local interaction rules: 1) attraction (cohesion rule), 2) collision avoidance (separation rule), and 3) velocity matching (alignment rule) between the boids located within a certain radius. When properly applied, these 3 local rules create a collection of autonomous agents that produce realistic flocking behavior.

This behavior is interesting in that not only must the individual behavior of the birds be evaluated, but also the overall flocking behavior needs to be considered.

Based on the boids model, local control laws for a collection of mobile agents that result in self-organization have been investigated. A collection of agents, like birds in a flock, must be able to align their velocities, to move with a common speed, and to achieve the desired interagent distances while avoiding collisions with each other.

Watts and Strogatz [17] introduced a network model called a small-world network that was capable of interpolating between a regular network and a random network using a single parameter. A small-world is a network with a relatively small characteristic length. In a small-world network, any two nodes can be linked using a few steps despite the large size of the network.

The small-world model of Watts and Strogatz has led researchers working in many different fields to study the topological properties of complex networks [6].

Dept. of Computer Science, National Defense Academy of Japan,
Yokosuka, Hashirimizu 1-10-20, Japan
e-mail: {g45074, nama } @nda.ac.jp

These properties include degree distribution, characteristic length, clustering coefficient, robustness to node failure, and search issues. The researchers who have most contributed to this effort are in such diverse fields as statistical physics, computer science, economics, mathematical biology, communication networks, and power networks.

## 2 Research on Flocking Behavior

Several researchers working in the area of statistical physics and complexity theory have addressed flocking and schooling behavior in the context of non-equilibrium phenomena in many degree of freedom dynamic systems, as well as the self-organization of a system of self-propelled particles [4][14].

Watts and Strogatz [17][16] introduced and studied a simple tunable model that can explain the behavior of many real world complex networks. Their small-world model creates a regular lattice and replaces its original edges with random edges based on some probability. Due to the short paths between distant parts of the network, which cause high-speed spreading of information that may result in fast global coordination, it is conjectured that compared to a regular lattice of the same size, dynamic systems coupled in this way would display enhanced signal propagation and global coordination.

The model starts with a ring of $n$ nodes, each connected by undirected nodes to its nearest neighbors to a range $k$. Shortcut links are added rather than rewired between randomly selected pairs of nodes, with some probability per link on the underlying lattice. Thus, there are typically short cuts.

Vicsek et al. [15] proposed a simplified model, that turns out to be a special case of boids model [12][13] where all agents move with the same speed and only follow an alignment rule. Each agent's heading is the average of its nearest neighbors' ones, plus some additive noise.

Such a system reduces computational effort, because each rule does not always need to be computed. This would also reduce the computational effort required by a flocking animal. Such a hierarchical decision tree was implied in [15]. B. L. Partridge showed that in a conflict between following the visual sense or the lateral line sense, schooling fish followed the visual sense.

R. Olfati-Saber et al. [11] theoretically established the stability properties of an interconnected closed loop system by combining results from classical and non-smooth control theory, robot navigation, mechanics, and algebraic graph theory. Stability is shown to rely on the connectivity properties of the graph that represents agent interconnections, in terms of not only asymptotic convergence but also convergence speed and robustness with respect to arbitrary changes in the interconnection topology. Exploiting modern results from algebraic graph theory, these properties are directly related to the topology of the network through the eigenvalues of the Laplacian of the graph.

Olfati-Saber[8] also demonstrated a phase transition phenomenon in the algebraic connectivity of flocking behavior on small-world networks. Algebraic connectivity of a graph is the second smallest eigenvalue of its Laplacian matrix and represents the speed required to solve consensus problems in the network. Hovareshti et al. [3] made the same conclusions for a discrete-time consensus algorithm.

Such an algorithm is possible using local control action agents that exploit the network properties of the underlying interconnection among the agents. Network connectivity affects the performance and robustness properties of a system of networked agents. A consensus protocol is an iterative method that provides the group with a common coordination variable[11]. However, local information exchange limits the speed of convergence of such protocols. A reasonable conjecture is that a small-world network should result in good convergence speed for the self-organization consensus problems due to the low average pair-wise path length that should increase the speed of information diffusion in the system.

We are interested in self-organization and group behavior about independent and autonomous agent. They are decentralized and doesn't share information. This paper examines the conjecture using simulations that show the emergence of flocking behavior on small-world networks. As well, the emergence of flocking behavior is directly associated with the connectivity properties of the interconnection network. To achieve coordination, individual reactive autonomous agent doesn't need to share information but only to refer others action. Finally, it will be shown that small-world networks are robust to arbitrary switching of the network topology.

## 3 Emergence of Flocking Behavior

The following method for simulating the flocking behavior of such animals that form flocks to avoid air, land, or sea obstacles is accurate and efficient enough to run in real time. This method modifies boids model [13] as follows:

(1) Cohesion: steer to move towards the average position of the neighboring flockmates
(2) Separation: steer to avoid crowding the neighboring flockmates
(3) Alignment: steer towards the average heading of the neighboring flockmates

Each agent is an individual and its behavior determines how an agent reacts to other agents in its local neighborhood. Agents outside of the local neighborhood are ignored. A flock often consists of multiple local interactions of each agent. A flock is defined as an area of multiple local interactions.

Each anget reacts just neighbors behavior and determines to behave. But distant relations are ignored or neglect. The flocking behavior emrgenet from such multiple local interaction in each neighboring agents. Cohesion behavior gives an agent, which is an outlined triangle located in the centre of the diagram, the ability to approach and form a group with other nearby agents. This rule causes each active flock member, which is represented by an outlined triangle located in the center of the diagram, to try to orient its velocity vector in the direction of the centroid (average spatial position) of the local flock. The degree of locality of the rule is determined by the sensor range of the active flock member, represented by the light colored circle.

Separation behavior gives an agent the ability to maintain a certain separation distance from others. This can be used to prevent agents from crowding together and colliding with each other. To compute steering for separation , a search is first made to find other agents within the specified neighborhood. This might be an exhaustive search of all agents in the simulated world, or might use some sort of spatial partitioning or caching scheme to limit the search to other local agents.

Alignment behavior gives an agent the ability to align itself with, that is, head in the same direction or speed as, other nearby agents . Steering for alignment can be computed by finding all agents in the local neighborhood (as described above for separation), averaging together the velocity, or, alternately, the unit forward vectors, of the nearby agents. This average is the desired velocity, and so the steering vector is the difference between the average and the agent's current velocity, or, alternately, its unit forward vector. This behavior will tend to turn the agent so that it is aligned with its neighbors.

The flocking algorithm works as follows: For a given agent, centroids are calculated using the sensor characteristics associated with each flocking rule. Next, the velocity vector the given agent should follow to carry out the rule is calculated for each of the rules. These velocity vectors are then weighted according to the rule strength and added together to give an overall velocity vector demand. Finally, this velocity vector demand is resolved into a heading angle, pitch attitude, and speed demand, which are passed to the control system. The control system then outputs an actuator vector that alters the motion of the agent in the appropriate manner.

# 4 Dynamic Analysis of Emergence

A theoretical analysis of the emergent dynamics of flocking behavior given in [1],[2] was performed. Each agent recognizes two physical values: 1) the distance to its nearest flockmates and 2) the relative velocity of its flockmates. Agent $i$ sees a neighboring agent $j$ in its visual sensor range. Agent $i$ can recognize the vector $\mathbf{d}_{ij}$, that is, the position vector to the neighbor agent $j$ and can calculate the center position vector $\mathbf{D}_i$ of the neighboring flockmates. Agent $i$ also recognizes vector $\mathbf{v}_{ij} = d\mathbf{d}_{ij}/dt$, that is, the relative velocity vector and can calculate the average relative velocity vector $\mathbf{V}_i$ of the neighboring flockmates. The center position vector $\mathbf{D}_i$, its unit vector $\mathbf{e}_{\mathbf{D}_i}$, the average relative velocity vector $\mathbf{V}_i$ and its unit vector $\mathbf{e}_{\mathbf{V}_i}$ are:

**Fig. 1** $\phi_{csi}$: the potential energy in Eq.(4).

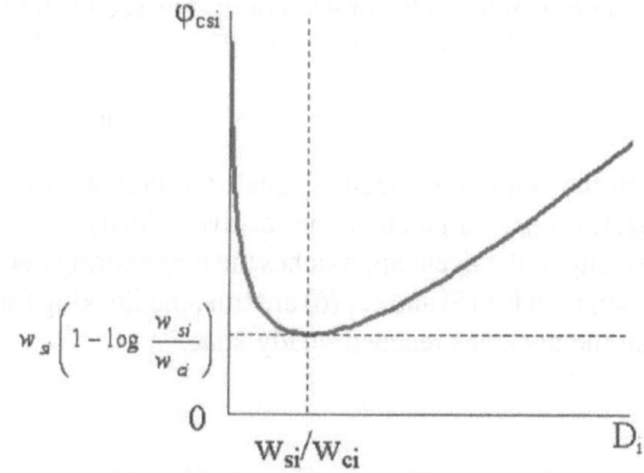


$$\begin{cases} \mathbf{D}_i = \dfrac{1}{n_i}\sum_j^{n_j}\mathbf{d}_{ji}\,,\ D_i = |\mathbf{D_i}|\ ,\ \mathbf{e}_{\mathbf{D_i}} = \dfrac{\mathbf{D_i}}{D_i} \\ \mathbf{V}_i = \dfrac{1}{n_i}\sum_j^{n_j}\mathbf{v}_{ji}\,,\ V_i = |\mathbf{V_i}|\ ,\ \mathbf{e}_{\mathbf{V_i}} = \dfrac{\mathbf{V_i}}{V_i} \end{cases} \tag{1}$$

A linear combination of the cohesion force vector $\mathbf{F}_{ci}$, separation force vector $\mathbf{F}_{si}$, and alignment force vector $\mathbf{F}_{ai}$ are used to define the flocking force vector $\mathbf{F}_{fi}$.

$$\begin{aligned} \mathbf{F}_{fi} &= \mathbf{F}_{ci} + \mathbf{F}_{si} + \mathbf{F}_{ai} \\ &= \left(w_{ci} - \frac{w_{si}}{D_i}\right)\mathbf{e}_{\mathbf{D}_i} + w_{ai}\mathbf{e}_{\mathbf{V}_i} \end{aligned} \tag{2}$$

where coefficient $w_{ci}$, $w_{si}$, and $w_{ai}$ are positive. The first term of Eq.(2) is the resultant force of the cohesion force vector $\mathbf{F}_{ci}$ and the separation force vector $\mathbf{F}_{si}$. The resultant force vector $\mathbf{F}_{csi}$ relates position between agents.

$$\mathbf{F}_{csi} = \left(w_{ci} - \frac{w_{si}}{D_i}\right)\mathbf{e}_{D_i} \tag{3}$$

The potential energy $\phi_{csi}$ of $\mathbf{F}_{csi}$ is given by the following equation.

$$\phi_{csi} = w_{ci}D_i - w_{si}\log D_i \tag{4}$$

Fig.1 shows $\phi_{csi}$, the size of the force vector $\mathbf{F}_{csi}$ , and the potential energy $\phi_{csi}$. The potential energy $\phi_{csi}$ has a local minimum at

$$D_i = \frac{w_{si}}{w_{ci}} \tag{5}$$

At this point, the force vector $\phi_{csi}$ equals the zero vector. When the distance $D_i$ from the center of neighbors is less than the value of right side of Eq.(5), the force vector $\mathbf{F}_{csi}$ becomes repulsive. Otherwise, the force vector $\mathbf{F}_{csi}$ becomes attractive.

If $w_{si}$ is smaller or $w_{ci}$ is larger, then the absolute value of the distance $D_i$ from the center of neighbors becomes shorter. The second term in Eq.(2), the alignment term, aligns the velocity of each agents. For this term, if

$$\mathbf{V}_i = \mathbf{0} \tag{6}$$

then the velocity of agent $i$ equals the neighboring flockmates' velocity. As show Fig.1, if $w_{ai}$ is larger, then the relative velocity vector $\mathbf{v}_{ij}$ goes to zero faster, and the velocity of the agent approaches the neighboring flockmates' velocity more quickly. When both Eq.(5) and Eq.(6) are true, the flocking force vector $\mathbf{F}_{fi} = \mathbf{0}$. This implies that the flock has reached steady-state.

## 5 Synchronization in Small-World Networks

The small-world model of Watts and Strogatz [17] caused a tremendous amount of interest among researchers working in multiple fields on the topological properties of complex networks. Due to the short paths between distant parts of the network, which result in high speed spreading of information that may result in fast global coordination, it is conjectured that, compared to regular lattices of the same size, dynamical systems coupled in this way would display enhanced signal propagation and global coordination.

In most engineering and biological complex systems, the nodes are dynamic, that is, "real-life" engineering networks are interconnections of dynamic systems. The same applies to biological networks, including gene networks and coupled neural oscillators. From the perspective of systems and control theory, the stability properties of a system of dynamic agent networks are a topic of interest.

In networks of dynamic agents, "consensus" means to reach an agreement regarding a certain quantity of interest that depends on the state of all agents [10]. A "consensus algorithm" is an interaction rule that specifies the information exchange between an agent and all of its neighbors on the network. Olfati-Saber [8] demonstrates a phase transition phenomenon in algebraic connectivity of network and show good convergence speed for consensus problems on small-world networks.

Each agent has cohesion, separation and alignment (CSA) rules that were originally introduced by Reynolds [13]. Each agent inputs the relative velocity and position of neighboring agents in its visual range and computes its steering and driving acceleration at that time. In other words, each agent has a local-directed link to other agents and the emerging flocking behavior. However, if agents are dintant each other, flocking behavior cannot emerge.

A theoretical framework for the design and analysis of flocking algorithms for mobile agents was developed by Olfati-Saber [9] as a consensus problem. They demonstrated that flocks are networks of dynamic systems with a dynamic topology. This topology is a proximity graph that depends on the states of all the agents and is determined locally for each agent. The notion of a state-dependent graphs was introduced by Mesbahi [5] in a context independent of flocking.

To construct a small-world network, Watts and Strogatz [17] introduced the following model (WS model) about a regular network that consists with $N$ nodes in which every node is bidirectorically connected to its $k(= (0,N])$ neighbors. Every link rewires with probability $p_{WS}(= [0,1])$ by changing one of the endpoints of a link uniformly at random. No self-loops or repeated links are allowed and the rewired links are calls shortcuts. Each node of the WS model always has $k$ undirected links.

Newman, Moore and Watts [7] sugest another small-world network model (NMW model) other than WS model. The model also start a regular network that consisted $N$ nodes. And each node bidirectorically connects with $k(= (0,N])$ neighboring nodes. Shortcut links are added rather than rewired between randomly selected pairs of nodes, Its probability is $p_{NMW}(= [0,1])$. Each node of the NMW model has, on average, $k + kp_{NMW}$ undirected links.

Thus, each agent is allowed to have an additional external directed link based on the boids model. The agent decides to whom it will connect using the external link based on a given probability during every simulation step. Using a stochastic external link has the advantage that the agent keeps a low throughput and allows for flock behavior to emerge. It also limits ergodically both the temporal average of degree (the number of links) an agent has and the ensemble average of degree the group of agents has.

A simulation was performed where the state of the emerging flocking behavior depended on the probability of the external directed link $P$. In the group of $N$ agents, an agent $i = 1,2,....,N$ has local undirected links with $n_i = [0,N]$ neighbors in its visual range and additional external link at a probability of $P = [0,1]$ in each simulation step. If the probability of an external link $P = 1$, then the group of agents always forms a complete graph. If $P = 0$, then each agent has only a local link. Thus, on average, each node has $n_i + (N - n_i)P$ links.

## 6 Simulation Results

A simulation was performed where the state of the emerging flocking behavior depended on the probability of the external link $P$. If $0 < P < 1$, then the group of agents are in a small-world network. At initial state (time step = 0), 100 agents are randomly deployed. Most of the agents are too far apart to be locally linked. Thus, each agent has a random initial velocity. The probability for an external link was set to $P = 1$, $10^{-3}$ or $10^{-4}$ for the simulation.

**Fig. 2** Snapshot of simulation. $P = 10^{-3}$, time step = 200.

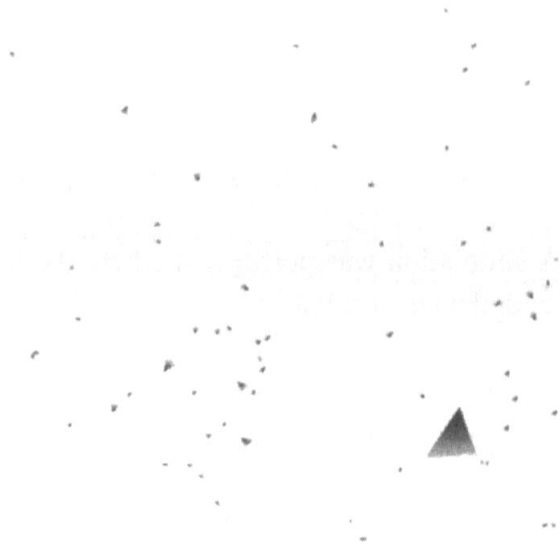

**Fig. 3** Snapshot of the simulation. $P = 10^{-4}$, time step = 200.

## *6.1 Flocking of Agents*

Each agent flocks together based on the cohesion and separation rules. The standard deviation of agents' position, $SD(x,y,z)$ shows the magnitude of agents' flocking and distancing. Each agent adjusts its velocity using the alignment rule. The standard deviation of agents' velocity, $SD(v)$ shows the magnitude of agents' velocity adjustments. Once flocking behavior occurs and steady-state is reached, both $SD(x,y,z)$ and $SD(v)$ become constant.

Fig.2 shows a snapshot of the simulation where each agent has a probability for external link $P = 10^{-3}$ at a time step of 200. It shows emergent flocking behavior. Fig.4 shows the standard deviation of agents' position over time, $SD(x,y,z)$. When

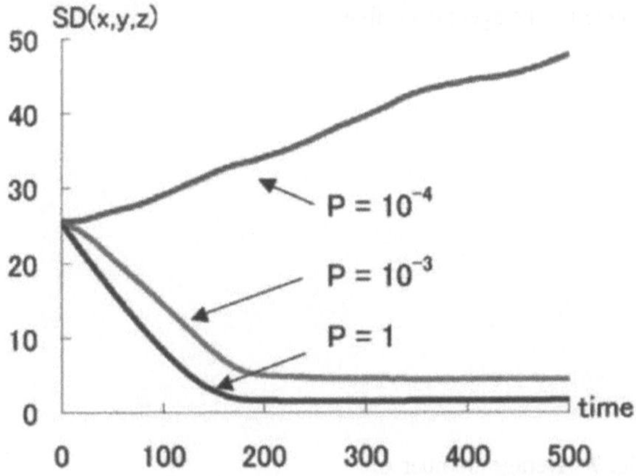


**Fig. 4** The standard deviation of agents' position over time, $SD(x,y,z)$.

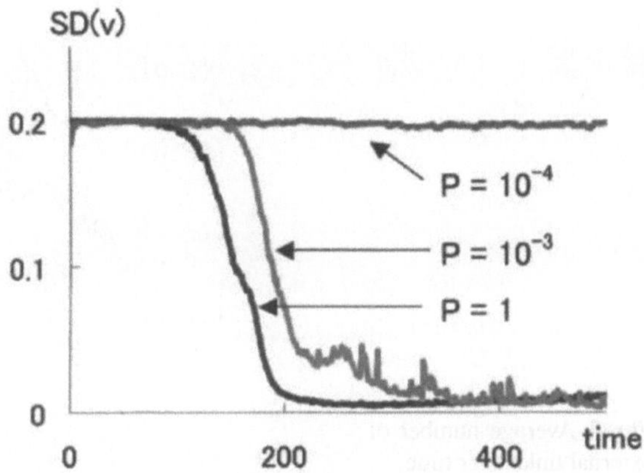


**Fig. 5** The standard deviation of agents' velocity over time, $SD(v)$.

$P = 1$, $SD(x,y,z)$ decreases monotonically at first but becomes constant at a time of 200. For $P = 10^{-3}$, the $SD(x,y,z)$ is similar to the graph for $P = 1$ except that the magnitude is different.

Conversely, when $P = 10^{-4}$, $SD(x,y,z)$ increases monotonically. Fig.3 shows that the agents disengage from the flock and, thus, flocking behavior cannot occur. This implies that a group of agents cannot form a flock unless there are enough external links ($P > 10^{-3}$). Fig.5 shows the standard deviation of the agents' velocity over time, $SD(v)$. This figure shows that the phase transition and emergent flocking behavior change dramatically between the time of 100 and 200 for cases $P = 1$ and $10^{-3}$. However, in the case of $P = 10^{-4}$, $SD(v)$ is constant. This implies that flocking behavior cannot occur.

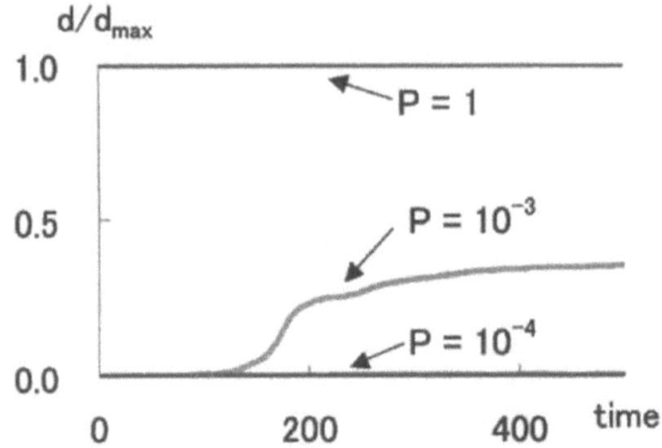


**Fig. 6** The normalized mean degree for the agent over time.

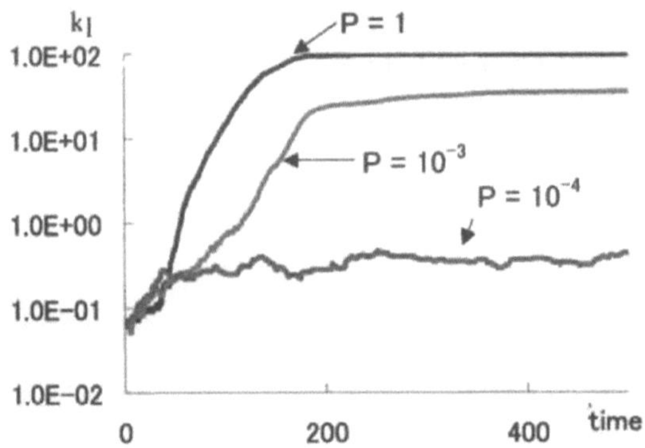


**Fig. 7** Average number of local links over time.

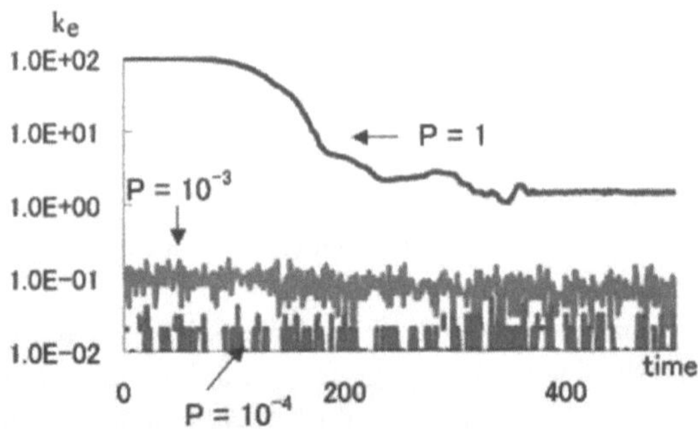


**Fig. 8** Average number of external links over time.

## *6.2 The mean degree of agents' network*

The mean degree $d$ is defined as the average number of links that an agent rewires to other agents. All agents link each other when $P = 1$, in other words, the agents' network is a complete graph, and the mean degree $d$ is a constant value $d_{max} = n - 1$, where $n$ is the number of agents. Figure6 shows the mean degree of the agents' network $d$ normalized by $d_{max}$. Figure7 shows the average number of local links and Figure8 shows the average number of external links.

In the case where $P = 10^{-4}$, there are only a few local and external links (Fig.7, Fig.8). The normalized mean degree $d/d_{max}$ remains at about 0.0. Since each agent does not have enough external link to effectively follow the cohesion rule, flocking behavior cannot emerge (Fig.6).

In the case where $P = 1$, the mean degree $d$ is the maximum value $d_{max}$. Thus, the normalized mean degree $d/d_{max}$ remains at 1.0 (Fig.6). At time = 0, the agent's links consist of a few local links and most external links(Fig.7, Fig.8). When agents are close by and proceed to recognize each other in their visual range at time between 100 to 200, the ratio between local and external links reverses.

In the case where $P = 10^{-3}$, there are always few local and external links at time = 0 (Fig.7, Fig.8). Thus, this case is similar to the case where $P = 10^{-4}$. The normalized mean degree $d/d_{max}$ is also approximately 0.0 (Fig.6). Initially, the number of external links is small, but increases starting at time = 100, before leveling off at time = 200. At this point, the result resembles the case where $P = 1$. Therefore, the graph of the normalized mean degree $d/d_{max}$ proceeds to increase at time = 100 and becomes constant after time = 200. This means that a phase transition occurs between the regions where flocking behavior is emerging and where flocking behavior is stable.

In the case where $P = 1$, all agents always link directly each other. Thus, the agents can aggregate, and flocking behavior emerges. In the case where $P = 10^{-4}$, each agent never has enough links and, thus, the agents are dispersed. In the case where $P = 10^{-3}$, all agents always have enough links to flock together. However, the agents as a group undergo a state transition from dispersing to aggregating.

## 7 Conclusion

Networked multi-agent systems are comprised of many autonomous interdependent agents found in a complex network. An example of such a system is a mobile sensor network. A common challenge in networked multi-agent systems is to decentralize group formation among the spatially and logically extended agents. Even in cooperative multi-agent systems, efficient team formation is difficult due to limited local information available to the individual agents.

This paper showed how flocking behavior can emerge and converge to stability on a small-world network using a few external links, in addition to the local links among the self-driven agents. A model for distributed multi-agent group formation in a networked multi-agent system based on local information was presented. As well, empirical results were obtained for improving flocking formation performance. Analyzing network characteristics of self-driven agents in a large-scale network requires both a suitable analysis method and a means to reduce the amount of data that must be collected. Thus, this paper described a novel technique that allows easy observation of the shift in the network-wide patterns over time. To illustrate this technique and its potential application, the simulation and analytical results were presented. In boids model, each neighbor agent forms a directed network that allows flocking behavior to occur. The topology of the network is important for robustness, since each agent needs to recognize not only its neighboring ones but also distant agents. In this manner, the agents can be connected or reconnected to create a flock.

This paper was inspired by the flocking behavior of network mobile agents. It is known that a class of decentralize and local control laws for the collection of mobile agents results in self-organization. This is possible due to local control action resulting from an exploitation of the network properties of the underlying interconnection among agents. Network connectivity affects the performance and robustness properties of the system of networked agents.

## References


1. Aoyagi, M., Namatame, A.: Dynamics of emergent flocking behavior. In: Proc. of 7th International Conference on Celluar Automata for Research and Industry, ACRI 2006, pp. 557–563 (2006)
2. Aoyagi, M., Namatame, A.: Network dynamics of emergent flocking behavior. International Transactions on Systems Science and Applications **3**(1), 35–43 (2007)
3. Hovareshti, P., Baras, J.: Consensus problems on small world graphs: A structural study. Technical report, Institute for Systems Research (2006)
4. Levine, H., Rappel, W.J., Cohen, I.: Self-organization in systems of self-propelled particles. Phys. Rev. E **63**(1), 017,101 (2000)
5. Mesbahi, M.: On state-dependent dynamic graphs and their controllability properties. IEEE Transactions on Automatic Control **50**, 387–392 (2005)
6. Newman, M.: The structure and function of complex networks. SIAM Review **45**, 167–256 (2003)
7. Newman, M.E.J., Moore, C., Watts, D.J.: Mean-field solution of the small-world network model. Phys. Rev. Lett. **84**(14), 3201–3204 (2000)
8. Olfati-Saber, R.: Ultrafast consensus in small-world networks. In: Proceedings Proc. of American Control Conference, pp. 2371–2378 (2005)
9. Olfati-Saber, R.: Flocking for multi-agent dynamic systems: Algorithms and theory. IEEE Transactions on Automatic Control **51**, 401–420 (2006)
10. Olfati-Saber, R., Fax, J.A., Murray, R.M.: Consensus and cooperation in networked multi-agent systems. In: Proceedings of the IEEE, vol. 95, pp. 215–233 (2007)
11. Olfati-Saber, R., Murray, R.M.: Consensus problems in networks of agents with switching topology and time-delays. IEEE Transactions on Automatic Control **49**, 1520–1533 (2004)
12. Reynolds, C.W.: Flocks, herds, and schools: A distributed behavioral model, in computer graphics. In: SIGGRAPH '87 Conference Proceedings, pp. 25–34 (1987)
13. Reynolds, C.W.: Steering behaviors for autonomous characters. In: Proceedings of Game Developers Conference, pp. 763–782 (1999)
14. Toner, J., Tu, Y.: Flocks, herds, and schools: A quantitative theory of flocking. Phys. Rev. E **58**(4), 4828–4858 (1998)
15. Vicsek, T., Czirók, A., Ben-Jacob, E., Cohen, I., Shochet, O.: Novel type of phase transition in a system of self-driven particles. Phys. Rev. Lett. **75**(6), 1226–1229 (1995)
16. Watts, D.: Small Worlds: The Dynamics of Networks Between Order and Randomness. Princeton University Press (1999)
17. Watts, D.J., Strogatz, S.H.: Collective dynamics of 'small-world' networks. Nature **393**(6684), 440–442 (1998)

# Evaluation of Mass User Support Strategies in Theme Park Problem

Yasushi Yanagita and Keiji Suzuki

**Abstract** The theme park problem is one of exercise to research on the mass user support in order to solve social problems and it is suggested as the problem that simplified on a large scale dynamic scheduling. In this paper, it is suggested the theme park models which used the general idea of the complex network such as small-world networks and scale-free networks to realize the complicated environment such as real problems. It is compared with the effectiveness and efficiency of the models and the strategies to introduce the multi-agents which are applied the traversal strategies that can support the various situations to coordinate the institutions using instantly. In addition, it is shown that possibility of the mass user support by the traversal strategies for large-scale personal schedule adjustment in the suggestion model.

## 1 Introduction

Over the past few years, a considerable number of studies have researched to realize ubiquitous societies. Those research's goals are not only to optimize individual life utility but also to support a social system made up of a group of individuals. It's called mass user support. Mass user support research is to solve various social problems. In particular, it aims to develop a social coordination mechanism that increases social welfare without reducing individual utility. How to increase social welfare and individual utility is researched from the viewpoint of theory and com-

Yasushi Yanagita,
Yasushi Yanagita, Graduate School of System Information Science, Future University - Hakodate, Kameda-Nakano 116-2, Hakodate, Hokkaido, 041-8655, Japan, e-mail: g2107034@fun.ac.jp

Keiji Suzuki,
Keiji Suzuki, Graduate School of Information Science and Technology, Hokkaido University, North 14 West 9, Kita-ku, Sapporo, Hokkaido, 060-0814, Japan, e-mail: suzuki@complex.eng.hokudai.ac.jp

plicity by terms of game theory and multiagent systems. It is necessity to study in practical mass user support model to apply practical problems.

The "Theme Park Problem" is suggested as one of exercise to solve those problems. The theme park problem is consists of two kinds of element, a spatial segment and a software agent. The spatial segment is one component of the theme park, and it may be one of several types, i.e., attractions, roads, plazas, entrances, and exits. The software agent represents a visitor to the theme park, and it has individual preferences regarding each attraction. The objective of the theme park problem is to develop an algorithm that dynamically coordinates agents' visiting behavior in order to reduce congestion and increase the individual visitor's satisfactions. In other words, the theme park problem is a dynamic coordination problem which needs to coordinate many individual behaviors and to optimize individual and social satisfaction by using distributed information.

The general research on the theme park problem [1] aim to develop heuristic algorithms which are called coordination algorithms. The coordination algorithm may be able to improve the efficiency of the whole theme park by coordinating a lot of individual behaviors. The agents decide a destination by the defined coordination algorithm based on congestion information of each attraction and preference that was given every agent. Then, the agents go around each attraction in the theme park. It is aimed to show the following effects to apply the coordination algorithm. First, agents can go around each attraction in the theme park more smoothly while dealing with the various environments and situations. Secondly, it is coordinated that the whole theme park by distributing loads of the queue which occurs because agents are crowded to each attraction so that the agents can use each attraction effectively.

This paper aims to achieve the above purposes. In addition, it is shown that the effectiveness, the efficiency, and the versatility of the coordination algorithm in the complicated model setting such as the real problems. In this research, the theme park is modeled by applying the general idea of the complex networks to realize the complicated environment such as the real problems. The experiment is carried out by the multi agents who applied the coordination algorithm in each theme park model which contained different network structures. It is compared with the results of the experiments. It is shown the influence, the factor and the environment that each model setting affects the coordination algorithm by the experiments. It is shown that what kind of network structure is desirable to build the model settings so that the agents go around the theme park more efficiently. Finally, this research aims to show what kind of the theme park model to evaluate the effectiveness of the coordination algorithm qualitatively.

Moreover, it is analyzed the results of the experiment. It is shown that the model setting condition which doesn't affect the coordination algorithm by network structures. Accordingly, it is carried out the comparison experiment to develop the more versatile coordination algorithm which doesn't depend on the network structures in the model setting condition. The comparison experiment is carried out by the coordination algorithms which are used by the general research and the suggestion algorithm which is developed based on data provided by the additional experiment in the model setting condition.

# 2 Modeling by Complex Networks

## 2.1 Theme Park Model

This paper can be applied the complex network to the theme park model to extend each model setting of the general research's theme park model. The theme park is defined as a directed graph in which the spatial segments are represented as nodes and each segment is connected by directed edges. The segments have one of five types defined as A(Attraction), R(Road), P(Plaza), En(Entrance), and Ex(Exit). The visitor agents transit these segments according to the directed edges. The simulation proceeds while the agents go around each attraction in the theme park. However, because the complex network which applies in this study is an undirected graph, the theme park model changes a directed graph into an undirected graph. Similarly, the spatial segments in the theme park represent it only at one En/Ex segment and the others are A segments. Figure 1 shows the general research's graph representation of the theme park and this research's one.

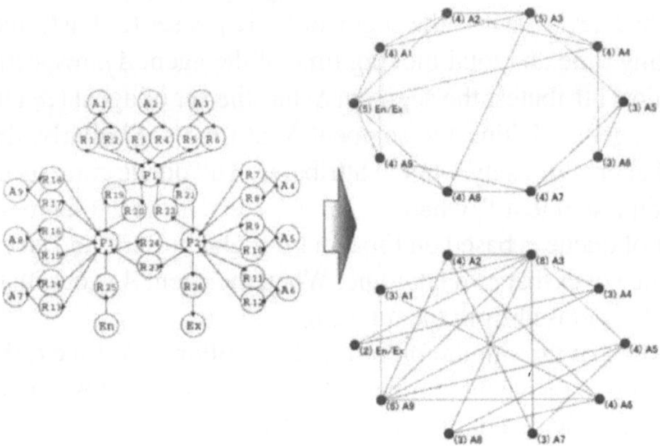

**Fig. 1** The general research's graph representation of the theme park and this research's one. The numbers in parentheses in the right figures are the number of edges of the node

The theme park model in this study is constructed from $N$ spatial segments a particular kind of service for the visitor agents and $M$ edges denoted as roads which connect each segment and $n$ visitor agents.

The segment $S_i (1 \leq i \leq N)$ has three static attributes, $T_i, c_i$, and $st_i$. Each type of segment is characterized by these parameters. The parameter $T_i$ represents one of two types defined as one En/Ex segment and the others are A segments. The parameter $c_i$ represents the service capacity of segment $S_i$, which means the maximum number of visitors that can be served in the segment $S_i$ at once. The parameter $st_i$ is the service time for the visitor agents. The agents require $st_i$ times to receive its service in the segment $S_i$. Similarly, the edge denoted as roads $R_k (1 \leq k \leq M)$ also has two static attributes, $c_k$ and $st_k$. The agents require $st_k$ times to move through the edge $R_k$.

In the theme park, $n$ visitor agents visit A segments through the edges. The visitor agent $A_j (1 \leq j \leq n)$ has the preference value $p_{ji}$ which represents the agent $A_j$'s preference degree for the segment $S_i$. For example, the higher value of $p_{ji}$ indicates the agent $A_j$ strongly hopes to visit the segment $S_i$. The En/Ex segment is set to 0. Each A segment is set to the value of $p_{ji}$ depending on the number of the attractions at random with no duplication every agent. The popularity of attraction segments dose not exist in this case.

The dynamical definition of the model is as follows. Let $t$ be the time step of the simulation. The simulation is iterated until $t$ reaches the termination time $t_{max}$. The agent $A_j$ has five dynamical attributes, $cs_j, pt_j, vt_{ji}, wt_j$, and $mt_j$,. $cs_j$ represents the current segment where the agent $A_j$ is on at time $t$. The agent $A_j$ belongs to only one segment any time, and starts from the En/Ex segment at the beginnings of the simulation. The past time $pt_j$ represents how long the agent $A_j$ spends in the segment $S_i$. This variable is increased by one when the time step of the simulation proceeds. The visiting time $vt_{ji}$ is the number of times the agents $A_j$ has visited the segment $S_i$. If the agent $A_j$ has already visited the segment $S_i$, $vt_{ji}$ is set to 1. Otherwise, if the agent $A_j$ has never visited the segment $S_i$, $vt_{ji}$ is set to 0. $wt_j$ and $mt_j$ represent the total waiting time and total moving time of the agent $A_j$, respectively.

As dynamical attributes, the segment $S_i$ has the set of agents $a_i$ and the queue $q_i$. $a_i$ consists of agents visiting the segment $S_i$ at time $t$. Similarly, the edge $R_k$ also has the set of agents $a_k$ as dynamical attributes. The queue consists of agents which desire to visit the segment $S_i$ when the service capacity of the segment $S_i$ is full. The priority order of queue is based on First-In First-Out buffers, and an earlier agent has prior admittance over that of a later one. When the agent $A_j$ goes inside the segment $S_i$, the agent $A_j$ is erased from the queue $q_i$.

The procedure of the simulation proceeds as follows. At time $t$, the agent $A_j$ acts in turn according to the agent number. The agent $A_j$ can choose the next segment to transit to if the following condition is satisfied.

$$pt_j \geq st_{cs_j} \tag{1}$$

The condition indicates that the service of current segment $S_i$ has finished. Otherwise, the agents $A_j$ spends more time in the current segment until the service has finished.

Next, suppose that the agent $A_j$ satisfies the above condition and chooses the segment $S_l$ as the next one. In the simulation, the next segment is indicated by the coordination algorithms explained later. If the following condition is also satisfied, the agent $A_j$ can transit from the current segment $S_i$ to the next segment $S_l$.

$$|q_l| = 0 \; and \; |a_l| + 1 \leq c_l \tag{2}$$

or,

$$|q_l| \neq 0,\ |a_l|+1 \leq c_l,\ and,\ the\ priority\ of\ the\ agent\ A_j\ is\ the\ first\ in\ q_l \tag{3}$$

The notation $|\cdot|$ means the cardinality of elements. However, if the segment $S_l$ is not a destination but a passing point to reach the destination, the agent $A_j$ transits to the next edge connected to the destination without spending the service time $st_l$ in the segment $S_l$ and without standing in the queue $q_l$. If the segment $S_l$ is the destination, the agent $A_j$ transits to according to the above condition (2) or (3). In addition, the agent $A_j$ updates the visiting time $vt_{jl}$ to spend the service time $st_l$ in the segment $S_l$.

When the agent $A_j$ moves to the next segment after satisfies of the condition (2) or (3), the total wait time of the agent $A_j$ is update as

$$wt_j = wt_j + (pt_j - st_{cs_j}) \tag{4}$$

Moreover, if the $cs_j$ is an edge, the total moving time is updated as

$$mt_j = mt_j + st_{cs_j} \tag{5}$$

After updating, the agent $A_j$ moves to the segment $S_l$ by changing the current segment $cs_j$ to $S_l$, and $pt_j$ is reset to zero. If the agent doesn't satisfy condition (2) or (3), the agent $A_j$ is added to the queue $q_l$ and waits until the segment $S_l$ is available for the agent $A_j$. Similarly, if the $A_j$ moves to the edge $R_l$, it is performed by the similar processing.

## 2.2 Evaluation

If $t$ reaches termination time $t_{max}$, the simulation is stopped. The evaluations $WT$, $MT$ and $P$ of the agent $A_j$ are calculated as follows.

$$WT = \frac{1}{n}\sum_j wt_j \tag{6}$$

$$MT = \frac{1}{n}\sum_j mt_j \tag{7}$$

$$P = \frac{1}{n}\sum_j\sum_i p_{ji} \cdot vt_{ji} \tag{8}$$

$WT$ and $MT$ represent the average of the total waiting time and the total moving time of the agent $A_j$, respectively. $P$ represents the average of the total $A_j$'s satisfactions according to the preference $p_{ji}$ and the visiting time $vt_{ji}$. Otherwise, it doesn't have a meaning that the agent $A_j$ visits the segment $S_i$ many times because if the agent $A_j$ has already visited the segment $S_i$, all $vt_{ji}$ are set to 1.

In this study, the evaluation $AFT$ is calculated as follows when the simulation has just been finished completely because all agents have visited all attractions.

$$AFT = \frac{1}{n}\sum_{j} T_{when\ A_j\ has\ finished\ the\ simulation} \quad (9)$$

$AFT$ represents the average of the total times when $A_j$ has finished the simulation.

## 2.3 Problem Settings

The theme park model setting applied the experiment premises the common setting condition which has a few uncertain elements. The theme park model setting is set as each element setting which is able to compare the experiment results more easily. That is, the total number of nodes in the graph $N$ are set to uniformity, and the total number of edges in the graph $M$ is set as $M = 2N$. Moreover, the service capacity of En/Ex segment and each road are set to $\infty$. The service capacity of all attractions are set to 1. To assume the real theme park, it is hard to happen to the situation which is crowded so that visitors cannot move between each attraction. In addition, the general theme park has the space large enough so that visitors can go around each attraction in the theme park more smoothly. Thus, it is appropriate that the service capacity of each edge $c_k$ is set to $\infty$. Because it is general that the entrance and the exit are common in the real theme park, it is made the segment En and the segment Ex the same segment En/Ex. The service time of each edge $st_k$ and each attraction $st_i$ is given the value according to the normal distribution to assume the real theme park. The normal distribution has the probability density function which is represented by equation (10). However, $\mu$ is the mean, the median, and the mode. $\sigma^2$ is the variance. The normal distribution represents as $N(\mu, \sigma^2)$.

$$f(x) = \frac{1}{\sqrt{2\pi}\sigma}\exp\left(-\frac{(x-\mu)^2}{2\sigma^2}\right) \quad (10)$$

The service time of each edge $st_k$ denoted as the moving time is given as $N(\mu, \sigma^2) = (50, 20)$. $st_k$ is given about between 10 to 90, and $\mu$ is 50. Similarly, The service time of each attraction $st_i$ is given as $N(\mu, \sigma^2) = (30, 10)$. $st_i$ is given about between 10 to 50, and $\mu$ is 30. Each element setting is represented in Table 1.

**Table 1** Each element setting applied the experiment

| Segment | $st$ | $c$ |
|---|---|---|
| En/Ex | 0 | $\infty$ |
| Edge(Road) | $10 \sim 90$ | $\infty$ |
| A | $10 \sim 50$ | 1 |

In addition, each experiment is carried out as the number of agents $n = 400$, the termination time $t_{max} = 10000$. $ST$ represents the simulation trials. The simulation is proceeded as $ST = 100$, and all results are shown as averages of $ST$ trials. The preference values $p_{ji}$ of the agent $A_j$ were randomly set in each trial.

## 3 Coordination Algorithm

For mass user support, a coordination algorithm provides the agents with theme park information and tries to control a sequence of agents' action for optimizing individual utility and social welfare. It is constructed that simple reactive coordination algorithms with no specific prediction and reservation system to simplify the dynamic complicated problem. The coordination algorithm informs each agent of the next segment after the service on current segment has finished. The algorithm can utilize the information about $c_i, c_k, st_i, st_k, p_{ji}, q_i, cs_j$, and $vt_{ji}$. It is assumed the situation that the algorithm obtains the information $p_{ji}, cs_j$, and $vt_{ji}$ from visitors' PDAs or cellular phones. Measures $q_i$ by observing each attraction's condition. $c_i, c_k, st_i$, and $st_k$ are given by the theme park setting.

For the theme park problem, four kinds of coordination algorithms are prepared in this study. Algorithm G and CA are suggested by the previous research, and Algorithm PE and CE are suggested by this research to increase the efficiency and the versatility. Each agent always follows the information provided by these algorithms.

**Greedy Algorithm (G)**

1. If the destination of the agent $A_j$ is not determined or the agent $A_j$ reaches the destination, this algorithm chooses the unvisited attraction whose preference is highest value among the unvisited attraction segments. If no unvisited attraction exists, En/Ex segment is set to the destination. The shortest path from the current segment to the destination is found by Warshall-Floyd's algorithm.
2. Inform the agent of the next segment that are on the shortest path.

This algorithm merely makes the agent select the most preferred A segment. Namely, each agent independently visits with unconscious of the others and theme park information aggregated from whole agents' behavior. This algorithm might represent no-coordination case.

**Congestion-Avoidance Algorithm (CA)**

This algorithm utilizes an alternative way to choose the destination segment for it of algorithm G. This algorithm chooses the unvisited attraction whose expected wait time $w_i$ is lowest value among the unvisited A segments. The value of $w_i$ is calculated with the simple equation $w_i = |q_i| \times st_i/c_i$ based on the current time data. Other procedure is same as algorithm G. This algorithm uses a part of information aggregated by whole agents' behavior, and this type of assist for a real theme park is easy to realize because it merely broadcasts congestion information or expected wait time.

**Priority-Emphasis Algorithm (PE)**

To determine the destination, this algorithm calculates the following value $pe_{ji}$ for each unvisited attraction, and selects the segment which has the highest value of all unvisited attraction.

$$pe_{ji} = \frac{st_i/c_i \times p_{ji}}{|q_i| + 1} \tag{11}$$

It is shown that the attraction which has a higher value of $st_i/c_i$ tends to be crowded by the additional experiment. In particular, the latter the simulation proceeds, the more it is crowded. To improve the efficiency of the agents and the whole theme park, this algorithm chooses the attraction which has a higher value of $st_i/c_i$ and $p_{ji}$, and a lower value of $|q_i|$, preferentially. Other procedure is also same as algorithm G.

**Cost- Emphasis Algorithm (CE)**

To determine the destination, this algorithm calculates the following value $ce_{ji}$ for each unvisited attraction, and selects the segment which has the highest value of all unvisited attraction.

$$ce_{ji} = \frac{p_{ji}}{|q_i| \times st_i/c_i + mc_i} \tag{12}$$

To improve the efficiency of the agents and the whole theme park, this algorithm chooses the attraction which has a higher value of $p_{ji}$, and a lower value of $|q_i| \times st_i/c_i$ and $mc_i$ denoted as the waiting cost and the moving cost. $mc_i$ represents the moving times from the current segment to the destination. This algorithm determines the segment which costs the lowest value of the waiting cost and the moving cost as the destination. Other procedure is also same as algorithm G.

## 4 Computer Simulation

### *4.1 Experiment 1*

In the experiment 1, the computer simulation is executed with several model settings that are combinations of the five complex networks {complete graph, regular graph, small-world network, random graph, scale-free network} and each node setting $\{N = 10, 20, \ldots, 70\}$. This simulation is proceeded to use the algorithm CA. The other detailed parameter settings in this experiment obey problem settings in section 2.3. It is shown the influence, the factor and the environment that each model setting affects the coordination algorithm by the experiments. In addition, it is shown that what kind of network structure is desirable to build the model setting so that the agents go around the theme park more efficiently. To execute the experiment 2, it is shown that the model setting condition which doesn't affect the coordination algorithm by the network structures. Figure 2 shows the results in Experiment 1.

## 4.2 *Experiment 2*

In the model setting condition which was provided from the experiment 1, the comparison experiment is carried out to develop the more versatile coordination algorithm which doesn't depend on network structures. The comparison experiment is carried out by the coordination algorithm G and CA which are used by the general research and the suggestion algorithm PE and CE which is developed based on data provided by the additional experiment in the model setting condition. The other detailed parameter settings in this experiment obey problem settings in section 2.3. Figure 3 and Figure 4 show the results in Experiment 2.

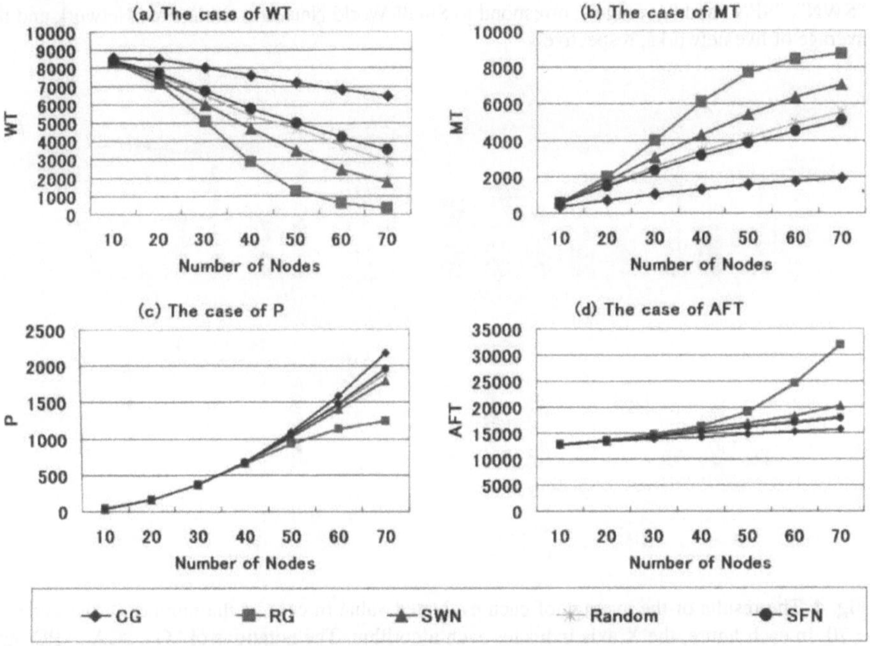


**Fig. 2** The results of the average of each evaluated value. In each figure, the X axis indicates the number of nods. The Y axis indicates the average of each evaluated value. The notations of "CG", "RG", "SWN", "Random" and "SFN" correspond to the network Complex Graph, Regular Graph, Small-World Network, Random Graph and Scale-Free Network, respectively

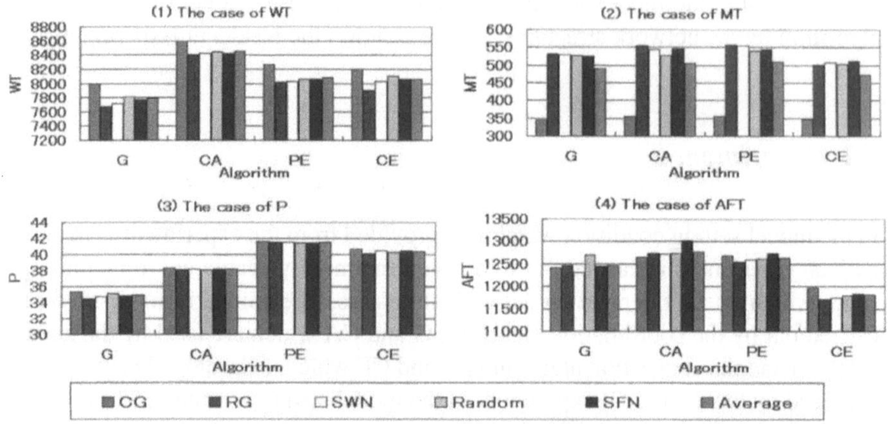


**Fig. 3** The results of the average of each evaluated value in case of the number of the nodes N = 10. In each figure, the X axis indicates each algorithm. The notations of "G", "CA", "PE" and "CE" correspond to the algorithm Greedy, Congestion-Avoidance, Priority-Emphasis and Cost-Emphasis, respectively. The Y axis indicates the average of each evaluated value. The notations of "SWN", "SFN" and "Average" correspond to Small-World Network, Scale-Free Network and the average of five networks, respectively

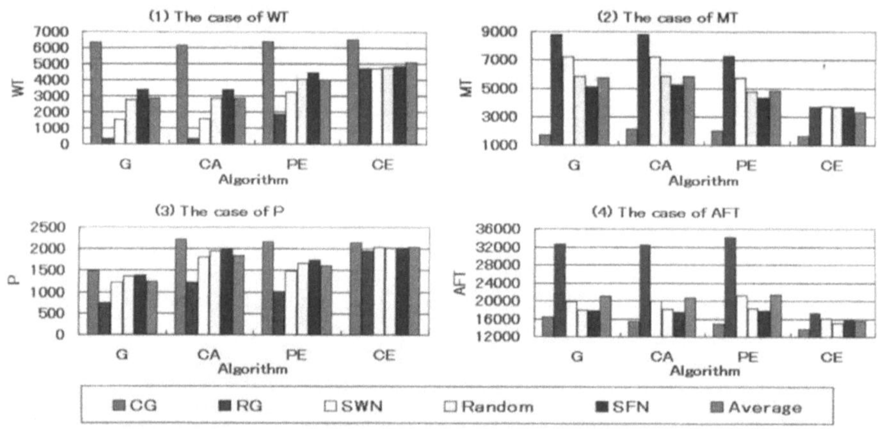


**Fig. 4** The results of the average of each evaluated value in case of the number of the nodes N = 70. In each figure, the X axis indicates each algorithm. The notations of "G", "CA", "PE" and "CE" correspond to the algorithm Greedy, Congestion-Avoidance, Priority-Emphasis and Cost-Emphasis, respectively. The Y axis indicates the average of each evaluated value. The notations of "SWN", "SFN" and "Average" correspond to Small-World Network, Scale-Free Network and the average of five networks, respectively

# 5 Discussion

## 5.1 Discussion about Experiment 1

Figures 2 (a) and 2 (b) show the average of the total waiting time WT and the moving time MT of the agent $A_j$, respectively. According to the increase of the number of nodes, WT gradually decreases and MT gradually increases in all networks, antithetically. These results mean that the number of attractions which each agent wants to visit increases according to the increase of the number of nodes since the number of agents is always defined as n = 400. Thus, each agent spreads each attraction easily. Because each segment's queue is also spread easily and each agent can go around each attraction without spending more times, WT decreases. To go around more attractions, agents spend MT instead of decreasing WT, and MT increases. It seems the above reasons why WT gradually decreases and MT gradually increases, antithetically. Next, let discuss WT's decreasing rate and MT's increasing rate in each network. It seems that the network which has these late lower hardly depends on the number of nodes and the model setting, and hardly affects the coordination algorithm.

Figure 2 (c) shows the average of the total satisfactions P of the agent $A_j$. According to the increase of the number of nodes, P gradually increases in all networks similarly. Agents can visit a lot of attractions according to the increase of the number of attractions and the above reasons why WT gradually decreases and MT gradually increases, antithetically. However, only the regular graph gradually decreases that increasing rate from the number of nodes N = 40. The average path length of the regular graph gradually increases according to the increase of the number of nodes. The agents have to spend the more moving time to reach the destination in the regular graph. In addition, when the number of nodes is less than 40, it seems that each result isn't so much of a difference in each network.

Figure 2 (d) shows the average of AFT of the agent $A_j$. According to the increase of the number of nodes, AFT gradually increases in all networks similarly. Only the regular graph sharply increases those increasing rate from the number of the nodes N = 40. As the results of Figure 2 (c), this result can be explained by that case.

## 5.2 Discussion about Experiment 2

According to the results of experiment 1, when the number of nodes is less than 40, each result isn't so much of a difference in each network because the average path length isn't so much of a difference in each network. In particular, the number of nodes N = 10 is the lowest of a difference in each result. It is assumed that the number of nodes N = 10 as the model setting condition which doesn't affect the coordination algorithm by network structures. In contrast, It is assumed that the number of nodes N = 70 as the model setting condition which affects the coordination algorithm very much by network structures. It is carried out the comparison experiment to develop the more versatile coordination algorithm which doesn't depend on network structures in each network as N = 10 and N = 70, respectively. Figure 3 and Figure 4 show the average of each evaluated value of each coordination algorithm in each network in case of N = 10 and N = 70, respectively. In case of N = 10, each result isn't so much of a difference in each network. In case of N = 70, each result is so much of a difference in each network, antithetically. It is shown several effectiveness of the suggestion algorithm PE and CE in comparison with the general research's algorithm G and CA. In particular, the suggestion algorithm CE shows highly versatile effectiveness in case of N = 10 and N = 70, respectively.

## 6 Conclusions

This paper researches the theme park problems for mass user support. It is shown that the effectiveness, the efficiency, and the versatility of the coordination algorithm in the complicated model setting such as the complex network. To assume the real problems, it is desirable for the theme park model to build small-world networks and scale-free networks according to the results of experiment 1. According to the results of experiment 2, it is desirable to combine various information such as the visitor's preferences, the attraction's congestion and the moving cost to develop the effective coordination algorithm. To evaluate the effectiveness of the coordination algorithm qualitatively, it is desirable to build the simple network model with a little number of the nodes according to the results of these experiments.

## References


1. Hidenori Kawamura, Koichi Kurumatani, Azuma Ohuchi: "Modeling of Theme Park Problem with Multiagent for Mass User Support", Multi-Agent for Mass User Support, International Workshop, MAMUS Acapulco, Mexico, August 2003, LNAI 3012, Revised and Invited Papers pp. 48-69, 2003.
2. Koichi Kurumatani: "Mass User Support by Social Coordination among Citizens in a Real Enviroment", Multi-Agent for Mass User Support, International Workshop, MAMUS Acapulco, Mexico, August 2003, LNAI 3012, Revised and Invited Papers pp. 1-17, 2003.
3. Takashi Kataoka, Hidenori Kawamura, Koichi Kurumatani, and Azuma Ohuchi: "Distrbuted Visitores Coordination System in Theme Park", First International Workshop, MMAS 2004, Kyoto Japan, Revised Selected and Invited Papers, pp335-349(2004).
4. Takashi Kataoka, Hidenori Kawamura, Koichi Kurumatani, and Azuma Ohuchi: "Effect of Congestion Reduction with Agent's Coordination in Theme Park Problem", Proceedings of 4th IEEE International Workshop WSTST's 05(Soft Computing as Trandisciplinary Science and Technology), pp245-254(2005).
5. Watts, D.J. and Strogatz, S.H.: Collective dynamics of 'small-world' networks, *Nature*, Vol.393, pp.440-442(1998).
6. Barabási, A.-L. and Albert, R.: Emergence of Scaling in Random Networks, *Science*, Vol.286, pp.509-512(1999).

# Agent-based simulation to analyze business office activities using reinforcement learning

Yukinao Kenjo, Takashi Yamada, and Takao Terano

**Abstract** This paper attempts to clarify team behavior in cooperative organizations by agent-based simulations. We focus attention on both the roles of manages and the initiatives of staffs, and then model them using agent-based model concepts. This enables us to investigate phenomena in organizations at micro-level and macro-level. Besides, we formulate the task processing of each member in real organizations as learning for maze problem. The advantages of applying maze problem for our simulation model are as follows: It is possible to describe agents who acquire skills by reinforcement learning and to represent environmental uncertainty by changing block placements dynamically. Several computational experiments show what the whole organization behaves from microscopic points of view. At the same time, the authors confirm that the ability to adapt environments under uncertainty is different from the characters of organization.

## 1 Introduction

In recent years, globalization and fierce competitions have made companies change their forms and styles. Therefore researchers need to offer new organization theo-

Yukinao Kenjo
Department of Computational Intelligence and System Science, Tokyo Institute of Technology (Now at Sony Corporation), 4259 Nagatsuta-cho, Midori-ku, Yokohama, Kanagawa 226-8502, Japan, e-mail: yukinao.kenjo@gmail.com

Takashi Yamada
Department of Computational Intelligence and System Science, Tokyo Institute of Technology, 4259 Nagatsuta-cho, Midori-ku, Yokohama, Kanagawa 226-8502, Japan e-mail: tyamada@trn.dis.titech.ac.jp

Takao Terano
Department of Computational Intelligence and System Science, Tokyo Institute of Technology, 4259 Nagatsuta-cho, Midori-ku, Yokohama, Kanagawa 226-8502, Japan e-mail: terano@dis.titech.ac.jp

ries meeting these trends. This paper attempts to clarify organizational behavior in corporate bodies through an agent-based simulation approach.

Attempts to analyze organizational phenomena computationally began in the 1960s [7] and have evolved into the computational organization theory in recent years [3, 4]. For example, the garbage can model provides an operational explanation of "not necessarily rational decisions" made in group decision making situations [6]. It can be said that the garbage can model is a good work of explaining not necessarily rational decision making that may occur in group decision situations. However, the garbage can model is highly abstract and only touching the surface of phenomena occurring due to the complex group interactions within actual organizations.

Agent-based simulation has been researched in recent years by a number of fields beginning with social science, and its usefulness is recognized [2, 8]. Especially, agent-based simulation of organization is one of the most challenging fields and several models have been developed[1]. In this study, we will focus on two local organizational elements, the role of the superior and the independence of the subordinate. We believe that this makes it possible to carry out both micro and macro analyses of complex organizational phenomena.

The relationship between superior and subordinate is especially important amongst the elements that compose organizations. "Indifference index" and "identification index" are proposed from organizational theory in order to detail the relations above in the proposed model. On the one hand, indifference index represents conformity with orders given by a superior. A high indifference index represents a passive attitude within the organization. On the other hand, identification index indicates agreement between purpose and values of an individual, and those of the organization. A high identification index becomes the motivation to actively work within the organization. Introduction of these indexes makes it possible to implement a simulation that handles a variety of organizational forms observed in actual organizations.

Furthermore, the simulation is conducted assuming that dealing with maze problems by reinforcement learning is a task process. In other words, the task processing of every member within the actual organization is formulated as a maze problem learning. In this way, the following two points are seen as advantages of introducing maze problem learning: Firstly, reinforcement learning allows expression of agents acquiring skills and abilities. This is based on observations of actual society such as actively working members of the organization becoming able to process that work in a short period of time. Secondly, dynamically changing maze blocks allows expression of environmental uncertainty. This corresponds to, for example, situations in which the need arises to learn new work processing methods due to such happenings as technical innovation.

The rest of this paper is organized as follows: The next section explains the details of our simulation model. Section 3 presents some of the highlighted results taken from our proposed model. Finally, Section 4 discusses some concluding remarks.

---

[1] For more details, see Chang and Harrington JR [5].

## 2 Setup

Our simulation model deals with hierarchical team observed in general industrial organizational forms to analyze what group form is applicable in response to environmental change.

### *2.1 Definition of objects*

The main constituents of the proposed model and their attributes are defined in Table 1. "Organization Object" represents an organization and "Member Object" include members in the organization. "Desk Object" is individually owned by members and correspond to the place of task processing. "Task Object" represents all work content processed by members of the organization, and is referred to as a unit task in this study.

- Attributes of *organization objects* are related to the organization as a whole.

  Structure (organizational structure) manages authority structure of the organization.
  Finished (finished task group) retains finished tasks.

- Attributes of *member objects* play an important role in distinguishing members.

  Role stands for top, middle, or bottom level.
  State is a present condition.
  Intention means intent in regard to working in the affiliated organization.
  Interest is an attentiveness in the work content of the unit task handled.
  Identification (identification index) represents identification with superior. This is derived from intention attributes of the two.
  Indifference (indifference index) means how eagerly an agent is going to deal with the presently held tasks. That is derived from interest attributes and attraction attributes of Task.
  Knowledge (knowledge of unit task) retains work content knowledge of unit task. This includes Q-table of reinforcement learning.

- Attributes of *desk objects* are inherent to each member.

  Buffer (task group waiting for processing) retains a list of tasks presently held by members.
  Phone (calling flag) is used when called by its superior.

- Attributes of *task object* identify each task.

  MazeId means the id number labeled in each maze problem corresponding to this unit task.
  MazeListId is an ID of generated task (maze string).

**Table 1** Definition of objects

| Object | Attribute |
|---|---|
| Organization | structure (int[]) |
| | finished (Set<Task>) |
| Member | role (String) |
| | state (String) |
| | intention (Character[]) |
| | interest (Character[]) |
| | identification (double) |
| | indifference (double) |
| | knowledge (Map<mazeId, Strategy >) |
| Desk | buffer (List<Task>) |
| | phone (boolean) |
| Task | mazeID (integer) |
| | mazeListID (integer) |
| | attraction (Character[]) |

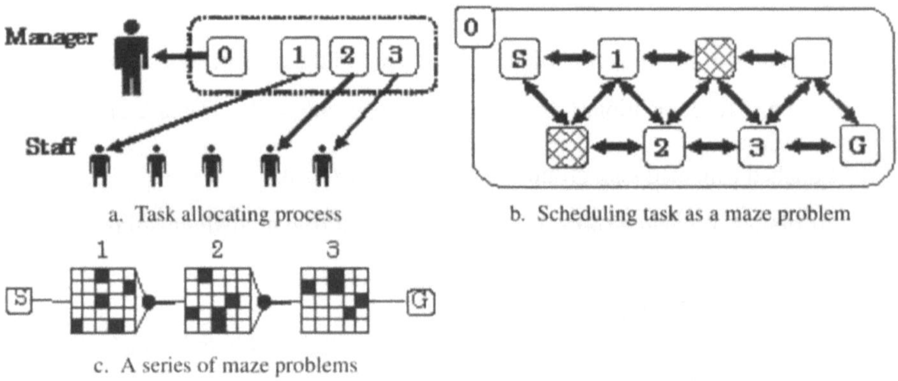


**Fig. 1** Processing flow of a maze list

Attraction is used to derive indifference attributes of a member. This represents the attraction of the work content of this unit task, set as a factor to induce members' interest.

## *2.2 Task processing and reinforcement learning*

This part of the section explains the main procedure of the proposed model:

1. Task Generation
   A task is generated within the organization at fixed periods. The generated task is piled on to a given managers desk. This task includes several unit tasks to

be handled by the manager's subordinates and is determined by the following algorithm.

a. A manager is chosen arbitrarily at fixed periods.
b. Only agents without subordinates are taken from the agent group corresponding to the subordinates under the manager chosen in the previous step.
c. Each agent taken in the previous step is drawn according to predetermined selective probability.
d. The unit task, handled by agents chosen in the previous step along with tasks from him handled by agents right up to managers chosen in the first step, is included in the generated task.

Since unit tasks are set as mazes, a generated task consists of two or more arranged maze strings. Unit tasks handled here are work content held by each agent, and signify unit tasks that must be processed by that agent.

2. Division and distribution of tasks
Tasks generated are broken down into partial tasks by the manager and allocated to the appropriate staff member who is called up to the desk (Figure 1a). The manager begins to process the single unit task left to him from the original task once all partial tasks have been allocated. This unit task, like the others, is a maze problem that must be processed by the manager, but is treated here as the manager solving the management scheduling problems of his subordinates (Figure 1b).
3. Learning
In this model, each agent carries out an appropriate action from the following three decision making rules at a certain time, namely interaction with others through 1) the decision making rule to manage the work of subordinates as a manager, 2) the one to follow the directions of the superior, or 3) the non-interactive one to process your own work. Although the agent at mid-level, namely middle management, is caught between the management of subordinates and directions from upper-level agents, behavioral rules as a staff member take precedence in this case. Interaction with others is given priority with regard to prioritization of each behavioral rule, and placed in staff-manager-individual order.
Their exists a unit task to be handled by each agent which is processed by learning maze problems. Maze problems are used extensively as the benchmark for reinforcement learning. The proposed model employs the particularly basic Q learning. The renewal formula for the Q value in Q learning is expressed as

$$Q(S_t, a_t) \leftarrow Q(S_t, a_t) + \alpha(r_t + \gamma \max_{a \in A} Q(S_{t+1}, a) - Q(S_t, a)), \tag{1}$$

where $S_t/a_t$ is the agent's condition/behavior, and $Q(S_t, a_t)$ is the value of selecting $a_t$ in regard to condition $S_t$. Moreover, $\alpha$, $\gamma$, $r_t$ are learning rate, discount rate, and compensation respectively. While agents processing unit tasks are to solve the maze step by step, the behavior selection technique employed is the $\varepsilon$-greedy method. This $\varepsilon$ value makes use of the indifference index which will be explained in the next section and is defined as

$$\varepsilon = 1 - indifference \tag{2}$$

where *indifference* takes the real value in [0, 1].
Therefore, an agent with a low indifference index becomes an exploratory agent. Unit task completion signifies arrival at the maze's goal, acquisition of compensation and a return to the starting point. In other words, processing one unit task is equivalent to moving through one level of maze problem learning.

4. Reports
Agents who have completed unit tasks or partial tasks given by their superior go to the superior's desk to report. They join the end on the line if other agents are waiting at the desk to report. Staffs called up during task distribution are to join the front of the line.
5. Task completion
Generated tasks and partial ones received from superiors are completed when unit tasks included have been processed. Therefore, tasks passed on to subordinates are explained as part of the critical path for the scheduling problem maze to be processed (Figure 1b). In other words, the given maze string is completed when a single path has been laid to the goal (Figure 1c). Partial tasks are reported to the superior upon completion, and generated tasks are registered on the processed task list when completed.

Incidentally, maze strings are produced by complex processes during generation of the above tasks. This results from modeling that dictates that the higher the level in the organization the more prolonged the job handled is to be, and the lower the level the shorter. The manager cannot reach process completion even if they have completed their own unit tasks until all partial task process reports are received from lower levels. Therefore, the higher the level of the manager who is given a task, the greater the number of agents involved and hence the longer the period. Moreover, choosing a manager ad lib induces the election of lower level managers in a pyramidal structure. As a result, lower level managers frequently handle short-term jobs, while high level managers incidentally handle long-term jobs.

## *2.3 Indifference index and identification index*

This study employs two significant indexes from organizational theory knowledge embedded as a distinguishing index of members composing the organization. The one is indifference index and the other is identification index. Indifference index is an indication of compliance with orders from superiors, and identification index is an indication of the extent of identification of the member's purpose and values against that of the organization. Members are classified into the following four types in accordance with the values of two indexes [9]:

type 1 passive instrument type

The passive instrument type which has both a high indifference index and identification index is seen as an agent that is loyal to organization orders while also sharing common purpose and values. Although they carry out given tasks, there is little self-initiated action. Additionally, motivation does not become a problem as they share common purpose and values with the organization.

type 2 alienated worker type

The alienated worker type has a high indifference index and low identification index. Although drawing the same line with the organization in terms of purpose and values, it is referred to as a public official or bureaucrat type as orders are followed in behavioral terms.

type3 problem solving type

The problem solving type which has a low indifference index and high identification index is not simply a loyal follower of orders, but rather makes decisions through problem identification from an organization point of view based on shared purposes and values.

type4 non-contributor type

The non-contributor type which has both a low indifference index and identification index not only shares no common purpose or values, but is also in-compliant to orders and can not be expected to act organizationally. Not very common in stably function actual organizations.

In this way, even though each constituent of the organization has several factors governing their environment, they are broken down by these two indicators.

The identification attributes and indifference attributes correspond to identification index and indifference index and are derived from interest/attraction attributes (Table 1). In this model, identification index is the consistency of intention attributes with superiors, and indifference index is the inconsistency of the attraction attributes of presently held tasks with one's interest attributes. The length of these intention attributes and interest/attraction attributes match respectively, and are expressed by an ad lib character string containing defined quantity characteristics. This is a tag from Axelrod's model "The Dissemination of Culture" [1], where a tag has number quantity properties for each length quantity characteristic.

Although identification index was conventionally an indicator of each member against the organization, it is used in the proposed model in terms of identification with the direct superior for greater micro-analysis.

In this model, identification attributes and indifference attributes are represented as real numbers in the domain [0, 1], and are derived from

$$identification = \frac{1 - distance(staff.intention,\ manager.intention)}{length}, \tag{3}$$

$$indifference = \frac{distance(staff.interest,\ task.attraction)}{length} \tag{4}$$

where $distance(tag1,\ tag2)$ is a function that seeks the hamming distance of $tag1$ and $tag2$ which have the same characteristics.

**Table 2** Parameters

| Parameter | | Value |
|---|---|---|
| Structure | | 13 |
| | (top level) | (1) |
| | (middle level) | (3) |
| | (bottom level) | (9) |
| Intention | Features | 30 |
| | Traits | 2 |
| Interest | Features | 10 |
| | Traits | 3 |
| Maze size | | 5 |
| Block rate | | 0.2 |
| Generate span | | 5 |

**Table 3** Identification and indifference values

| Agents | Identification | Indifference |
|---|---|---|
| Member 1 | 1.00 | 0.60 |
| Member 2 | 0.53 | 0.80 |
| Member 3 | 0.50 | 0.60 |
| Member 4 | 0.30 | 0.90 |
| Member 5 | 0.27 | 0.50 |
| Member 6 | 0.43 | 0.60 |
| Member 7 | 0.73 | 0.70 |
| Member 8 | 0.53 | 0.70 |
| Member 9 | 0.57 | 0.50 |
| Member 10 | 0.50 | 0.60 |
| Member 11 | 0.50 | 0.70 |
| Member 12 | 0.50 | 0.80 |
| Member 13 | 0.60 | 0.60 |

## *2.4 Environmental uncertainty*

In addition, the proposed model handles an environmental uncertainty as the variability of maze problems, namely we suppose that an organization under an increasing environmental uncertainty has more information to deal with in order to adapt its changes. We assume that how much there is environmental uncertainty is equivalent to the extent that the blocks in a maze are moved. This means that it could be possible to express environmental uncertainty by changing the characteristics of a task because a task for an organization is considered as the one with external information. In simulation experiments, we express this environmental uncertainty as the transfer of a block within the maze to a section without blocks at regular intervals.

## 3 Results

From here, we will present two computational results regarding the relations between the degree of environmental uncertainty and the performances of each members. All the parameters for this experiment are shown in Table 2. An organization where one top-level, three mid-level, and nine bottom-level members are active on three levels from the top-down is used as the organizational structure. In other words, this is a simple structure in which each agent has three subordinates. The agent name index is prioritized in term of width from the top of the organization down. With this parameter condition *Member 1*'s subordinates are *Member 2, 3,* and *4* and *Member i*'s subordinates are *Member* $3i-1$, $3i$, and $3i+1$ ($i = 2, \cdots, 4$). The length of the intention attribute tag associated with the identification index is 30, and the length of the interest/attraction attribute tag associated with the indifference index is 10. Maze problems of $5 \times 5$ in size with a block ratio of 0.2 are treated, and tasks (maze strings) are generated every five steps. The identification and indifference of each agent are shown in Table 3.

### 3.1 Basic experiment: under low uncertainty conditions

First, we implemented a simulation assuming a low level of environmental uncertainty, namely all the blocks in a maze were determined randomly but unchanged.

Panels a and b of Figure 2 are time series plots of how long it takes each member to finish his/her task allocated: While panel a show results of nine bottom-level members and panel b does those of three mid-level members. The vertical axis is set as the number of steps involved from when the task is given by the superior until the report is completed, and follows variations during each step. The superior and subordinate's id are given in the two legend brackets. Also, Figure 2c represents the state ratio of each member after 1000 steps. Moreover, the left panels of this figure are the results with the parameters in Table 3 and the right ones are those in case of $\varepsilon$-greedy method.

An apparent fact from Figure 2a is that *Member 5*'s task processing efficiency is strikingly poor due to a low identification index. Although unprocessed tasks are piling up on *Member 5*'s desk, he/she does not actively process tasks since he/she has a low identification with his/her superior, *Member 2*. Other agents seem to be carrying out routine work because their task processing cycle is broadly flat. Since it took *Member 7* and *Member 12* few steps to complete each task, we can consider them as high performers in this environment.

When checking the performances of mid-level members shown in Figure 2b, the behavior of *Member 2* is quite different from those of other two members. This is because *Member 2* has two different subordinates, a high performing *Member 7* and a poorly motivated *Member 5*. If a given task to the *Member 2* has a sub-task for the *Member 5*, then the *Member 2* has to be kept waiting for the completion of the *Member 5*. Conversely, tasks handed to the *Member 7* return quickly because he/she has a high task processing ability. This misfit is represented as this kind of rectangular wave.

Those situations are also observed from Figure 2c, namely the distinguishably low motivation of *Member 5* is represented by this result. More than half of all work time is spent shirking, or avoiding responsibility. In short, the *Member 5* has a strong 'non-contributor' tendency. Furthermore, this panel shows that time allocated to the management of subordinates by middle management *Members 2, 3*, and *4* is more distinguishable than expected, and it can be seen that they have almost no free time. When reconsidering Figure 2b with this in mind, the task processing period is ever-increasing on a management level even though each agent is task learning. The cause is a short task generation cycle, and it can be said that this result reflects the high-load of middle management in the organization.

The above result successfully clarified a part of organizational activity, but it is difficult to tell whether maze problem learning is functional because the $\varepsilon$ value is relatively large. Therefore, we implemented another experiment using the $\varepsilon$-greedy method, the value of $\varepsilon$ was set to zero. Which means, the agents are trying to solve their maze exploratively.

*Member 5* is again the bottom level agent as was the case with the right panel of Figure 2a. However, an improvement due to learning can be seen in the task processing efficiency of agents excluding *Member 5*. This processing efficiency arises solely from the identification index as only knowledge utilization is relied on without any exploration. It can be said that *Member 5* with his low identification index has few learning opportunities, as processing improvement cannot keep pace with task generation. Additionally, agents were more exploratory when attacking maze problems in the first experiment than this one as the $\varepsilon$ value is set by the method described in the proposed model. This is represented by the fact that the work time of most agents is longer in the left panel of Figure 2c results than those in the right panel of the same figure. This comparative experiment shows that exploratory attributes work negatively when there is no maze variation and low uncertainty, resulting in more time tied down by work.

## *3.2 Advanced experiment: increases in uncertainty*

Next, we will describe an experiment in order to analyze the effects of environmental uncertainty. In other words, an experiment was conducted where maze blocks varied in a given cycle. The parameters used in this experiment are the same as those of the previously mentioned basic experiment, only that the task generation cycle is doubled, assuming an organization of reduced load. Furthermore, an element is added in which a maze block held by each member changes position every 500 steps during the simulation period. The simulation result is shown in Figure 3.

The time series performances of bottom and mid-level agents ( Figure 3a, b) became more complex and fluctuated. Especially, it is worth noting that the behavior of *Member 7*, a high performer in the previous experiment, is much different, namely

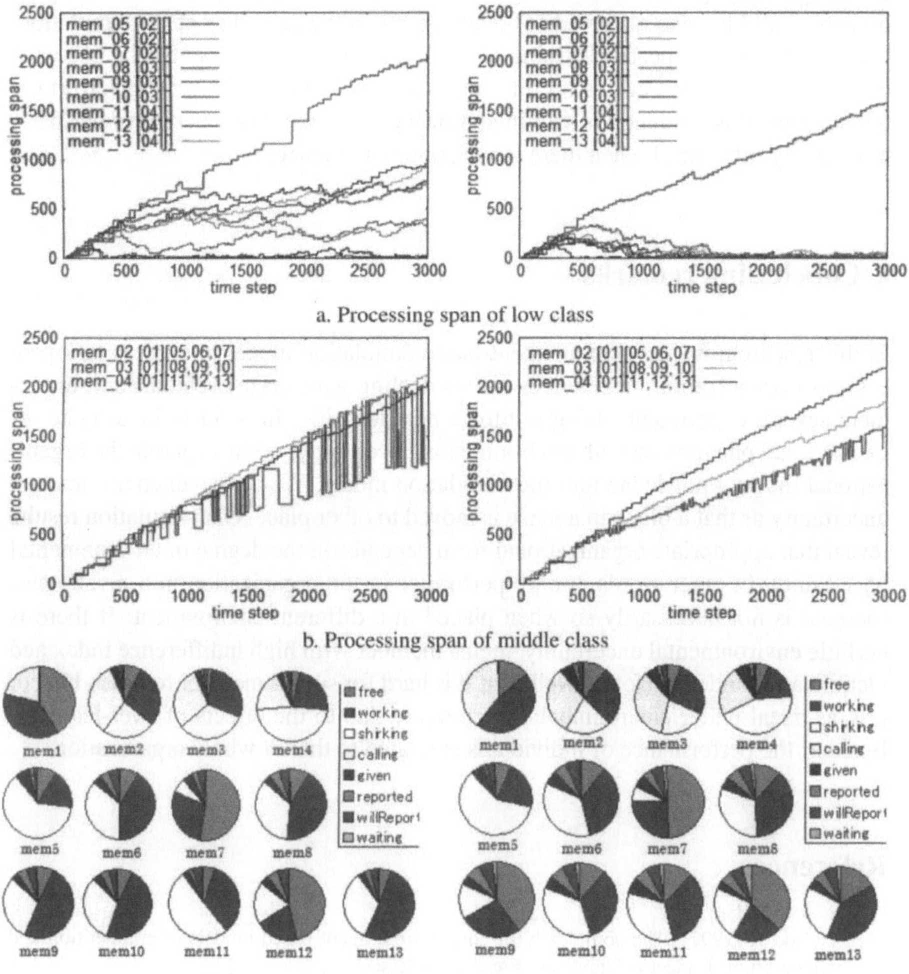


a. Processing span of low class

b. Processing span of middle class

c. Task processing situation of each member after 1000 steps

**Fig. 2** Comparative performances (left panel: non-$\varepsilon$-greedy method, right panel: $\varepsilon$-greedy method)

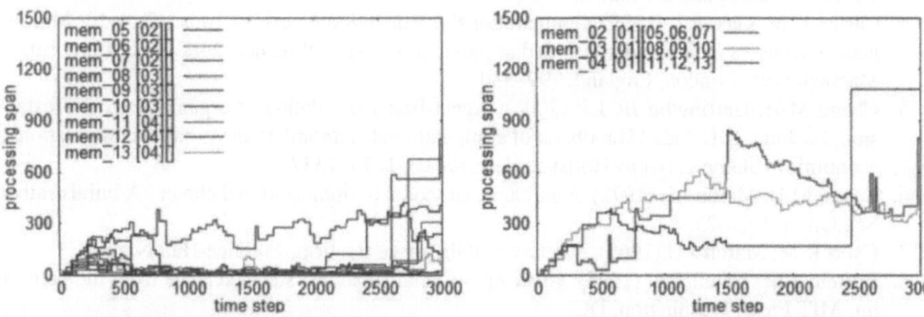


a. Performance of the low class

b. Performance of middle class

**Fig. 3** Comparative performances after adding maze change parameter his/her processing time increases rapidly around the 2700 step mark. These phenomena could be also observed for other agents in the later half of the simulation. This is because a repeated learning led to over-learning and thus he/she becomes confused when facing even a bit different maze. In short, the higher the performer, the stronger this tendency is within stationary organizations. Therefore, he/she is more likely influenced when there is variation in the maze.

## 4 Concluding remarks

In this research, we develop an agent-based simulation model to express task processes of organization members as agents dealing with maze problems and implement several experiments using reinforcement learning. In order to investigate organizational phenomena with environmental uncertainty, we incorporate the organizational theory knowledge into our simulation model. We define an environmental uncertainty as that a block in a maze is moved to other place. Our simulation results reveal that appropriate organizational form depends on the degree of environmental uncertainty. In other words, a high performer in the organization in a given environment is not necessarily so when placed in a different environment; If there is no/little environmental uncertainty, then a member with high indifference index and identification index performs well. But it is hard for such a member to finish his/her task as usual under environmental uncertainty due to the effects of over-learning. Besides, the performance of individuals is related to that of whole organization.

## References

1. Axelrod R (1997) The complexity of cooperation: agent-based models of competition and collaboration. Princeton University Press, New Jersey
2. Axelrod R, Cohen M D (2000) Harnessing complexity: organizational implications of a scientific frontier. Free Press, New York.
3. Carley K M (1995) Computational and mathematical organization theory: perspective and directions. Compu. and Math. Org. Th. 1:39–56
4. Carley K M, Gasser L (1999) Computational organization theory. in Weiss, G. (ed.): Multiagent systems - a modern approach to distributed artificial intelligence. MIT Press, Cambridge, Massachusetts London, England, 299–330
5. Chang M-H, Harrington JR J E (2006) Agent-based simulation of organization. in Tesfatsion, L., Judd, K.L. (eds.) Handbook of computational economics: agent-based computational economics volume 2, North-Holland, Netherlands, 1273–1337
6. Cohen M D, March J G (1972) A garbage can model of organizational choice. Administrative Sci. Quart. 17:1–25
7. Cyert R M, March J G (1963) A behavioral theory of the firm. Prentice-Hall NJ
8. Epstein J M, Axtell R L (1996) Growing artificial societies: social science from the bottom up. MIT Press, Washington, DC
9. Takahashi N (1992) An evaluation of organizational activation. Omega: Int. J. Manag. Sci. 20:149–159

# Fundamentals of Agent-Based and Evolutionary Approaches

# A Model of Mental Model Formation in a Social Context

**Umberto Gostoli**

Department of Economics, Università Politecnica delle Marche
Piazzale R. Martelli, 8, 60121 Ancona – Italy
u.gostoli@univpm.it

**Abstract.** This paper presents a model of learning in a context of a relatively large population interacting through random bilateral matching to play a bilateral game in strategic form. While the theory of learning in games commonly assumes that the players can observe only the strategies chosen by their opponents, in this paper is introduced the additional assumption that the players are characterized by phenotypic traits observable by the other players with whom they interact. The extension of the traditional framework allows to introduce a more sophisticated and cognitively plausible expectations' formation model than the ones proposed so far. In particular, this paper introduces a new model of the induction process through which the agents build mental models that take the form of lexicographically structured decision trees.

**Keywords:** Theory of Learning in Games, Categorization, Social Stereotyping, Fast and Frugal Heuristic Theory, Self-Organization, Data Mining.

## 1 Introduction

Any dynamic or repeated game is characterized by a particular information structure that defines what the players know before the game starts and what they can observe during the stages, or periods, of the game. Moreover, any game is, implicitly or explicitly, characterized by a certain information processing algorithm that defines and represents the players' cognitive skills. The equilibria reached, as well as the dynamics that leads to them, depend on the information structure of the game and on the decision process through which the players compute the information at their disposal.

It is commonly assumed that the players know the payoff matrix of the game, can observe in every period the action of their opponent and, moreover, that all the players know that the other players have the same prior knowledge and observational skills, that all the players know that the other players know and so on. These are the so-called prefect and common knowledge assumptions that come from the traditional static game theory, whose formal application to the structure of the game can lead to technical and philosophical problems (Binmore, 1997).

While the information available to the players has been considered by the means of formal assumptions, less attention has been paid to the explicit formal definition of the cognitive process through which this information is computed by the players to reach a decision in order to the action to undertake. However, as behavioural game theory shows, the differences between the outcomes of experiments and the predictions of game theory are often due to the unrealistic computational power and rationality assumptions of the latter (Camerer, 2003).

The theory of learning in games is the branch of the literature that introduced models of players less than perfectly rational and that explicitly formalized, together with the information structure of the game, the decision process that underlies the player's action. While game theory tells us which Nash equilibria a particular game has, the theory of learning in games gives us models that determine the path through which a certain equilibrium is reached. In these models, the equilibrium is the outcome of a process in which less than fully rational players grope for optimality over time (Fudenberg and Levine, 1999).

The approaches to learning in games can be divided in two big classes: reinforcing and forecasting learning models. The first class includes models where the players choose a strategy on the basis of its past performance. The imitation models belong to this class: the players are endowed with the skill to asses their neighbours' success and to associate this performance to a particular strategy, unsophisticated but not trivial cognitive skills. The second class is represented by models where the players are endowed with the cognitive skills to develop a forecasting model of the behaviour of other players and, so, to choose the strategy that is the best reply to their opponent's expected action.

The *fictitious play* models belong to this last class. In general, in these models players are endowed with the basic skill to observe and memorize their opponent's action. This cognitive skill allows the agents to keep track of the relative frequencies with which each strategy is played. These relative frequencies represent the players' expectation about the strategies' distribution of the population.

In the model proposed in this paper the information structure of the game is extended to include the players' skill to observe their opponent's phenotype, represented by a string of three binary attributes. Given the information processing skills that characterize the fictitious play models, this extension would allow the players to develop expectations conditional on their opponents' phenotypes.

Of course, this extension is not new in the field of the theory of learning in games. Previous papers have introduced models where the agents are endowed with a visible tag composed by one binary attribute (Axtell, Epstein and Peyton

Young, 1999) or by three binary attributes (Hoffmann, 2006). These works demonstrated that the agents' capacity to develop expectations conditional on their opponents' type leads to the formation of social classes. In other words, in these models we can observe the endogenous emergence of social stereotyping.

The main difference between the papers mentioned above and this one is the cognitive process through which the information at the agent's disposal is processed. While in the former papers the agents develop conditional probabilities over opponents' play conditional on the particular type they are playing against, according to the classical fictitious play algorithm, in this paper is proposed a model of information processing algorithm inspired by a recently introduced decision process theory: the Fast and Frugal Heuristic Theory (FFHT).

The FFHT proposes a lexicographic decision process, that is, a process through which the attributes are looked up in a particular order of validity by the means of a decision tree. Figure 1 shows a lexicographic decision tree with which is possible to classify instances, identified by three binary attributes, in the two classes A and B. In front of the situation 0-1-1, the agent would classify it as A: the third attribute (identified with the number 2) has value 1, so it has to consider the second node, containing the first attribute (identified with the number 0). Its decision process does not need to proceed further because, being this attribute's value equal to 0, the decision tree ends with a leaf, containing in this case the class A. The model presented in this paper is based on the assumption that the agent's mental model is structured as a decision tree.

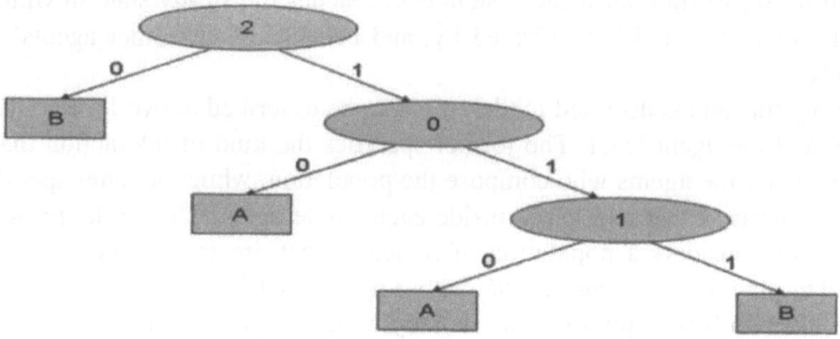


**Figure 1**

The FFHT is based on the bounded rationality paradigm: because of mind's limitations, real decision makers have to use approximate methods to handle most tasks. The function of these methods, that we will call *heuristics*, is to make reasonable, adaptive inferences about the social and physical world given limited time and knowledge. In this situations, indeed, real agents cannot perform the cognitively demanding multiple linear regression, but, more plausibly, they look up only few available cues in a fixed order of validity (Gigerenzer and Goldstain, 1996).

The cues relevant in the decision-making process and the order in which they have to be considered can be learned in three main ways: they could be genetically coded, they can be learned through cultural transmission or they can be learned from the agent's experience. In the model presented in this paper, the number and the kind of cues the agents can detect are genetically determined but the decision tree by which they are hierarchically ordered to form their mental model is the result of an induction process that the agents perform on their experiences' database.

According to FFHT, in an evolutionary context, decision-making strategies are selected for their accuracy, frugality and speed, measures that relate the decision process to the environment in which the decision has to be taken. In fact, the FFHT introduces a new concept of rationality called *ecological rationality*: a heuristics, or mental model, is rational from an ecological point of view if it allows the agents to take effective decisions in the environment in which the agents live. This means that in order for a given heuristics to be rational it has to match the information structure of the environment in which that particular heuristics is used.

In a social context, however, the environment of each agent is represented by other agents. So, the mental model of each agent is rational from an ecological point of view if it matches the mental model of the other agents with whom it interacts. The main aim of this paper is to find out if a social system formed by agents that make heuristic inferences about their opponent's behaviour can reach an equilibrium, that is, a situation where the mental model that each agent develops through its experiences allows it to make ecologically rational decisions given the mental models developed by the other agents with whom it interacts. The model presented in this paper is a model of *co-adaptation* of mental models with which we try to find out if the system ever reaches the steady state in which each agent's mental model is confirmed by, and because of, the other agents' mental models.

Being this an agent-based model, it has to be described at two levels: the population and the agent level. The former specifies the kind of interaction that takes place among the agents who compose the population, while the latter specifies the decision process that take place inside each single agent. The model presented in this paper considers a population of $N$ agents that are in each period randomly paired to play the *Stag Hunt Game*, whose payoff matrix is shown in Table 1. This game has two Nash equilibria: the socially optimal equilibrium S – S and the risk-dominant equilibrium H – H. In each period the agent has to take a decision about which strategy to play: it will play S if it expects its opponent to play S, it will play H otherwise. This game has been chosen because it represents the typical coordination problem present in many social contexts: if the agents trust each other, they can coordinate on the social optimum, otherwise they will be trapped in the sub-optimal equilibrium.

Table 1: the Stag Hunt Game

| | S | H |
|---|---|---|
| S | 2,2 | 0,1 |
| H | 1,0 | 1,1 |

To understand how expectations form and evolve, we have to describe the model at the level of the agent. Regarding the prior knowledge of the agents, this model adopt the limited knowledge paradigm that characterizes the models adopted by the theory of learning in games: the agent knows only its own payoff matrix and does not know what its opponent gets from the interaction.

Each agent is characterized by phenotypic traits represented by three binary attributes. This means that the population is composed by eight different phenotypes: 000, 001, 010, 011, 100, 101, 110, 111. Moreover, each agent can observe the phenotypic traits of its opponent, together with the action its opponent performs. So, after each game, the phenotypic traits and the action of the opponent is stored in the agent's memory M, its size being a parameter of the model. An example of an agent's experiences database could be the matrix shown in Table 2.

Table 2: an agent's experiences database

| Attribute 1 | Attribute 2 | Attribute 3 | Strategy |
|---|---|---|---|
| 0 | 1 | 0 | S |
| 1 | 0 | 0 | S |
| 1 | 1 | 1 | H |
| ... | ... | ... | ... |
| 1 | 0 | 1 | H |
| 1 | 1 | 0 | S |
| 0 | 0 | 0 | H |
| ... | ... | ... | ... |
| 0 | 1 | 1 | H |
| 0 | 1 | 1 | H |
| 0 | 0 | 1 | S |

This matrix shows that in its last game, the agent met an opponent whose phenotype was 001 and who played strategy S. Of course its own phenotype and action have been stored at the same time in the experiences database of its opponent. We call *s* the agents' memory size, that is, the number of experiences the agent can store in its memory. When the number of periods $p = s$, the agent's memory is full new experiences are stored according to the FIFO system: the oldest experience in the database is discarded to make room for the last experience.

Having defined the agent's gathering and storing cognitive skills, we come to the information processing algorithm that underlies the agent's decision. We call $l$ the number of periods that constitutes the agent's *learning period*, its length being another parameter of the model. For the first $l$ periods the agent has not yet developed a mental model that allows it to forecast its opponent's move, so it makes a random forecast and chooses the action that is the best response to this forecast. We have to notice that, in general, to best respond to a random expectation is not the same of making a random choice: in the first case the agent never chooses strongly dominated strategies, an event that, instead, is possible in the second one.

At the end of the learning period the agent builds, through an induction process performed on its experiences database, a mental model that allows it to forecast, from its opponent's phenotype, the strategy its opponent will play. An example of an agent's mental model can be the decision tree of Figure 1, interpreting the letters A and B in the rectangles as, respectively, the two strategies S and H. If an agent with this mental model meets an opponent whose phenotype is, for example, 011, the agent's forecast of its opponent's move would be S. The mental model represents, in fact, a set of hierarchically organized behavioural rules of the kind *If/Then* that maps each phenotype to a strategy to be chosen when meeting an agent with that phenotype.

From the cognitive point of view, we have to distinguish two different processes: the *induction process* through which the agent develops its mental model on the basis of its experiences, and the *decision process*, that takes place when the agent forecasts, with its mental model, its opponent's move. In other words, with the induction process the agent makes the mental model, with the decision process it puts it into use. While the decision process takes place in each match, the induction process can also be performed once every given number of periods $r$, this number being another parameter of the model.

In this model, this cognitive process is modelled by the means of data mining techniques. Let's consider the database of Table 3, containing 20 instances. The database is formed by three independent variables that define each instance (V1, V2, V3) and one dependent variable representing the class to which the instance belongs (C). The first and the third independent variables are binary variables: they can have either value 0 or 1. The second independent variable can have three values: 0, 1 or 2. The instances can belong either to class 0 or to class 1. The first step of the data mining process is to find out the first independent variable of the decision tree. In order to do so, we have to compute the *average information value* for each independent variable. According to the information theory, the information value of a database is 1 minus the *entropy* of the database. The database's entropy is defined as the number of bits required to specify the class of an instance given that it belongs to the database. The entropy value can go from 0, if we the information that an instance belongs to the database is all we need to classify the instance, to 1, if the fact that the instance belongs to the database do not gives us any useful information about its class.

Table 3: a database

| | V1 | V2 | V3 | C |
|---|---|---|---|---|
| 1 | 0 | 2 | 1 | 1 |
| 2 | 0 | 1 | 0 | 1 |
| 3 | 0 | 2 | 0 | 1 |
| 4 | 1 | 2 | 1 | 0 |
| 5 | 0 | 1 | 1 | 0 |
| 6 | 1 | 0 | 1 | 0 |
| 7 | 1 | 2 | 1 | 1 |
| 8 | 0 | 0 | 0 | 1 |
| 9 | 0 | 0 | 0 | 0 |
| 10 | 1 | 0 | 1 | 1 |
| 11 | 1 | 1 | 0 | 0 |
| 12 | 1 | 1 | 1 | 0 |
| 13 | 1 | 1 | 0 | 0 |
| 14 | 1 | 2 | 0 | 1 |
| 15 | 0 | 1 | 0 | 1 |
| 16 | 1 | 1 | 1 | 0 |
| 17 | 0 | 1 | 0 | 1 |
| 18 | 0 | 1 | 0 | 1 |
| 19 | 0 | 2 | 0 | 1 |
| 20 | 1 | 0 | 1 | 1 |

If we call $N$ the total number of instances in a database composed by a variable that can be classified in $r$ classes and $n_i$ the number of instances belonging to class $i$, the entropy $E$ of a database is given by (1):

$$E = -\frac{n_0}{N}\log_r\frac{n_0}{N} - \frac{n_1}{N}\log_r\frac{n_1}{N} - \ldots - \frac{n_i}{N}\log_r\frac{n_i}{N} - \ldots - \frac{n_r}{N}\log_r\frac{n_r}{N} \qquad (1)$$

The decision tree we get at the end of this data mining algorithm will be the decision tree that maximizes the number of instances that are correctly classified or, in other words, that minimizes the number of the exceptions. This decision tree can also be defined as the theory that best describes the database or, alternatively, the most likely theory given the database.

From a general point of view, the model described above is a feed-back process: the agents' experiences determine the agents' mental model which determine the agents' behaviour which determines the agents' experiences and so on. The main aim of the simulations, whose results will be presented in the following section, is to find out if this process ever reaches the steady state where the agents' behaviour produces experiences that confirm the agents' mental model.

## 4 Simulations' Results

In this section are presented the results of a particular simulation that will serve us as example, results that are, however, representative of the outcome of simulations based on the social and cognitive model presented above. In fact, making the agents in their learning period random forecasts, different simulations will produce results that are different in their details, but, nevertheless, we can see that they share general statistical features.

In the simulation, the population is formed by 1000 agents who, in each period, are randomly paired to form 500 couples to play a one-shot stag hunt game. At their birth the agents are assigned a randomly generated phenotype and a memory, which is randomly drawn from a normal density function having average 40 and variance 4. The learning period for each agent is ¼ of her memory. This means that each agent will form a mental model after its memory has been filled for ¼ of its size. Then, the mental model will then be updated after every game (r = 1).

The graph in Figure 2 shows the *forecasting performance* of the population of agents, measured as the number of agents that, in each period, forecast correctly their opponent's choice. From the beginning of the game till the end of the learning period this performance will be around 50%, that means that half of the population has a correct expectation about the opponent's behaviour, because of the random guess the agents make during the learning period. Then, as the agents develop and update their mental model, we expect this performance to increase until it reaches, eventually, the equilibrium, a state where 100% of the agents have correct expectations about their opponent's strategy.

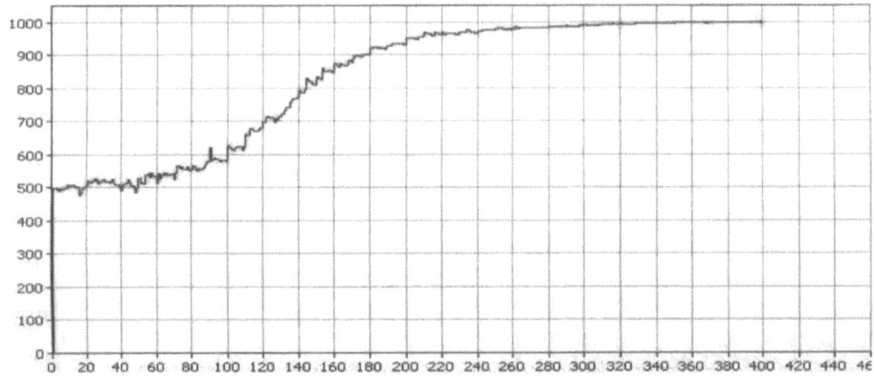


**Figure 2**: forecasting performance

Even if the exact timing of the forecasting performance growth changes from one simulation to the other, the common feature of all simulations is that it always reaches 100%. This means that the system always reaches the mental model equilibrium, where the expectations of each agent of the population consistently match the other agents' expectations. After period 350, the experiences the agents make are consistent with their mental model and, consequently, the evolution of their mental models reaches a steady state. At this point, for each agent its mental model is the 'true' mental model because it allows it to forecast perfectly its opponent's behaviour in every game.

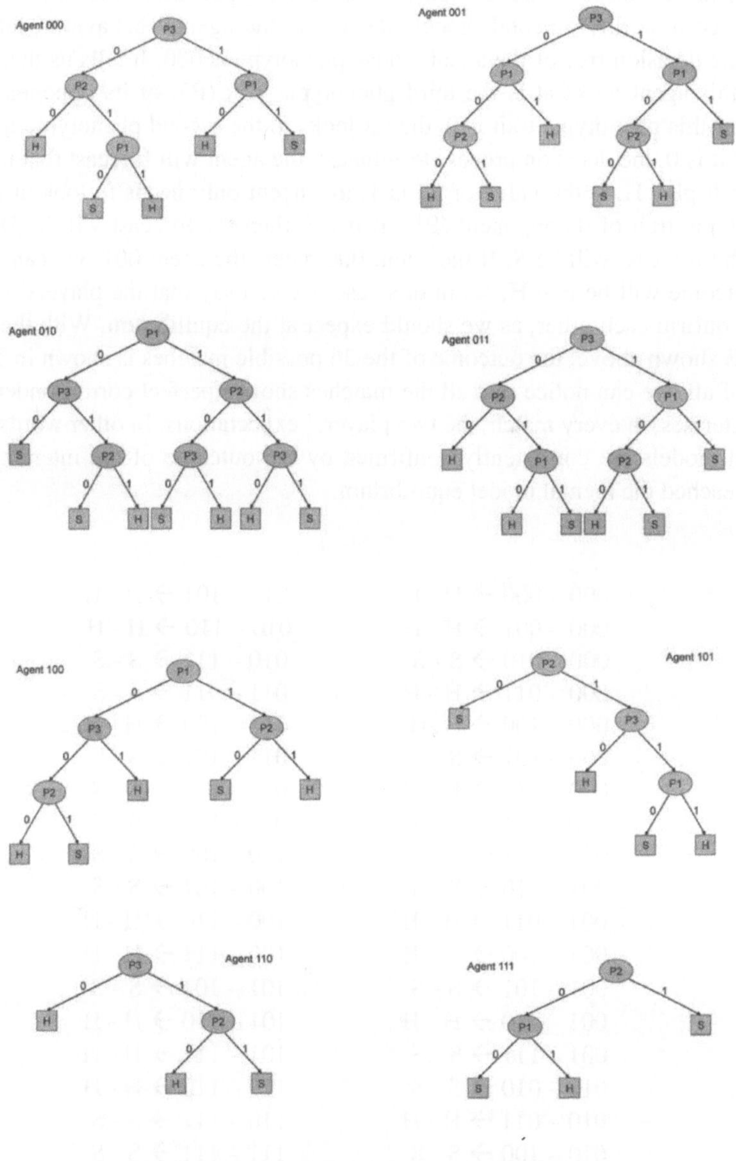


**Figure 3**: mental models at the equilibrium

The mental models that characterize the equilibrium are a peculiar character of each simulation because they depend on stochastic fluctuations that take place during the learning period, when each agent makes random forecasts. The eight decision trees of Figure 3 show the mental models of a sample of eight agents, one for each phenotype, after the system has reached the mental model equilibrium. The agents use these mental models to forecast its opponent's behaviour.

To see how these mental models determine the agents' behaviour, let's consider the decision tree of the agent whose phenotype is 000. It tells us that the fist thing this agent looks at is the third phenotypic trait (P3) of its opponent. If the value of this phenotypic trait is 0, then it looks to the second phenotypic trait (P2) and, if it is 0, the decision process terminates: the agent will forecast that its opponent will play H. If the value of P3 is 1, this agent only needs to look at the first phenotypic trait of its opponent (P1): if it is 0 then the forecast will be H, otherwise the forecast will be S. If the agent 000 meets the agent 001, we can see that the outcome will be H – H, so, in this case, we can say that the players' expectations confirm each other, as we should expect at the equilibrium. With the mental models shown above, the outcome of the 36 possible matches is shown in Table 4. First of all, we can notice that all the matches show a perfect correspondence that characterizes, in every match, the two players' expectations. In other words, all the mental models are consistently confirmed by the outcome of the interaction: we have reached the mental model equilibrium.

Table 4

| | |
|---|---|
| 000 – 000 → H - H | 010 – 101 → H - H |
| 000 – 001 → H - H | 010 – 110 → H - H |
| 000 – 010 → S - S | 010 – 111 → S - S |
| 000 – 011 → H - H | 011 – 011 → S - S |
| 000 – 100 → H -H | 011 – 100 → H - H |
| 000 – 101 → S - S | 011 – 101 → S - S |
| 000 – 110 → H - H | 011 – 110 → S - S |
| 000 – 111 → S - S | 011 – 111 → S - S |
| 001 – 001 → S - S | 100 – 100 → S - S |
| 001 – 010 → S - S | 100 – 101 → S - S |
| 001 – 011 → H - H | 100 – 110 → H - H |
| 001 – 100 → H - H | 100 – 111 → H - H |
| 001 – 101 → S - S | 101 – 101 → S - S |
| 001 – 110 → H - H | 101 – 110 → H - H |
| 001 – 111 → S - S | 101 – 111 → H - H |
| 010 – 010 → S - S | 110 – 110 → H - H |
| 010 – 011 → H - H | 110 – 111 → S - S |
| 010 – 100 → S - S | 111 – 111 → S - S |

Secondly, 19 of the 36 matches are characterized by the S – S strategy equilibrium, while the remaining 17 matches are characterized by the H – H equilibrium.

This means that, at the population level, in each period around 53% of the agents play the strategy S and around 47% play the strategy H. This represents the strategy equilibrium that determine the overall efficiency reached by the system in this particular simulation. While in all the simulations the system reaches the perfect internal coherence among the mental models of the agents, the level of efficiency reached by the population is path-dependent and so varies from one simulation to the other.

Third, we can notice that this equilibrium is characterized by a particular average payoff distribution: while the agent with the phenotype 111 plays the strategy S in 6 of the eight possible matches it can have, with an average payoff of 1.75, the agent with the phenotype 110 plays S only in 2 matches, with an average payoff of 1.22. Again, the average payoff distribution among the agents' types depends on the particular simulation run.

# 5 Conclusions

While the specific results of the simulation presented in the previous section depend on variables and parameters determined through stochastic algorithms and, consequently, change from one simulation to the other, we can point out some general characteristics of the dynamics of the social and cognitive model presented in this paper.

First of all, the mental model equilibrium, and the consequent strategy equilibrium, are reached in *all* the simulations, after around 300 periods. This means that, given the assumptions of the model, the emergence of a social stereotyping system, that is, a set of socially formed and evolved beliefs that tend to confirm and strengthen each other, is statistically a very likely phenomenon. These beliefs are not true from an *absolute* point of view: at the beginning of the simulation the phenotypes attached to each agent do not have any influence on the agent's behaviour. However, by playing repeatedly the game, the agents develop beliefs about their opponents' behaviour that become true from a *social* point of view: they are true because the agent's opponent holds beliefs that make them true. In other words, the model presented in this paper suggests a mechanism for the endogenous emergence of social conventions, defined as the equilibrium reached by a system of beliefs that evolves until it reaches a state of internal coherence.

We have to notice that, even if the particular social convention that the system develops depends on stochastic events, or historical accidents, nevertheless, once it gets established it is in the interest of each single agent to follow it. In other words, even if would be in the interest of each agent to change the convention to reach the socially optimal equilibrium, the situation where all the agents choose S, an agent that would decide not to follow the convention would be worse-off in a system where the other agents follow it. This makes the convention a steady state from which it is almost impossible to escape, unless subsets of the population decide collectively to change it.

Secondly, the strategy equilibrium that characterizes the social convention is not the social optimum but it is not the worst social outcome, represented by the risk-dominant equilibrium, neither. In fact, the simulations show that the proportion of interactions S – S goes from 40 to 60% of the total. However, we have seen that each equilibrium is characterized also by an unequal average payoff distribution among the various phenotypes. This fact would have important consequences in an evolutionary setting, where the phenotypes having the higher average payoff would tend to grow in the population, a dynamic that would lead to the socially optimal outcome, where all the agents play Stag.

Finally, from a general point of view, these simulations show that, if we give up the perfect and common knowledge paradigm that characterizes the classical game theory to embrace the bounded rationality paradigm of the theory of learning in games, the dynamics and the equilibrium eventually reached by the system depend crucially on the assumptions about the cognitive skills of the agents, cognitive skills that, in order to build models that have some positive or normative value, need to be empirically justified. One of the aim of this paper has been the proposal of an agent that is one step closer to the cognitive sophistication of real agents than the agent that have populated the fictitious play models so far. The additional assumption adopted in this paper is that the agents' behaviour is based on the mental model that takes the form of a decision tree through which the agents analyse the input in a lexicographic way.

## References


1. Axtell RL, Epstein JM, and Young PH (2001) The Emergence of Economic Classes in an Agent-Based Bargaining Model. In: Durlauf S, Young HP (eds) Social Dynamics. MIT Press, Cambridge, Mass.
2. Binmore K (1987) Modeling Rational Players - Part I. Economics and Philosophy 3: 179-214.
3. Camerer C (2003) Behavioural Game Theory. Princeton University Press, Princeton
4. Fudenberg D, Levine DK (1998) A Theory of Learning in Games. MIT Press, Cambridge, Mass.
5. Gigerenzer G, Goldstain DG (1996) Reasoning the Fast and Frugal Way: Models of Bounded Rationality. Psychological Review 103: 650-669
6. Hoffmann R (2006) The Cognitive Origins of Social Stratification. Computational Economics 28: 233-249
7. Shannon C (1948) A Mathematical Theory of Communication. Bell System Technical Journal 27: 379-423, 623-656

# A Thought on the Continuity Hypothesis and the Origin of Societal Evolution

Kazuhisa Taniguchi

## 1 Introduction

The purpose of this paper is to discuss the continuity hypothesis in Witt [10] and to consider the origin of societal evolution. The continuity hypothesis which is addressed in Witt [10] is an assumption that evolution of life and society are continuing and evolutionary change continues beyond the range of what can be explained by Darwin's theory of evolution. In addition to this point, Witt [10] has claimed that the assumption gives the ontological basis of evolutionary economics.[1] According to the continuity hypothesis, societal evolution and organic evolution must be differently separated from each other, because, since these evolutions are separated, it can be insisted that there is continuity between them. So where is the continuing point of organic evolution and societal evolution? Where can be found the range which can not be explained by Darwin's theory of evolution? This is the first question of this paper and the author holds that the rhetoric in the continuity hypothesis is artful and it attracts readers, though some parts of the rhetoric seem odd.

The birth of Homo sapiens with an identical anatomical structure to ours is dated at around 100,000 years ago, however around 30-60,000 years ago, that is to say a mere 1500-3000 generations ago, something happened to Homo sapiens living in Africa, following which Homo sapiens spread out the globe. This step is referred to as "Great Leap Forward"[2]. At this time there was

Faculty of Economics, Kinki University
3-4-1 Kowakae, Higashiosaka, Osaka 577-8502 Japan

[1] Prof.Witt wrote, "Somewhere in the history of the humankind there is, thus, a point where the power of Darwinian evolutionary theory for explaining (economic) behavior ends. But evolutionary change continues beyond that point - only with different means and in other forms. I call this the assumption of an "ontological continuity of evolution" which sets the frame for approach to evolutionary economics in this book." U. Witt [10] p.3.

[2] "Great Leap Forward" is quoted from Diamond [2].

a large evolutionary step in Homo sapiens society, the final result of which is our existence today. So what actually happened during this great leap forward? What made this leap possible? For those familiar with mainstream economics, this kind of question is recognized as not usually addressed in the study of economics and seems to be beyond the scope of economics. It is, however, a fundamental question when one looks for the origins of societal and economic evolution. Hayek gave us the following words:

> If we wish to free ourselves from the all-pervasive influence of the intellectual presumption that man in his wisdom has designed, or ever could have designed, the whole system of legal or moral rules, we should begin with a look at the primitive and even pre-human beginnings of social life.[3]

However, just as Homo sapiens' muscles and voice have not left behind any traces, the important phenomena of customs, conducts and rules do not remain as evidence as do fossils and stone implements. And so, in addition to economics, investigation into this type of problem relies on research into cultural anthropology, physical anthropology, evolutionary biology and other disciplines; however philosophical enquiry is also vital. Hayek has the followings words on the subject:

> The facts about which we know almost nothing are the evolution of those rules of conduct which governed the structure and functioning of the various small groups of men in which the race developed. On this study of still surviving primitive people can tell us little. Though the conception of conjectural history is somewhat suspect today, when we cannot say precisely how things did happen, to understand how they could have come about may be an important insight.[4]

With such viewpoints, this paper asks: "how could this have happened?" "In order for the great leap forward to occur what must have happened?" This is the second question of this paper. The author assumes that the answer is the emergence of rules within groups of Homo sapiens, and that this is also the origin of societal and economical evolution. The origin and cause of the evolution of society are therefore considered, and at the same time we give thought to the meaning and significance of the continuity hypothesis.

## 2 The Continuity Hypothesis

*"The Evolving Economy"*[5] consists of a long introductory chapter and 19 published papers. The continuity hypothesis is described in the introductory chapter and four papers published earlier. In describing the continuity hypothesis, Prof. Witt first discusses Darwin's theory of evolution; he indicates that the main concept of economic evolution was obtained from the theory

---

[3] F. A. Hayek [5] p.73.

[4] F. A. Hayek [6] p.156.

[5] U. Witt [10]

of Darwinism independently. Seeing 'domain-unspecific feature of evolution', he has defined evolution as follows.

> *Evolution is the self-transformation over time of a system under consideration.* In this definition, the term 'transformation' means a process of change governed by regularities. The prefix 'self-transformation' points to the endogenous sources and causes of novelty. Self-transformation can be split into two logically (and usually also ontologically) distinct processes: the *emergency* and the *dissemination* of novelty. It is in these two processes, I think, that we are faced with two characteristic, domain-unspecific feature of evolution.[6]

Following this introduction, he talks about the continuity hypothesis as the ontological basis of economic evolution. He begins his theory with the following typical example.

> Darwin sailed around the world with the Beagle and recorded many undiscovered species. Particularly, he observed the avian species in the Galapagos Islands in the Pacific Ocean almost unspoiled by humans. This had a strong impact on him and on his identification of species. However, what can we discover on the Galapagos Islands in these modern days less than 200 years after Darwin's visit? We discover not new biotic species but artificial materials such as cottage, roads and landing fields. No signs of any genetic program, which indicate the appearance of those artificial materials, can be discovered. How can such human artefacts be explained, if no sings are observed? What kind of evolution has been occurred there? Darwin's theory of evolution can never answer this question. Because evolution of human economic activities had been shaped by natural selection at the early stage of human history, but another form of evolution has become dominant as evolutionary process. Evolution occurs continuously beyond the scope of what Darwin's theory of evolution can explain.[7]

He described the continuity clearly as follows.

> It (Natural evolution) has therefore shaped the ground and still defines the constraints for man-made, or cultural, evolution. In this sense, there is, thus, also an ontological continuity despite the fact that the mechanisms and regularities of cultural evolution differ from those of natural selection.[8]

Darwin described the evolutionary process by natural selection as a tree of life.[9] Does the continuity hypothesis assume that it is possible to attach the tree of societal evolution to the tree of biological evolution as if it were a graft? Assume that the z-axis is a time axis, the x-axis is the diversity of biological species, and its phylogenic tree is drawn on the x-z-plane. Then, setting the diversity of economies on the y-axis, can a stereoscopic phylogenic tree be drawn, which connects the base of the human economic evolutionary tree continuously to the biologic evolutionary tree? Does Prof. Witt describe continuity in such sense?

---

[6] U. Witt [10] p.13.

[7] U. Witt [10] p.3. and pp.15-8.(Summary)

[8] U. Witt [10] p.15. The sentence in parentheses was written by the author.

[9] C. Darwin [1] pp.98-9.

The example of the Galapagos Islands used by Witt [10] to explain the continuity hypothesis is very interesting and effective; however, it also seems to be just odd. The number of endemic species discovered in the Galapagos Islands is so great because they have evolved uniquely in geographical isolation. Obviously, cottage and roads in the Galapagos Islands example did not evolve by themselves on the islands. This example might be fitting to explain the continuity hypothesis, if aliens who cannot distinguish the difference between animate and inanimate beings came down to earth, then the aliens might think a bird evolved into an air plain and a fish evolved into a submarine. However, this is the odd case clearly. By citing the example of cultural evolution in an uncivilized society without any contact with the outside world, the example of continuity hypothesis is given and the different sign of evolution such as diversity in languages should be discussed.[10]

## 3 In What Way did the Great Leap Forward Occur? -Emergence of rules of conduct in small roving bands-

Threats, warnings and such like, pressure on other parties, advice, instructions for seeking shelter and such forms of simple communication that take place between group members can also be observed in various animal species. However, as the ability to hypothesize, and expressions of temporal distinction between past, present and future all require a high level of intellect, it is possible to say that these polished skills appeared at a different time from the appearance of simple communication of volition. In the same way, it is easy to imagine that the innate factors giving rise to behavioural rules, such as escaping when sensing danger, arose at a different time from abstract rules, like justice. Having said this, as was already stated at the beginning, the emergence of rules and the evolution, just like language itself, do not leave traces in the way that fossils do. Hayek has written extensively on the origins of rule generation in "the extended order"[11], and thus we will proceed with our investigations here based on his writings.

Behaviour related to the senses of closeness and of fear is seen in advanced animals such as mammals and fowl, however finding these senses being common between different individuals requires quite a degree of mental ability. If agent A perceives the sense X (e.g. the feeling of closeness) and agent B perceives the sense Y (the same feeling of closeness), in order for agent A

---

[10] Darwin tells that the family tree of ethnic groups can be classification of languages. Using the example of language to show this viewpoint regarding classification might be effective. If a perfect family tree of humans is made, the systematic arrangement of ethnic groups gives the best classification to various languages currently spoken in the world. (Darwin [1] p.345.)

[11] F. A. Hayek [7] Chapter 2 and 8.

to recognize the signals put out by agent B it is necessary for agent A to observe these signals, whether they be facial expressions or gestures. According to Hayek, in order to recognize the behavioural patterns of others by one's own innate sensing, it is necessary to first of all understand one's own behavioural pattern as the mould of the behaviour pattern for the others. In addition it is necessary to have the ability to recognize the other agent's behavioural styles and patterns - that is Gestalt perception is required. Perception here is not the perception of special forms or shapes, but the ability to understand, even for a different agent, that a particular different situation belongs to a certain same kind of situations.[12] Initial simple rules of conduct most likely came about from the innate senses of closeness and fear. However several stages along the evolutionary process should have been taken before certain innate senses being common were understood.

Through understanding of another agent having similar senses and feelings by comprehending both one's own feelings and the changes of the other agent's expression or posture, this common understanding recognition spreads throughout a band and then band members become to be able to react to the situation according to a predetermined behavioural pattern. By the use of facial expression and gestures, or by the high level means of expression through language, the situation that, even though no one in the band knows the reason or cause for the situation being indicated, any one in the band understands the meaning of the signal and knows how to respond to the situation is raised. With repetition, this kind of behaviour in a particular situation becomes more routinized, and rules of conduct concerning who plays what role when responding to a given situation emerge. For example, rules regarding respect at a first meeting require previous knowledge of the sense of what ill feeling is like, and requires knowledge of by what kind of gestures ill feeling is expressed. As a result of this, it is possible, with a smiling expression, to show that there is no intent to do harm to the other agent. For another example, the rules surrounding reciprocity and gift giving require understanding of the sense of the other agent's happiness, and it is necessary to understand what kind of gesture expresses this. Routine conduct is transmitted through generations by learning and education, and this gives solidity to rules of conduct and customs.

Where can we find the origin of this kind of rule and the ultimate origin of the societal evolution? Hayek believed that these origins lay further back than the appearance of Homo sapiens in the Homo genus range.[13]

> This (Cultural evolution) took place not merely after the appearance of *Homo sapiens*, but also during the much longer earlier existence of the genus *Homo* and its hominid ancestors.[14]

---

[12] F. A. Hayek [3] Section 2.

[13] Within the hominids, there is the Homo genus as well as the Australopithecus genus. Within the Homo genus, there are Homo sapiens, habilis and erectus. Within Australopithecus there are boisei, robustus, ethiopicus, afanensis and others.

[14] F. A. Hayek [6] p.156. The parentheses were added by the author.

However, directly after this, Hayek states that even if one looks back to the birth of the Homo genus to find the origins of cultural evolution, the most important point of cultural evolution is that it occurred only in humans.

> The most important part of cultural evolution, the taming of the savage, was completed long before recorded history begins. It is this cultural evolution which man alone has undergone that now distinguishes him from the other animals.[15]

Chimpanzees are capable of recognizing and distinguishing between uncles, aunts, nephews and nieces. The hominids is said to have forked off from the animal anthropoid ape 5,000,000-7,000,000 years ago and also lived in small-scale blood related bands.[16] Within these small bands there were no strangers. The size of the band was at most numbered somewhere in the tens, and no bands would have passed the hundred mark.[17] These bands were likely made up of no more than five generations, comprising parents, grandparents, aunts and uncles, brothers and male cousins, children and grandchildren. If the individual member can not recognize the others as one's family through the dead member's relations in the family, there is a limit to the number of individuals within any given band. According to Diamond [2], Homo sapiens lived in small blood related groups until around 30,000-50,000 years ago. The great leap forward occurred around 50,000 years ago. Given that the number of people within the small groups did not increase, for Homo sapiens to spread out the globe it would have been necessary for the number of small groups to increase, and it is thus thought that Homo sapienses began its spread across the globe as small, blood related bands.[18] At some stage in the existence of the small blood related bands the unique behavioural patterns and rules of Homo sapiens arose and began their evolution. There were probably unique behavioural patterns and rules in Homo species other than Homo sapiens; however they were replaced by the sapiens species through the process of natural selection.[19] It is probable, looking at ourselves nowadays, to imagine that countless peculiar unrecognizable rules should have been arisen at this time.

Hayek strongly emphasizes the distinction between the rules of conduct controlling individual band member's conduct and the order within the band composed of these members. Rules of conduct control the conduct of individual members within a band; however natural selection works upon the whole group as composed by all the individual members.[20]

---

[15] F. A. Hayek [6] p.156.

[16] Footprints that show upright walking are from 3,600,000 years ago.

[17] According to Hayek [6](p.160), around 15-40 individuals. Diamond [2] puts this figure at 80.

[18] Homo sapiens left Africa and crossed the Bering Strait around 15,000 years ago.

[19] Neanderthal man is also classed as Homo sapiens; however they were exterminated around 30,000 years ago.

[20] F. A. Hayek [4] p.66.

As a result, there are two different paths when rules of conduct are transmitted through small bands. One path is through vertical transmission. In other words, the original primitive rules are transmitted through generations while being filtered by learning and education within the small band. Those rules must have been added with various elements such as accompanying manners to gradually become customs within the small band. Some particular customs may have become dominant and important, and others may have been abandoned. Needless to say, different small bands have developed different rules and customs. If adaptability to natural environments differed among small bands according to rules and customs, the band with the advantageous customs would have been the one to survive.[21] It is easily conceivable that such a small band could have expanded successively from children's generation to grandchildren's generation. In addition, where there happened to be an artificial conflict among bands, the survival rate of the bands may have differed depending on how they coped with the conflict, since conflicts vary in type and character.[22] Provided that bands that successfully adapted not only to natural environments but also to artificial environments left numerous descendants and increased the number of the bands exponentially, the bands'governing rules and customs could have been rapidly diffused.

The other path for transmission of rules of conduct is through horizontal one. Rules of conduct of groups that are better able to adapt to environments can spread to other groups through interaction among groups. The customs, which have been spread into the group, are propagated at an accelerated pace by being transmitted to the descendants of the group. At the same time, it is easy to assume that smaller or weaker bands may have been absorbed by larger or stronger ones. Convergence and dispersion of bands must have repeatedly occurred over a long period of time. In this way, rules that have better adoptability to environments have expanded beyond small-scale bands to spread into larger bands as common rules. If the shared common rules were advantageous to survival, they would have generated a synergistic effect, and then have further contributed to the maintenance and enlargement of the groups. It is still unknown whether vertical transmission or horizontal transmission is dominant. In the former, a small band has acquired better adaptability to environments than other bands because of certain advantageous customs, and the group's descendants increased in number to finally replace other groups and to spread their customs (vertical transmission). In the latter, many small-scale bands simultaneously had various customs and they learned those different customs among the bands to finally share some of them (horizontal transmission). Considering the fact that learning and education are provided mainly by generation overlapping, vertical transmission

[21] For example, whether they could find medical herbs and poisonous herbs and established eating habits using the knowledge.

[22] For example, either aggressive or peaceful, and whether 'tit for tat'strategy can be taken or not.

may have been popular.[23] Customs may have been, however, simultaneously and collaterally transmitted both vertically and horizontally.

In either case, a small-scale band society could survive provided that the members of the band followed certain rules of conduct. The rules of conduct were transmitted through continued existence of the band. Through processes of selection, rules were abstracted and generalized over an extended period of time to create adaptive customs which helped expand the social scale. It is difficult, here, to describe when and which rules were created and the relation between the rules and social forms, based on concrete traces.

Tribal society which emerged next to small-scale blood related bands, according to Diamond [2], came into existence not later than 13 thousand years ago. Each tribal society consisted of several hundreds of people. On the assumption that a tribal society simply grew from small-scale blood related bands, its ancestor band of several generations earlier had a clear memory of blood relationships among members. Considering the number of members of a group, the group must have been a society in which members put names with faces and the background of even visitors could easily be known and memorized within the group. Small related bands and tribal societies have in common the point that members are acquainted with one another and know each other's background. It is difficult to clearly define a division between the small related band period and the tribal society period since some tribal societies may have returned to small related groups. When a small related group grew to a tribal society consisting of several hundreds of people, rules and customs would have further evolved. For example, general arbitration rules may have been established. Such arbitration rules helped reduce unnecessary fights within a band and helped the society to survive maintaining a size of several hundreds of people. In such a society, concepts of private property and fairness and justice, which we have now, must have existed. It must rather be more reasonable to say that such a society has been emerged because concepts of private property and justice grew earlier among the people. Expansion of society and evolution of rules may have occurred both synergistically and repeatedly. It is supposed, however, that in principle the rules which allowed expansion in the size of society evolved before social expansion took place.

The emergence and diffusion of many rules helped the gradual, or even rapid, increase in population to lead to the advent of a large-scale society with lots of people. Then, evolution from a tribal society consisting of several hundreds of people to a larger chiefdom society, where a meeting of several thousands of people could be held, emerged. According to Diamond [2], the chiefdom society was born approximately 7,500 years ago. It was a large-scale society consisting of several thousands of strangers and must have had highly-evolved rules. It is supposed that the basic moral rules we have now may have been created at this time. It may therefore be that the advent of

[23] K. Taniguchi [8]

the chiefdom society was the end of the period of the great leap forward of Homo sapiens.

It can be said that those abstract rules are similar to a compass which helps a group of ships determine which way to go when the ships sail through dense fog. The ships in the group have, in the meantime, to sail avoiding internal breakdown among them though the final destination of sailing remain unclear. So, each ship takes a definite course by the compass. It is the luck of the draw whether the ships can reach their intended destination. However, they can avoid collision against one another by sailing in the same direction. And the final destination of the ships remains unknown.

## 4 Conclusion - Reconsideration of the Continuity Hypothesis

The continuity hypothesis is an assumption that organic evolution and social evolution are continuous and evolutionary change continues beyond the range of what can be explained with Darwin's theory of evolution. As the author wrote previously, according to the continuity hypothesis, societal evolution and organic evolution must be differently separated from each other. Since these evolutions are differently separated from each other, it can be asserted that there is continuity between them. With respect to stereoscopic grafting of evolutionary trees, a branch of the social evolutionary tree is a new branch and its growth is independent from a branch of the organic evolutionary tree. At present, however, social evolution is not independent of biotic evolution and they are inseparable. Hayek points out that there are not only two separable categories, i.e., the natural and the artificial, but there is also another category which belongs neither to the natural nor the artificial. The phenomena of this category has been created through human activity; and yet neither by design nor intentionally.[24] Nobody can predict how our socioeconomic activities may affect those phenomena. As described in Taniguchi [8], instances of societal evolution affecting biotic evolution can be seen here and there. According to this, it is impossible to consider social evolution and biotic evolution separately at present. Furthermore, Hayek argues that a decisive change from animals to human beings lies in the restriction of human innate responses by cultural determinations. We have been inclined to follow learned rules both unconsciously and customarily taking them as human innate instincts. Rules have gradually replaced instincts. In addition, rules and instincts are intricately interrelated and cannot be strictly discriminated.[25] It can be said that biotic evolution and social evolution are co-evolving.

---

[24] F. A. Hayek [5] p.20.

[25] F. A. Hayek [7] p.17.

As seen in this paper, the origin of social evolution can be traced back to the time of the great leap forward where Homo sapiens was bestowed with the first blush of dawn. At that time, primitive forms of property, liberty and justice were born.[26] It might be conceivable, therefore, that the consecutive point of the biotic evolution and social evolution, i.e., the grafting point of the evolutionary tree is the point where human beings made the great leap forward. Nonetheless, as we don't yet see the implementation of social evolution independent of biotic evolution, our time may be still included in the grafting point of the organic and social evolutionary trees. 50,000 years from the great leap forward to the present makes only 1 percent of the time elapsed from the divergence of human beings from anthropoids, or 0.00125 percent of four billion years of development of life on earth. As Hayek stated, we are subject to ignorance. It might be supposed that the biotic and societal evolutions cannot be divided at the stage of Home sapiens.

Then, where do we find the meaning and significance of the continuity hypothesis? This question gives rise to other questions; whether the two evolutions can be finally divided, even though it cannot be done by Homo sapiens, and when and how they can be divided. In addition, next one issue emerges; where can we find the next consecutive point. Taniguchi [8] stated that a consequence of biotic evolution operates as driving factor of social evolution. Assuming that evolution continues, it is suggested that a consequence of social evolution can create the next form of evolution. From this assumption, the continuity hypothesis provides for the possibility of continuous emergence of evolution by any vehicle other than organisms and our societies, as well as a motive for exploring the possibility of continuous emergence of evolution. Such an evolution would be made by a Super Homo sapiens who may already have existed somewhere.

## References


1. Charles Darwin. *On the Origin of Species - By Means of Natural Selection or the Preservation of Favoured Races in the Struggle for Life -*. Watts & Co., 1859.
2. Jared Diamond. *Guns, Germs, and Steel: The Fates of Human Societies.* W.W.Norton and Company, Inc., New York, 1997.
3. Friedrich A. Hayek. "Rules, Perception and Intelligibility". In *Proceedings of the British Academy.* 1962. Reprinted in F.A.Hayek, *Studies in Philosophy, Politics and Economics*, The University of Chicago Press, pp.43-65, 1967.
4. Friedrich A. Hayek. "Notes on the Evolution of Systems of Rules of Conduct". In *Studies in Philosophy, Politics and Economics*, pages 66–81. The University of Chicago Press, 1967.
5. Friedrich A. Hayek. *Law, Legislation and Liberty, Volume 1 - Rules and Order -*. The University of Chicago Press, 1973.
6. Friedrich A. Hayek. *Law, Legislation and Liberty, Volume 3 - The Political Order of a Free People -*. The University of Chicago Press, 1979.


[26] K. Taniguchi [9]

7. Friedrich A. Hayek. *The Fatal Conceit - The Errors of Socialism -*. The University of Chicago Press, 1988.
8. Kazuhisa Taniguchi. "Why Can the Mecca of the Economist Lie in Economic Biology? - Driving Factors of Social Evolution and the Popularisation of Economics". Presented in Tenth Annual Conference of the European Society for the History of Economic Thought, University of Porto, Portugal, April 2006.
9. Kazuhisa Taniguchi. "A Thought on the Origin of Social Evolution and Justice". Presented in Eleventh Annual Conference of the European Society for the History of Economic Thought, The Louis Pasteur University, Strasbourg, France, July 2007.
10. Ulrich Witt. *The Evolving Economy.* Edward Elgar, 2003.

# Modeling a small agent society based on social choice logic programming

**Kenryo Indo**

Department of Management, Faculty of Economics, Kanto-Gakuen University

200 Fujiagu-cho, Ota-si, Gumma-ken 373-8515, Japan

kindo@kanto-gakuen.ac.jp

**Abstract** This paper reports a computational modeling toolkit, with several non-trivial applications, for collective decision making in a small artificial society of three alternatives, and two, three agents, or possibly of larger scale models. Based on the standard axioms in voting and social choice, this modeling makes the axiomatic approaches in social sciences computationally verifiable and intellectually tangible by means of the logic programming.

## 1. Introduction

Logic has been a basic modeling tool in two independent academic fields: social choice theory and artificial intelligence. Social choice theory [2, 6, 7, 10, 15, 18] studies formal, but not computational, modeling of collective decision making based on preference orderings in the rational choice vein. Unlike other conventional economic analyses, where the basic tools are probability theory and calculus, this field utilizes discrete orderings and proving based on logic, especially after the celebrated theorem proved by K. J. Arrow [1].

This paper introduces the CMAS-SCLP (Computational Modeling for Agent Society based on Social Choice Logic Programming) approach. This approach provides both a toolkit for logic-based modeling and simulation for collective decision making using PROLOG—a programming language based on first-order predicate logic and on the Horn-clause resolution—to simulate theorem proving. PROLOG [5, 17] has been widely used in artificial intelligence, especially for expert systems and natural language processing.

Regarding agent-based simulation, parallel development of social science and artificial intelligence has been recently observed [19]. While focusing on the emergent properties in complex social systems, it has been observed that the agent-based approach using computer simulations is a promising option for interrelating social science to artificial intelligence. Contrastingly, the logic-based (or axiomatic) approach has not been intensively used in this regard. There is room for logic-based as well as computational approach. Recently, PROLOG has been applied to prove Arrow's general impossibility theorem [12]. In this study, we will attempt to further develop this approach. We present a collection of PROLOG codes that recursively model the basic components of social choice theory, including the axioms on the preferences, profiles, (preference) domains, social decision rules, winning coalitions, core, and so on.

This paper is organized as follows. Section 2 introduces the modules and the preference modeling of the CMAS-SCLP approach. In Section 3 we use the domain management module to specify the permissible orderings for the individual members of the society. Section 4 provides an automated proof of the Gibbard-Satterthwaite theorem [8, 14], a dictatorial result, and models preference aggregation, social choice function, and strategic manipulation. Section 5 argues for several domain conditions with regard to the simple majority rule. Section 6 demonstrates the simple game analysis, which relates Arrow's theorem to majority voting. Section 7 presents an application that proves some of the nontrivial results regarding the Dodgson metric. The last section concludes with our remarks.

## 2. Modeling in CMAS-SCLP

In accordance with the social choice theory, the system CMAS-SCLP enables preference (i.e., permissible ordering) based modeling and simulation for the collective decision making problem, which consists of two or three agents and three alternatives, accompanied by the following axiomatic analysis.

The system can be divided into six basic conceptual components called modules, though these are not explicitly PROLOG modules.

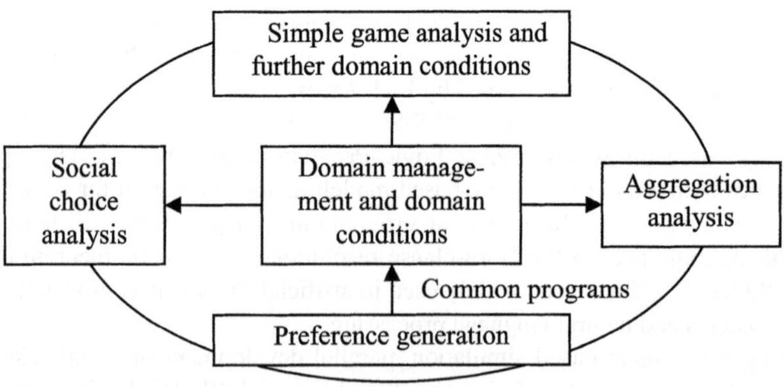


**Fig. 1.** The CMAS-SCLP modules: A collection of PROLOG codes as a toolkit for logic-based modeling of the individual preferences and collective decision making rules.

- Preference generation module (PG) gprf06.pl
- Domain management module (DM) cswf07.pl, dcdi06.pl, drsc06.pl
- Aggregation analysis module (AA) cswf07.pl
- Social choice analysis module (SCA) spcf06.pl
- Simple game analysis module (SGA) sged06.pl, power.pl, metric.pl
- Common programs–libraries: arithmetic, set theory, file I/O, and so on.

Each module—along with its program name—is treated as a separate functionality, to build and analyze the logical models in accordance with the social choice theory.

Figure 1 shows the modules of this approach and their interrelationships. For example, the module DM depends on the module PG because of the possible combination of individual orderings (i.e., the permissible orderings and profiles). Table 1 represents excerpts of some representative models. Throughout our modeling, the modules PG and DM form the basis of all the other modules.

**Table 1.** Some theoretical models in the modules

| Modules | Functionality and predicative models |
|---|---|
| Preference generation (PG) | Generating and analyzing orderings and default setting.<br>x: alternative, r: weak ordering, p: strict ordering |
| Domain management (DM) | Permissible profiles, domain conditions, and majority vote<br>j: agent, rr: permissible profile, r_j: the agent j's preference |
| Aggregation analysis (AA) | Modeling Arrow's axioms and the dictatorial result.<br>swf: social welfare function, majority: simple majority rule |
| Social choice analysis (SCA) | Proving the Gibbard-Satterthwaite theorem<br>scf: social choice function, sp: strategy-proofness |
| Simple game analysis (SGA) | Essential decomposability and coalitional power distributions.<br>gen_win: generating a simple game, core: the core |

In our modeling, a society consists of a set of agents (numbers), the alternatives (alphabets), and the possible preferences (orderings of alternatives). First, we must specify the required properties of the orderings listed in Table 2 using the PG module, which includes PROLOG clauses, as follows:

- r((X,Y), R). % Weak preference relation.
- p((X,Y), R) :- r((X,Y), R), \+ r((Y,X), R). %Strict relation.
- i((X,Y) :- r((X,Y), R), r((Y,X), R). %Indifference relation

In the above codes, r, p, and i are *predicate* symbols, X and Y are variables to be matched as alternatives, and R denotes a variable for an ordering. A neck ':-' represents the *if* relation that separates the head (LHS) and the body (RHS), which is read as a conjunction. \+ denotes NOT operator.

**Table 2.** Orderings and their required conditions

| Type of ordering | Complete | Reflexive | Acyclic | Transitive | Asymmetric |
|---|---|---|---|---|---|
| l: strict (or linear) | Yes | Yes | Yes | Yes | Yes |
| t: weak (or transitive) | Yes | Yes | Yes | Yes | Strict part |
| q: quasi-transitive | Weak part | Yes | Strict part | Strict part | No |
| o: complete | Weak part | Yes | No | No | No |
| binary | No | No | No | No | No |

# 3. Domain management

The DM module is the most basic component of the CMAS-SCLP approach, and it determines the orderings to be used later in the analyses for the model generated by the PG module. These analyses are based on the rather classical expertise of the social choice theorists.

| Show current domain | Change basic ordering type | Set the number of agents |
|---|---|---|
| ?- display_domain.<br>current: ACITZN<br>[base domain = l: linear]<br>Yes | ?- chdom(_->t: N), display_domain.<br>current: ABCFIJOSTWZnN<br>[base domain = t: transitive]<br>N = transitive<br>Yes | ?- make_n_agents(3).<br>Yes<br>?- model(A, B).<br>A = states: [a, b, c, d]<br>B = agents: [1, 2, 3]<br>Yes |
| Call an ordering | Show all strict binary relations | Generate profile |
| ?- r(B, [+, +, +]).<br>B = a, b<br>Yes | ?- p(B, [+, 0, −]), write(B: ''), fail.<br>(a, b): (c, b):<br>No | ?- rr(R).<br>R = [[+, +, +], [+, +, +], [+, +, +]]<br>Yes |
| Ordering expressions | Expressions with indifferences | Alphabetize the profile |
| ?- id_r(1: N, A, B).<br>N = 'A'<br>A = [+, +, +]<br>B = [a > b, a > c, b > c]<br>Yes | ?- id_r(22: N, A, B).<br>N = 'W'<br>A = [+, 0, −]<br>B = [a > b, a > c, c > a, c > b]<br>Yes | ?- name_domain('AIZ', R).<br>R = [[+, +, +], [−, −, +], [+,−, −]]<br>Yes |

**Fig. 2.** Command line sequences and results obtained using preference models. Left: Since the default domain is characterized by linear orderings, there are no indifference relations. Each ordering is symbolized by a number, sign profile, and alphabet as the synonym. For example, ordering 1 is represented by A or [+, +, +]. A list [S1, S2, S3] implies that each Sk represents the binary relation between ab, ac, and bc, respectively. Center: Changing the ordering type to transitive, where the ordering might have indifference relations. Right: Setting model and generating profiles of individual orderings.

**Table 3.** The orderings for three alternatives (additionally aliened from left to right). Each ordering is represented by (No.: binary [ab, ac, bc]: synonym respectively).

| Linear | Weak | Quasi-transitive | Acyclic | Cyclic |
|---|---|---|---|---|
| 1: [+, +, +]: A | 2: [0, +, +]: B | 5: [0, 0, +]: E | 4: [+, 0, +]: D | 7: [+, −, +]: G |
| 3: [−, +, +]: C | 6: [—, 0, +]: F | 11: [0, +, 0]: K | 8: [0, −, +]: H | 21: [−, +, −]: V |
| 9: [−,−, +]: I | 10: [+, +, 0]: J | 13: [+, 0, 0]: M | 12: [−, +, 0]: L | |
| 19: [+, +, −]: T | 14: [0, 0, 0]: O | 15: [−, 0, 0]: P | 16: [+, −, 0]: Q | |
| 25: [+,−, −]: Z | 18: [−, −, 0]: S | 17: [0, −, 0]: R | 20: [0, +, −]: U | |
| 27: [−, −, −]: N | 22: [+, 0, −]: W | 23: [0, 0, −]: X | 24: [−, 0, −]: Y | |
| | 26: [0, −, −]: n | | | |

Depending on his/her requirements, an analyst or planner of the artificial agent society can modify and perform the analysis in a more detailed manner using novel ideas. The classical theory primarily assumes weak or linear ordering as the preference model; sometimes, it uses quasi-transitive or acyclic ordering [1, 6, 10, 15].

The DM specifies a set of possible profiles, i.e., possible combinations of individual orderings as permitted by PG, the preceding module. A preference aggregation is a function from the set of possible profiles, i.e., the domain, to the set of orderings. By default, the system assumes the universal (or unrestricted) domain, which is later modified into restricted domains. For example, the sequence in which this system is used is depicted in Figure 2. Table 3 summarizes the orderings for the three alternatives generated by the system.

## 4. Preference aggregation and strategy-proofness

Kenneth J. Arrow proved his celebrated theorem, known as the general (im)possibility theorem [1]. His theorem can be interpreted as a dictatorial result of the aggregated preference in weak (and linear) orderings. The Arrow-type aggregation of individual preferences is known as the *Social Welfare Function* (SWF). Both Gibbard [8] and Satterthewaite [14], independently, proved another dictatorial result of voting procedures: more precisely, the *Social Choice Functions* (SCF) implied by strategy-proofness and citizens' sovereignty. That is, an SCF cannot be manipulated by a single voter, if and only if the voting procedure is dictatorial.

These two theorems commonly assume an unrestricted domain, and are logically equivalent (see [18]). The axioms for the SWF and SCF are simple enough to be modeled in our framework. With regard to Arrow's theorem, only the weak Pareto optimality (i.e., unanimity), the IIA condition (independence of irrelevant alternatives), and the unrestricted value (linear or transitive valued SWF) are needed. Indo [12] showed an automated proof of this theorem for the 2-agent and 3-alternative case using PROLOG. A similar program for the strategic-proof SCF is shown in Figure 3. Figure 4 presents the results of the simulation. We can provide the similar result for linear ordering in [12], of course, but possibly for more general orderings. The experiment is left to the reader.

```
oof(A,B):- member(A,B).
xof(Y,A):- agenda(A), oof(Y,A).
% social choice function
scf([],[],_).
scf([RR->C|F],[RR|L],X):- scf(F,L,X),axiom_scf(X,RR->C,F).
scf(F,X):-all_rr(L),scf(F,L,X).
% axioms for scf
axiom_scf(d(J,A),RR->C,_):- xof(C,A),dictatorial_scf(J,A,[RR-
>C]).
axiom_scf(d(J),RR->C,_):- x(C),dictatorial_scf(J,[RR->C]).
axiom_scf(p(A),RR->C,_):- xof(C,A),pareto_scf(A,[RR->C]).
axiom_scf(p,RR->C,_):- x(C),pareto_scf([RR->C]).
axiom_scf(sp,RR->C,F):- x(C),spx([RR->C|F]).
% recursive (cumulative) conditions for SCF
pareto_scf(A,F):- agenda(A),
   \+ (oof(PP->C,F), \+ (r_j(_,PP,P),\+ best(B,A,P)),C\=B).
dictatorial_scf(J,A,F):- j(J),agenda(A),
   \+ (oof(PP->C,F),r_j(J,PP,P),\+ best(C,A,P)).
cs_scf(A,F):- \+ (xof(X,A),\+ oof(_->X,F)).
manipulable(J,PP->X,QQ->Y,F):- dop((X,Y)),oof(QQ->Y,F),
   udrr(PP,QQ,_,(J,P,_)),oof(PP->X,F),p((Y,X), P).
% udrr: unilaterally deviated profile (omit)
manipulable(A,J,PP->X,QQ->Y,F):-
   xof(X,A),xof(Y,A), manipulable(J,PP->X,QQ->Y,F).
% spx: strategy-proofness in the recursion
spx([W|F]):-
   \+ manipulable(_,W,_,[W|F]),
   \+ manipulable(_,_,W,[W|F]).
spx(A,[W|F]):-
   \+ manipulable(A,_,W,_,[W|F]),
   \+ manipulable(A,_,_,W,[W|F]).
% Note: The non-recursive versions are trivial and omitted.
```

**Fig. 3.** A portion of PROLOG codes which implements the strategic-proof social choice function. To prove the Gibbard-Satterthwaite theorem, the axioms of sp (non-manipulability) are recursively cumulated in scf/3.

```
?- model(A,B).
A = states:[a, b, c]
B = agents:[1, 2]
Yes
?- display_domain.
current domain: ACITZ
[base domain=l:linear]
Yes
?- stopwatch((
  scf(F,sp),cs_scf(F),display_scf(F),nl,fail;true
  ),T).

scf:row  col ACITZN
--------------------
[+, +, +]=A  aaaaaa
[-, +, +]=C  bbbbbb
[-, -, +]=I  bbbbbb
[+, +, -]=T  aaaaaa
[+, -, -]=Z  cccccc
[-, -, -]=N  cccccc

scf:row  col ACITZN
--------------------
[+, +, +]=A  abbacc
[-, +, +]=C  abbacc
[-, -, +]=I  abbacc
[+, +, -]=T  abbacc
[+, -, -]=Z  abbacc
[-, -, -]=N  abbacc
% time elapsed (sec): 0.344

F = _G168
T = 0.344
Yes
```

**Fig. 4.** An automated proof of the Gibbard-Satterthwaite theorem, which states that any non-manipulable scf should be dictatorial for the unrestricted domain of two-agent and three-alternative, using a predicate scf/2 in Figure 3.

# 5. Domain conditions

The necessary and sufficient condition for avoiding the no winner result for pairwise majority rule (the so-called simple majority principle) is proved by A. Sen and P. K. Pattanaik (see [15], Chapter 7) by extending Inada's [11] conditions on transitive majority. They proved that excluding a certain type of Latin Squares (see Section 6) from the common admissible domain, which is expressed as the disjunction of three conditions—value restriction, extremal restriction, and limited agreement—is necessary and sufficient to ensure that majority vote works well.

```
?- restricted_domain(L,1,N),
scf(F,sp),cs_scf(F),\+ dictatorial_scf(_,F),display_scf(F).

scf:row  col ACITZ
--------------------
[+, +, +]=A  abbaa
[-, +, +]=C  bbbbb
[-, -, +]=I  bbbbb
[+, +, -]=T  abbaa
[+, -, -]=Z  abbac

L = [[+, +, +], [-, +, +], [-, -, +], [+, +, -], [+, -, -]]
N = 5
F = [([[+, +, +], [+, +, +]]->a),
 ([[-, +, +], [+, +, +]]->b),([[-, -, +], [+, +, +]]->b),
 ([[+, +, -], [+, +|...]]->a), ...]

Yes
```

**Fig. 5.** Generating an admissible domain with a non-dictatorial, non-manipulable (i.e., strategy-proof), and citizen sovereign SCF against the Gibbard-Satterthwaite theorem.

Moreover, Sen also tried another approach to the difficulty of democratic society from Arrow's proof, in his article "The impossibility of a Paretian liberal" ([15], Chapter 13). His pessimistic result was proved without the independence axiom for unrestricted domains. Impartial distribution of *Decisiveness*, the power to determine the social choice for a pair of alternatives, among individuals is not possible. Kalai and Muller [9] (and Blair and Muller [4]) proved that (essential) decomposability is necessary and sufficient for the existence of a nondictatorial and strategy-proof SCF, and also for a nondictatorial SWF, for any (individually) admissible domain.

The CMAS-SCLP system proves these theoretical results by directly generating both admissible domains and social decision rules. Figure 5 shows the experimentation that generates non-dictatorial and strategy-proof social choice functions for common admissible domains. The system is capable of enumerating all such SCFs and, of course, of verifying the Kalai-Muller decomposability, and other domain conditions.

## 6. Simple game analysis

A simple game is a coalitional structure, given a power distribution [10, 16]. Many domain conditions for transitivity and the social decision rule that use simple game analysis, have been proposed [6].

A coalition is a nonempty subset of the set of all agents. Formally, it was defined as a set of *monotonic* winning coalitions. A simple game is *proper* if no complement set of a winning coalition is winning. A simple game is *strong* if no complement set of a losing coalition is losing. A winning coalition is a subgroup of agents who are decisive if they unanimously agree, regardless of members who are not a part of the coalition. The unanimity of the winning coalition for each profile derives a *dominance relation* from the consequences of social decision making.

This scenario falls within the realm of the cooperative game theory, and it elegantly connects Arrow's dictatorial result to the dominance relation. The *core* of a game is a set of undominated alternatives, assuming a preference profile. The core of the game is said to be *stable* if it is nonempty for every profile (Condorcet's paradox is a case, in which the core is empty: a circular dominance relation, under pairwise majority). If there are *vetoers*, who are commonly found in every winning coalition, the game is stable and the dominance relation does not circulate. A *dictator* is a singleton set of vetoers, such that there is no cycle of dominance relations. A game is *weak* if there is a vetoer.

The set of conditions in Arrow's theorem can be interpreted as proper and strong simple games ([6], Chapter 3). The dictatorial result is equivalent to the fact that, in this case, only dictatorial games are stable. However, in this case the cycle occurs only for majority rule at the Latin square profiles. Figure 6 examines this result using the SGA module. Therefore, the possibility result ([15], Chapter 5, Theorem 1) can be obtained by excluding these profiles.

```
?- gen_win(W,[monotonic:yes,proper:yes,strong:yes]),
\+ \+ core([],_),verify_win, inspect_empty_core,nl,fail.

game:[[1, 2, 3], [1, 2], [1, 3], [2, 3]]
 + :[monotonic, proper, strong, not weak, essential]
 - :[]
[[+, -, -], [-, -, +], [+, +, +]]: =>ZIA
[[-, -, +], [+, -, -], [+, +, +]]: =>IZA
[[-, -, -], [+, +, -], [-, +, +]]: =>NTC
[[+, +, -], [-, -, -], [-, +, +]]: =>TNC
[[+, -, -], [+, +, +], [-, -, +]]: =>ZAI
[[+, +, +], [+, -, -], [-, -, +]]: =>AZI
[[-, -, -], [-, +, +], [+, +, -]]: =>NCT
[[-, +, +], [-, -, -], [+, +, -]]: =>CNT
[[-, -, +], [+, +, +], [+, -, -]]: =>IAZ
[[+, +, +], [-, -, +], [+, -, -]]: =>AIZ
[[+, +, -], [-, +, +], [-, -, -]]: =>TCN
[[-, +, +], [+, +, -], [-, -, -]]: =>CTN

No
?-
```

**Fig. 6.** Only majority rule is a monotonic, proper, and strong, simple game, but it has an empty core. And every profile that brings about an emptiness of the core (i.e., the cycle of dominance relations) under the pairwise majority decision is a Latin square.

## 7. An application: the Dodgson rule

This section investigates a particular method of counterfactual reasoning with respect to social decision making. We want to ascertain the consequences in case of occurrence of some reversals in individual orderings, or to what extent the current state is stable. The *minimal distance* (or metric) is the minimum number of preference reversals that lead to a different outcome, and we would then want to re-rank the alternatives according to this metric.

Social choice theorists know that the Dodgson rule employs this type of reasoning, which is based on the minimal distance from the empty core, using the pairwise majority rule (See [7], pp. 100–104). Because determining the winner of the Dodgson rule is known to be computationally awkward (NP-hard problem), it is not considered useful for practical voting [3]. However, using the toolkit, we can simulate the minimal distance and extend the rule beyond pairwise majority.

Figure 7 summarizes the experimental result of the Dodgson rule, using a simple game of the majority vote for three agents. It is well-known that the number of alternatives is greater than three, then the Dodgson rule fails to reproduce the Condorcet winner. The result shown in Figure 7 confirms the invariance. However, this would not be trivial if, in addition to the orderings, the decision rule could vary, i.e., the game is permitted to change from majority vote to another power distribution. By adding code for rule mining, we present the reader with a set of findings on a generalized form of the Dodgson rule with co-varying individual orderings and simple games (See Figure 8).

```
?- rr(Q),is_upper_diagonal(Q),
r_dodgson(D,Q),majority(Q->M),D\=M,name_domain(A,Q),
nl,write((profile:A,majorty:M,dodgson:D)),fail.

profile:TTC, majority:[+, +, -], dodgson:[+, +, 0]
profile:AII, majority:[-, -, +], dodgson:[-, 0, +]
profile:AZI, majority:[+, -, +], dodgson:[0, 0, 0]
profile:ZZI, majority:[+, -, -], dodgson:[0, -, -]
profile:AAZ, majority:[+, +, +], dodgson:[+, +, 0]
profile:CCN, majority:[-, +, +], dodgson:[-, 0, +]
profile:TCN, majority:[-, +, -], dodgson:[0, 0, 0]
profile:TNN, majority:[-, -, -], dodgson:[0, -, -]

No
?- r_dodgson(D,Q),majority(Q->M),
best(B,D),best(W,M),B\=W.

No
```

**Fig. 7.** While the Dodgson rule and the Condorcet rule (i.e., simple majority rule) differ with respect to the ranking for many profiles, the winners are the same, assuming the majority game in Figure 6.

```
% minimal rule that is not sensitive to the metric.
d-sensitive(-):[monotonic, proper, not weak]->[]
d-sensitive(-):[monotonic]->[strong]
d-sensitive(-):[strong]->[monotonic, not weak]
% minimal rule that is sensitive to the metric.
d-sensitive(+):[monotonic]->[proper, strong]
d-sensitive(+):[monotonic, not weak]->[strong]
d-sensitive(+):[strong]->[not weak]
% theorem for simple games (metric-irrelevant).
d-sensitive(-):[]->[proper, not weak]
d-sensitive(+):[]->[proper, not weak]
```

**Fig. 8.** The rule mining results of a generalized form of the Dodgson rule. Social ranking, recalculated based on the minimal distance from the empty core, changes over co-varying individual orderings, and simple games. This can be seen as a generalization of the Dodgson rule. In the above figure, each rule has the form "d-sensitive(T): LHS->RHS," where "d-sensitive(T)" represents whether the minimal distance is changed (T = "+") or not (T = "–"), the LHS is a list that can be read as a conjunction and RHS as a disjunction. These rules are minimal since there is no other rule where both side hands are subsumed.

# 8. Concluding remarks

This paper outlined the CMAS-SCLP approach to model axiomatic social choice problems computationally, albeit for a small-sized society, including several nontrivial results. For further details regarding the programs shown in this paper, the reader should refer to the author's website [13].

These codes might be incorporated into more advanced analyses of group decision making. Its pedagogical use as courseware is a potential contribution of this paper. Operating the theoretical objects computationally helps in learning elementary social choice theory, in both the classroom and self-education. Further, the author hopes that this approach will prove useful to researchers who are willing to make use of it and to communicate new research ideas.

However, the reader might wonder about extending the presented approach into larger-scale realistic models that include more than two agents, more than three alternatives, or even incomplete orderings. Almost all the theoretical results mentioned in this paper are ternary-based, i.e., they have to be proved for every triple of alternatives. The dictatorial results can be modeled by simple game analysis, regardless of the number of alternatives. Therefore, the answer is partially affirmative, at least for modeling flexibility. The modification needed for each module is straightforward, although the computation might be awkward.

Finally, the reader may doubt the utility of our approach in finding any new result that are not already known. Under the traditional axiomatic approach, no proof step brings about an emergent property out of the logical consequences of the given axioms, as long as the proof is correct. This is because of the completeness of the deductive system at all levels of objectivity. But it is not always true at the levels of (inter-)subjectivity, even for a small society as a set of axioms considered by a researcher, or a group of researchers. It can produce unforeseen consequences, as demonstrated in the counterfactual reasoning experiment in Section 7. Simulation experiment is, thus, beneficial in this respect. We want to seek appropriate restrictions on our modeling, in order to maneuver the simulation experiments into a set of significant observations without loss of rigor, while minimizing computer resource consumption. I believe that the CMAS-SCLP approach opens up such explorations for knowledge.

## References

[1] Arrow, K. J. (1963). *Social Choice and Individual Values.*, 2nd edition. Yale University Press. (Originally published in 1951 by John Wiley & Sons)

[2] Arrow, K. J., et al. (2002). *Handbook of Social Choice and Welfare*, Vol. 1. Elsevier.

[3] Bartholdi III, J., et al. (1989). Voting schemes for which it can be difficult to tell who won the election. *Social Choice and Welfare*, 6(2), 157–165.

[4] Blair, D. and Muller, E. (1983). Essential aggregation procedure on restricted domains of preferences. *Journal of Economic Theory*, 30, 34–53.

[5] Clocksin, W. F. and Mellish, C. S. (2003). *Programming in Prolog: Using the ISO Standard*, 5th edition, Springer.

[6] Gaertner, W. (2001). *Domain Conditions in Social Choice Theory*. Cambridge University Press.

[7] Gaertner, W. (2006). *A Primer in Social Choice Theory*. Oxford University Press.

[8] Gibbard, A. (1973). Manipulation of voting schemes: A general result. *Econometrica*, 41, 587–602.

[9] Kalai, E. and Muller, E. (1977). Characterization of domains admitting nondictatorial social welfare functions and nonmanipulable voting procedures. *Journal of Economic Theory*, 16, 457–469.

[10] Moulin, H. (1988). *Axiom of Cooperative Decision Making*. Cambridge University Press.

[11] Inada, K. (1969). The simple majority decision rule. *Econometrica*, 37, 490–506.

[12] Indo, K. (2007). Proving Arrow's theorem by PROLOG. *Computational Economics*, 30(1), 57–63. doi: 10.1007/s10614-007-9086-2.

[13] Indo, K. (2008). Logic programming for modeling social choice. http://xkindo/cog_dec/wp/mplsc.html. Accessed 15 May 2008.

[14] Satterthwaite, M. A. (1975). Strategy-proofness and Arrow's conditions: Existence and correspondence theorems for voting procedures and social welfare functions. *Journal of Economic Theory*, 1, 187–217.

[15] Sen, A. (1982). *Choice, Welfare and Measurement*, MIT Press.

[16] Shapley, L. S. (1962). Simple games: An outline of the descriptive theory. *Behavioral Science*, 67, 59–66.

[17] Starling, L. and Shapiro, E. (1994). *The Art of Prolog: Advanced Programming*, 2nd edition, MIT Press.

[18] Taylor, A. D. (2005). *Social Choice and the Mathematics of Manipulation*. Cambridge University Press.

[19] Tesfatsion, L. and Judd, L. (Eds.) (2006). *Handbook of Computational Economics: Agent-based Computational Economics*, Vol. 2. North Holland.

# Production, Services and Urban Systems

# Modeling and Development of an Autonomous Pedestrian Agent – As a Simulation Tool for Crowd Analysis for Spatial Design

**Toshiyuki Kaneda,** Nagoya Institute of Technology, (kaneda@nitech.ac.jp)
**Yanfeng He,** Hitachi Systems & Services, Ltd., (he_yanfeng@hotmail.com)

Abstract: At present, it is expected that pedestrian agent simulation will be applied to not only accident analysis, but also spatial design; ASPF (Agent-based Simulator of Pedestrian Flows) has already been developed as a simulator for such purposes. However, in the present version ASPFver.3 a pedestrian agent merely walks straight ahead and simply avoids other agents, and it had been impossible to analyze crowd flows on a large-scale space with a complicated shape, a function is required that enables an agent to walk along a chain of visible target 'waypoints' to each destination, as well as a function the agent keeps the direction to the target. The study introduces newly a target maintaining (Helmsman) function, a concept of waypoint, and update mechanism of targets, and develops the simulator ASPFver4.0 that models an autonomous pedestrian agent on ArtiSoc(KKMAS). The performances tests of these additional functions of ASPFver4.0 are shown. Especially, to successfully model pedestrians' shop-around behavior in a Patio-style shopping mall at Asunal Kanayama, Nagoya, ASPFver4.1 has been also developed by introducing an optimization function of routes by Dijkstra method, and implemented several parameters based on data for survey of the pedestrians' behaviors in this mall. Through the test of four simulation cases; (1) weekday case, (2) weekday double case, (3) holiday case, and (4) at time of a music event in holiday case, the performance of ASPFver4.1 was also verified. Due to a series of these version-ups, we can conclude that ASPF is now available for analyzing crowd flows and density in space with complicated shapes.

## 1. Research Background and Objectives

At present, it is expected that pedestrian agent simulation will be applied to not only accident analysis, but also spatial design; ASPF (Agent-based Simulator of Pedestrian Flows) has already been developed as a simulator for such purposes.[5-9] However, in the present version ASPFver.3 a pedestrian agent merely walks straight ahead and simply avoids other agents, and it had been impossible to carry out a simulation on a large-scale space with a complicated shape.[9] In particular, to successfully model pedestrians' behavior in a Patio-style mall and in order for an agent to reach their destination by moving from point to point, a function is required that enables an agent to move along a route made up of a chain of visible waypoints linking each target destination.[4,12]

The study introduces newly a target maintaining function, concepts of way-point, update of targets and optimization of routes, and develops the simulator ASP-Fver4 that models an autonomous pedestrian agent with a function to move toward a target destination on ArtiSoc(KKMAS)[i]. More specifically, characteristics of this simulator are described through the analysis of crowd density in a simulation of pedestrian behavior within the Patio-style mall at Asunal Kanayama, Nagoya.

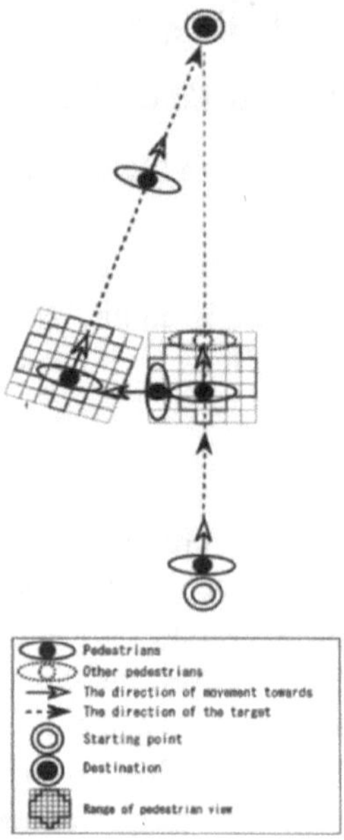


**Fig.1 Target maintaining function**

## 2. APSFver.4.0: Modeling an Autonomous Pedestrian Agent with a Function to Move toward a Target Destination

### *2.1 Improvement of Pedestrian Agent toward ASPFver4.0*

In ASPFver4.0, in order to install a movement to destination function into a pedestrian agent, a target maintaining (Helmsman) function, the concept of way-point and walking routes were introduced,

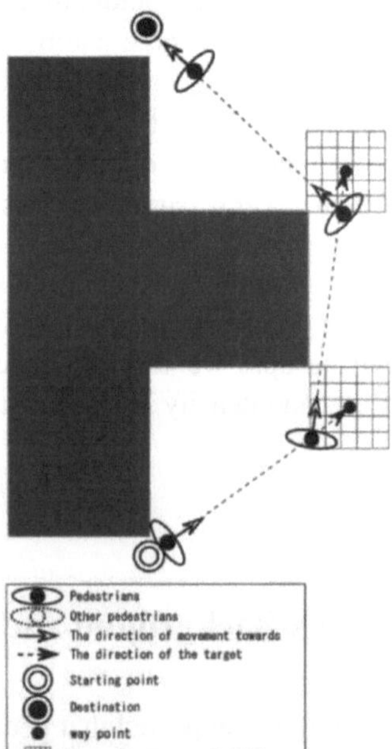


**Fig.2 Movement to destination**

and a target update algorithm at waypoints was implemented.[4, 12]

The target maintaining function (Fig. 1) refers to a function that firstly determines the direction of movement towards a given (visible) target and secondly by regularly reconfirming the location of the target corrects any difference of directions between the movement and the target; such differences may occur because of behavior to avoid other pedestrians and walls while moving towards the target. In addition, in a large-scale and complicated shaped space, there is no guarantee that a destination can be always confirmed visually; in this case, a list of waypoints that satisfy visual confirmation conditions from the starting point to the destination is given in advance and a pedestrian agent walks along the list of the waypoints (Fig. 2). The target mentioned above is either the final target destination or a waypoint. Update of the target means that the agent regularly confirms whether they are closing on their target and when they arrive in the neighborhood of the target, they update their target to the next one.

Furthermore, pedestrian behavior is affected by not only other pedestrians, but also walls and both these factors have different characteristics; therefore, version 4.0 was designed to allow the installation of wall agents at any location in the form of a unit of cells.

## *2.2 Structure of ASPF ver.4.0*

The general structure of the simulator followed the basic design of ASPF3.0. [7,8] The spatial scale is represented by 40 square cm cells and the time scale is set at one step per 0.5 seconds. In ASPFver4.0, with the introduction of a target maintaining function, target update and walls, in order to maintain the integrity of these new elements, additions and changes were made to the walking behavior rules.

Behavior rules were applied to an agent in the following order: (1) set a route; (2) maintain the target; (3) walking behavior rules; and (4) update the target (Fig. 3). The following parameters were set: confirmation to maintain a target was carried out every 10 steps, target update was confirmed every 2 steps by checking whether they were located within the 2 cells from a waypoint. Due to the introduction of a wall agent, walking behavior rules increased to a total of 36 rules comprising 14 new wall avoidance rules and 25 rules taken from the previous version, comprising 6 basic behavior rules, 8 slow-down rules, 4 avoidance rules, 3 high density flow rules and 1 pattern cognition rule (Fig. 4). Wall avoidance rules were designed to avoid a wall by changing the direction of movement. The existence of other agents was basically judged by using a relative coordinate system, but in order to standardize the differences in unit distances caused by each agent's individual progress, the existence of a wall agent was judged on an absolute coordinate system.

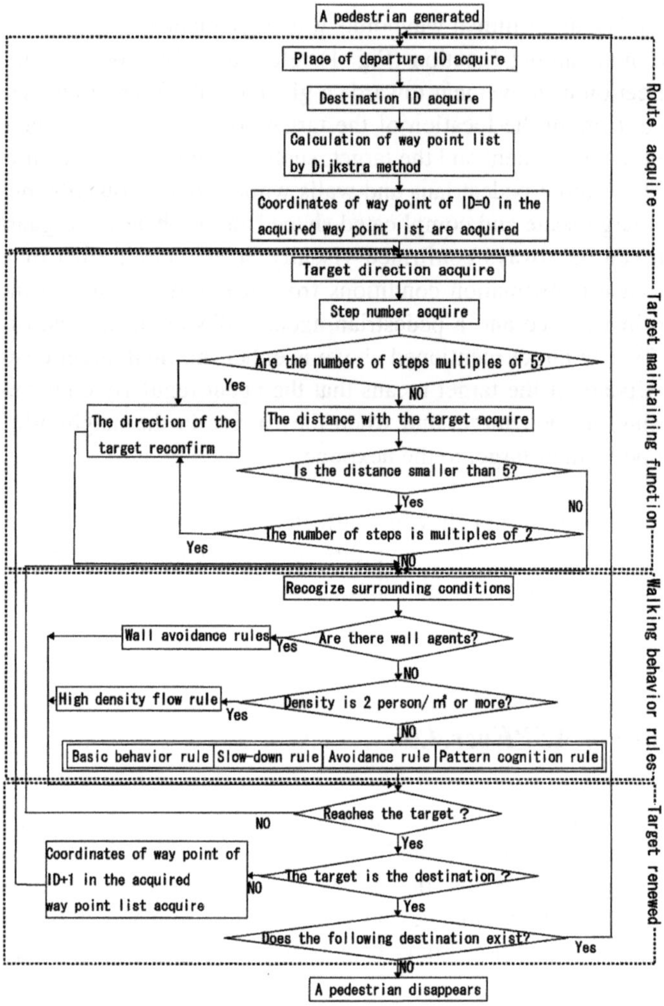


**Fig. 3 Pedestrian agent's algorithm**

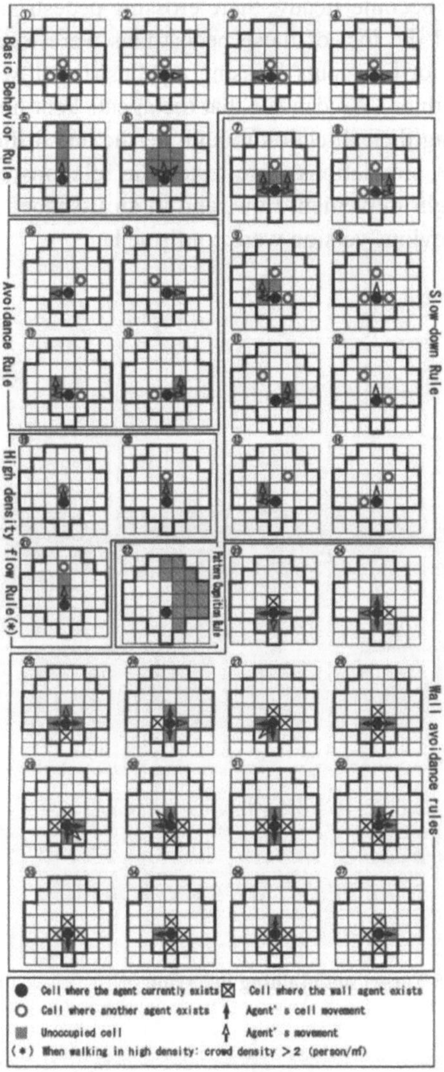


**Fig. 4 Pedestrian behavior rules**

## *2.3 Example of 'Crowd Flows-Crossing' Simulation*

Here, the target maintaining function was demonstrated. In a cross-shaped space, 40 cells in road width, opposing flows with a flow coefficient of 0.5 person/m · sec were generated from both the right and left sides, and after the number of pedestrians had become steady, three crossing pedestrians were generated from the lower part and the loci of these pedestrians were examined (Fig. 5). Target maintaining behavior by agents crossing a pedestrian flow from two different directions was confirmed. We can see that the agents kept holding their target destination by correcting their direction though he/she had drifted by the flows. Moreover, in this study, simulations were tested the relationship between density and speed with a straight movement flow, and in an L-shaped corridor, and the same results as in the previous version were confirmed.

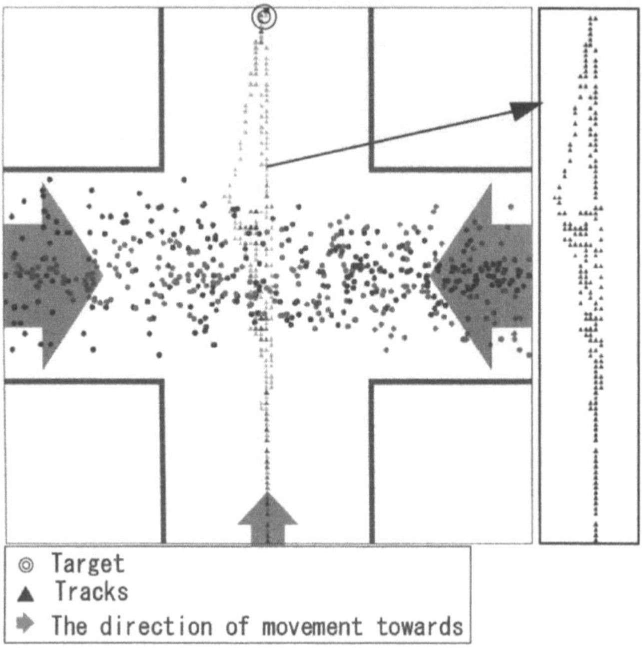


**Fig5 Target maintaining behavior**

# 3 ASPFver4.1: Autonomous Pedestrian Agent Simulation with an Introduction of a Route Optimization Function

## *3.1 Characteristics of ASPF ver.4.1*

By improving ASPFver.4.0, ASPFver.4.1 was developed, which deals with pedestrian behavior in the Patio-style mall standing next to Kanayama railway

Station, Asunal Kanayama, and an attempt was made to apply this updated version to the analysis of crowd density. Firstly, by using the data of pedestrian behavior research that was conducted in the previous year, an agent behavior algorithm shown in Fig. 6 was established. This version has the following characteristics: first, create a list of shop-around facilities from research data by attributes; then by using the Dijkstra method for movement between facilities, find and follow the shortest route. Moreover, in order to express the crowd density at the time of an event, an event routine was added; when an music event starts, the 80% of the agents within the facilities begins to gather in the event square and at the end of the event, these agents continue shop-around behavior again.

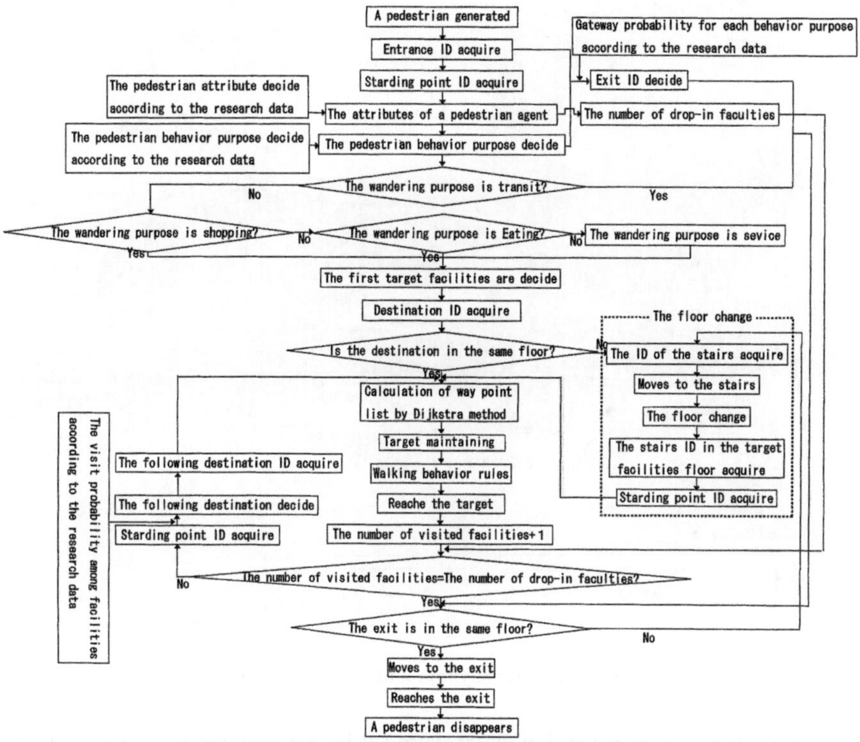


**Fig.6 Wandering action algorithm by ASPFver.4.1**

## *3.2 Crowd Density Analysis of Pedestrian Behavior in a Patio Style Mall*

This section describes a simulation experiment of an area of 150m x 100m on the ground floor of Asunal Kanayama. The target area covered 300 × 250 cells and comprised 11 gateways (10 gateways for the ground floor, 1 gateway for 2[nd] and 3rd floors), 19 stores for each store type (18 stores for the ground floor, 1 store for 2nd and 3rd floors). 105 waypoints (87 points for ground floor, 18 points for 2nd and 3rd floors) were set up: all the waypoints formed a route network and conformed to visual confirmation conditions. When any starting point and destination were given, one of 7,656 set shortest routes, found by using the Dijkstra method, was used.

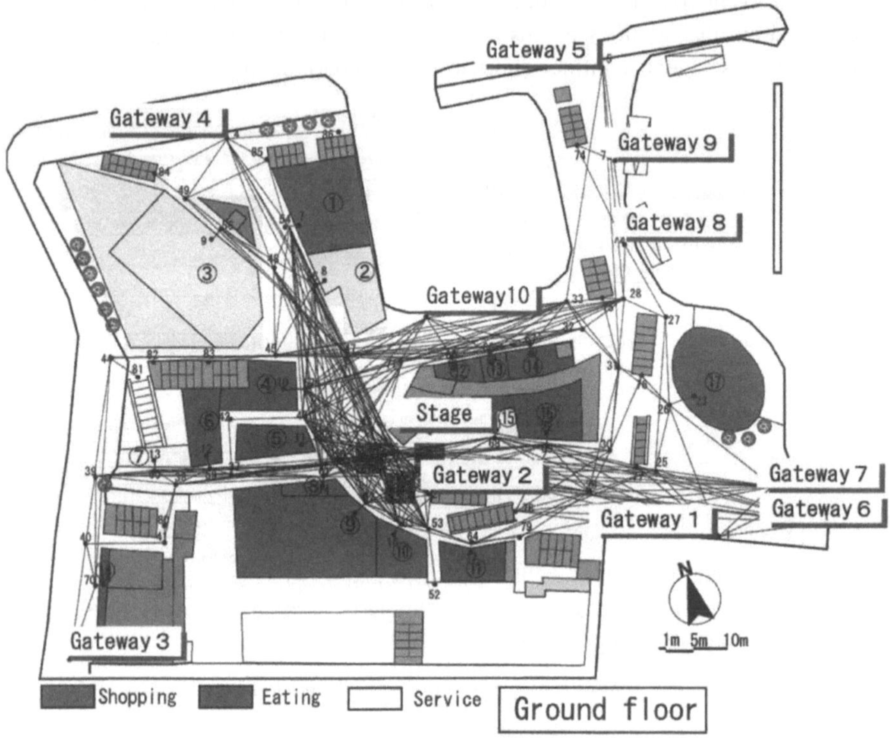


**Fig.7 Layout of Asunal Kanayama and waypoints and route network**

The attributes of a pedestrian agent were set by sex. Four categories of behavior purpose were set: shopping, eating and drinking, service and transit. For the number of drop-in faculties, 0 was given for the case of transit; for all other cases, based on research data, a random number value was obtained with a Poisson distribution using a minimum value of 1, a maximum value of 13 and an average value by attribute. From the research data, the pedestrian inflow rate at the entrances was obtained for weekdays and holidays.

Exit points were set for each behavior purpose. In the case of transit, the exit point was determined by a random number and for other purposes, 90% of pedestrians left the facilities from the same gateway as they entered, and 10% were determined by a random number. The first drop-in store was set on the same floor as the entry gateway for 90% of pedestrians and for all behavior purposes. The visit probability among facilities was set according to the research data.

In the simulation, as shown in Fig. 8, three areas were set for density measurement and a simulation was carried out for the following four cases: (a) weekday case; (b) weekday double case – the number of visitors on a weekday was doubled; (c) holiday case; and (d) holiday event case. In the simulation experiments, the value at the time of 600 steps, when the number of agents generally became steady for all cases, was measured.

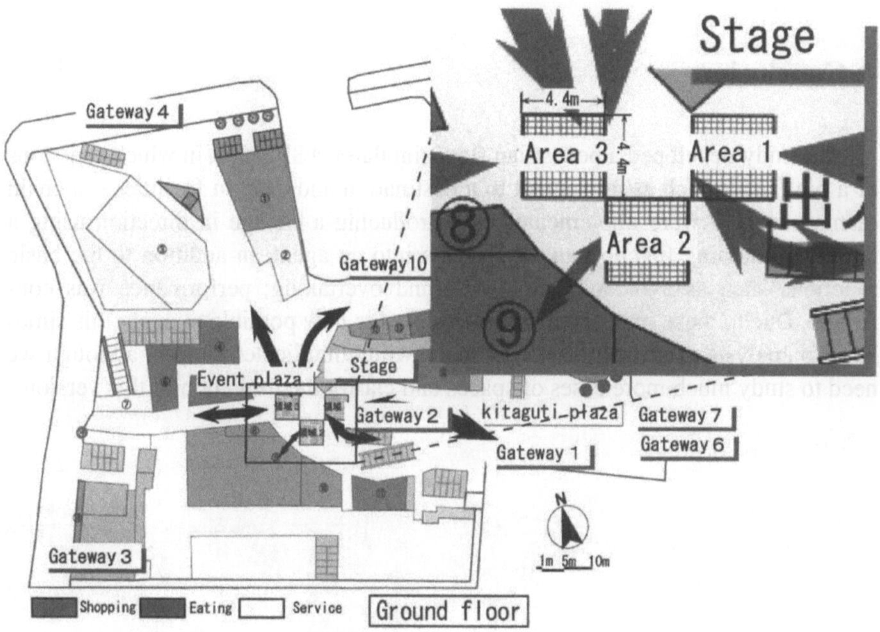


**Fig.8 Density measurement areas**

## 3.3 Analysis of Crowd Density Simulation Results

Fig. 9 shows the simulation results. In both the weekday and holiday cases, the density in the passage that leads from the event square to the west (Area 3) was relatively high (0.98 person/m$^2$ for both weekday and holiday). In the holiday case, the density in the passage that leads from the square to the square at the north entrance (Area 1) is high (1.6 person/m$^2$). This result conforms to the passing rate trends obtained from the previous year's research. In the weekday double case, as shown in Areas 2 and 3, the density increased in a different location from the weekday case. Apart from Area 1 of the weekday double case (2.22 person/m²); it was found that the density did not exceed 2.0.

Fig. 10 shows changes of density in Area 1 in the case of an event. After an event occurred, the density increased and after the end of the event, the density continued to increase for a time before declining to the same level as the holiday case. When the duration of the event time was set at 150 seconds, which is 1.5 times more than the norm, it was found that the density at peak time also went up. The conditions set were extreme; however, these results suggest that crowd control would be required at the time of an event.

# 4. Conclusion

The study developed a pedestrian flow simulator ASPFver4, in which functions of a high order such as movement to a destination and drop-in facilities – a chain of movement – were implemented by introducing a change in direction using a target maintaining function and waypoints into an agent, in addition to the basic functions such as avoidance, following and overtaking; performance was confirmed. Due to these improvements, we think it is now possible to apply this simulator to analysis of crowd density in space with complicated shapes, although we need to study much more cases of spaces and clarify the limitation of this version.

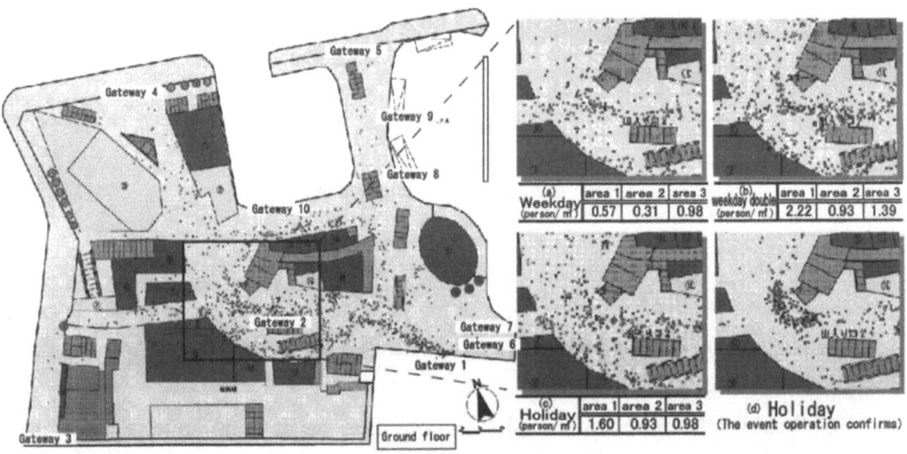


**Fig.9 Simulation results in four cases**

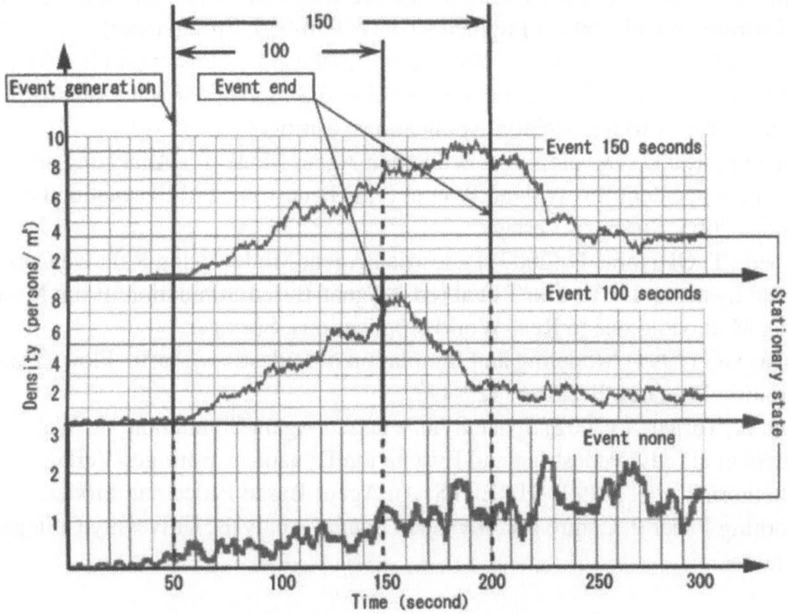


**Fig.10 Density rise at event (holiday area 1)**

## References


1. Antonini G, Bierlaire M (2007) A Discrete Choice Framework for Acceleration and Direction Change Behaviors in Walk Pedestrian. In: Waldau W et al (eds) Pedestrian and Evacuation Dynamics 2005. Springer-Verlag.
2. Batty M, DeSyllas J et al (2002) Discrete Dynamics of Small-Scale Spatial Events: Agent-Based Models of Mobility in Carnivals and Street Parades. Working Paper 56, Centre for Advanced Spatial Analysis. University College London.
3. Borgers A, Timmermans HA (1986) A Model of Pedestrian Route Choice and Demand for Retail Facilities within Inner-City Shopping Areas. Geographical Analysis 18: 115-128.
4. Haklay M, Thurstain-Goodwin MO et al (2001) “So Go Downtown ” :Simulating Pedestrian Movement in Town Centres. Environment and Planning B 28: 343-359.
5. Kaneda T, Yano H et.al (2003) A Study on Pedestrian Flow by Using an Agent Model - A Simulation Analysis on the Asagiri Overpass Accident, 2001--. In: Terano T, Deguchi H et al (eds), Meeting the Challenge of Social Problems via Agent-Based Simulation. Springer.

6. Kaneda T (2004) Agent Simulation of Pedestrian Flows. Journal of the Society of Instrument and Control Engineers 43/12: 950-955. (in Japanese)
7. Kaneda T, Suzuki T (2005) A Simulation Analysis for Pedestrian Flow Management. In: Terano T et al (eds) Agent-Based Simulation From Modeling Methodologies to Real-World Applications. Springer.
8. Kaneda T (2007) Developing a Pedestrian Agent Model for Analyzing an Overpass accident. In: Waldau.W et al (eds) Pedestrian and Evacuation Dynamics 2005. Springer-Verlag.
9. Kaneda T, Okayama D (2007) Pedestrian Agent Model Using Relative Coordinate Systems. In: Terano T et al (eds) Agent-Based Simulation: From Modeling Methodologies to Real-World Applications. Springer.
10. Lovas GG (1994) Modeling and Simulation of Pedestrian Traffic Flow. Transportation Research B 28B/6: 429-443.
11. Penn A, Turner A (2002) Space Syntax Based Agent Simulation. In: Waldau.W et al (eds) Pedestrian and Evacuation Dynamics. Springer-Verlag.
12. Schelhorn T et al (1999) STREETS: An Agent-Based Pedestrian Model. Working Paper 9, Centre for Advanced Spatial Analysis. University College London.

---

[i] Developed by Kozo-Keikaku Engineering Inc.

# Agent-Based Adaptive Production Scheduling – A Study on Cooperative-Competition in Federated Agent Architecture

Jayeola Femi Opadiji [1)] and Toshiya Kaihara [2)]

1) Graduate School of Science and Technology, Kobe University, Japan.
2) Graduate School of Engineering, Kobe University, Japan.
femi@kaede.cs.kobe-u.ac.jp and kaihara@cs.kobe-u.ac.jp

**Abstract.** An increasingly popular method of improving the performance of complex systems operating in dynamic environments involves modeling such systems as social networks made up of a community of agents working together based on some basic principles of social interaction. However, this paradigm is not without its challenges brought about by the need for autonomy of agents in the system. While some problems can be solved by making the interaction protocol either strictly competitive or strictly cooperative, some other models require the system to incorporate both interaction schemes for improved performance. In this paper, we study how the seemingly contradictory effects of these two behaviours can be exploited for distributed problem solving by considering a flexible job shop scheduling problem in a dynamic order environment. The system is modeled using federated agent architecture. We implement a simple auction mechanism at each processing center and a global reinforcement learning mechanism to minimize cost contents in the system. Results of simulations using the cooperative-competition approach and the strictly competitive model are presented. Simulation results show that there were improvements in cost objectives of the system when the various processing centers cooperated through the learning mechanism, which also provides for adaptation of the system to a stream of random orders.

## 1 Introduction

The need for very robust and flexible models for complex system simulation explains the growing trend in exploiting theories of agent interactions from the social sciences. These theories range from market-based theories in economics to social contract and culture formation theories in sociology. The main impact of these theories is in the construction of interaction protocols for a society of agents co-habiting in a specified environment. In order to employ such protocols however, it is necessary to clearly define the nature and behaviour of each type of agent existing in the environment, as well as the environmental conditions under which they exist.

One important area of application of Multiagent system (MAS) paradigm is in solving problems relating to the supply network of organizations. Problems such as manufacturing resources allocation, production system scheduling, distribution network planning, increase in complexity with increase in the size of a supply network.

## 1.1 Research Background

The background of our research stems from an interest in providing efficient algorithms for management of supply chains. The activities involved in a typical supply chain are depicted in Fig. 1. At the present stage of our work, we consider a problem at the operational level of the activity matrix in Fig. 1. This is the production scheduling problem. The ability of a production system to respond rapidly to changes that occur in its external environment is one of the factors that determine how flexible the supply network of an organization is. We therefore take a look at this problem with respect to changes in external orders placed for various goods manufactured by a production system.

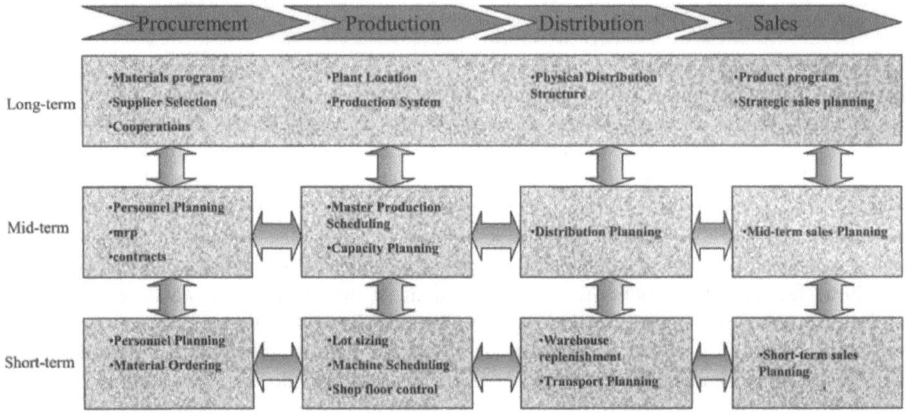


**Fig. 1.** Supply Chain Activity Matrix

At the operational level, machine scheduling is done to minimize a given cost objective or completion time objective both of which eventually have effects on the overall profit maximization objective of a supply network. Up until recently, most production scheduling problems have been addressed from a normative point of view where mathematical models employ solution methodologies such as branch and bound algorithms, dynamic programming techniques[1,2,7,8] etc. Also, meta-heuristic algorithms such as tabu-search, simulated annealing and genetic algorithm [3,4,5] have played important roles in solving a number of normative models. However, there are situations in which the production environment is stochastic and the performance of the system in terms of meeting predefined objectives strictly depend on the complex interaction of the various units that make up the system. For such systems, algorithms based on Multiagent architectures [11,15] are being developed to meet the numerous challenges resulting from uncertainties in the environment where the system is resident. Literature on market-based algorithms which form a great deal of research efforts in this area can be found in [6,12,13]. Details on auction based protocols can also be found in [9,10].

Our approach in this research work is to make use of the principles of competition in auction markets and cooperation in federated societies to provide the system with a good heuristic ability in responding to continuous changes in the external environment of the production system. The production system environment is plagued with fluctuating order volumes for the various goods manufactured by the system. The production system therefore has to schedule dynamically and find a way of adapting to orders coming in while keeping an eye on the global objective of the system, which in our case we assume to be production cost minimization.

In the next section, we describe in detail, the target production system and the scheduling environment scenario. This gives insight into how complex a dynamic scheduling problem can be. Next, the solution methodology applied to this problem is discussed. Here, we define the production environment as a social system and we give a brief description of how the agents are modeled. We also describe briefly, the interaction protocol we used in the system. In section 4, we present results from simulations carried out using hypothetical values as system parameters while in section 5, we give brief notes on the results presented in the preceding section. Conclusions drawn from the research work are given in section 6 with a pointer to future research directions.

## 2 Definition of Scheduling Problem

**Table 1.** List of Notations

| Terms | Notation | Terms | Notation |
|---|---|---|---|
| Set of task processing centers | $\boldsymbol{W}$ | Task Sequence | $\boldsymbol{S}$ |
| Task processing center | $w$ | Number of tasks in sequence | $\lvert \boldsymbol{S} \rvert$ |
| Set of machines in task processing center | $\boldsymbol{M}_w$ | Task | $j$ |
| Machine i in center w | $m_{wi}$ | Volume of task j | $v_j$ |
| Machine capability label | $L$ | Task release time | $r_j$ |
| Availability of machine at time t | $x(t)$ | Task due time | $d_j$ |
| Speed of machine i in w (volume/unit time) | $\omega_{wi}$ | Assumed speed vector for job centers (volume/unit time) | $\boldsymbol{\Omega}$ |
| Processing cost of machine i in w per unit time | $c_{iw}$ | Assumed processing cost vector at task processing centers | $\boldsymbol{\rho}$ |
| Set of orders | $\boldsymbol{Q}$ | Tardiness weight of order q | $k_q$ |
| Number of orders | $\lvert \boldsymbol{Q} \rvert = N$ | Tardiness of order q | $\tau_q$ |
| Order in time bucket i | $q_i$ | Total weighted tardiness | $Z$ |
| Payment for $q_i$ | $P_i$ | Profit generated from $q_i$ | $\pi_i$ |
| Processing cost of $q_i$ | $C_i$ | | |

The scheduling problem being considered is that of a Flexible Job shop, which is a special case of the classical Job Shop Scheduling problem [8]. We define our scheduling problem as follows:

- There exists a set of job types $\boldsymbol{o_i} \in \boldsymbol{O}$ $(i = 1,\ldots, n)$
- Each job type $\boldsymbol{o_i}$ is made up of a sequence of tasks denoted $\boldsymbol{S_i}$

- Each task sequence $S_i$ contains tasks $j_i$ $(i = 1,\ldots, n)$; therefore, a given order $y$ of type $i$ will be fulfilled by executing the task sequence $S_{iy}$.
- There exists a set of task processing centers $w_i \in W$ $(i = 1,\ldots, n)$
- There exists a set of parallel machines $m_{wj} \in M_W$ $(j = 1,\ldots, k_w)$ in each work center $w$, where $k_w$ is unique to each processing center.
- Each machine in a processing center has a unique speed $\omega_j$ at which it performs its task.
- There exists a dynamic order stream $Q$ which is random in both volume and type (i.e. processing sequence required).

The scheduling objective is to minimize the overall cost of processing all the given orders over a given period of time subject to profit feasibility constraint, i.e.,

$$\min(\sum_{i}^{N} \max C_i) \tag{1}$$

Subject to

$$\pi_i = P_i - C_i \geq 0; (for\ i = 1,\ldots, N) \tag{2}$$

The total processing cost of a given order is the sum of actual processing and the work in process inventory cost. We make use of the profit feasibility constraint because we assume that the manufacturer has a completion time flexibility advantage in which it can inform the customer of the earliest time an order can be fulfilled. This assumption is plausible because the customer can not force a manufacturer to complete a task when there is no available capacity. The constraint also means that a schedule which has been constructed for an order may be infeasible when the total cost of processing exceeds the payment to be made by the customer.

We therefore state the operational environment of the flexible job shop production system:

(a) Once an order is accepted, it must be fulfilled
(b) It is possible to reject orders if expected profit is negative
(c) Orders are fulfilled by executing a sequence of tasks.
(d) The environment does not permit preemption of tasks.
(e) A task cannot be started until the preceding task has been completed.
(f) Recirculation is permitted in the task sequence

## 3 Solution Methodology

We propose a federated-agent architecture representation of the flexible job shop as shown in Fig. 2. This model is made up of three types of agents, namely, controller agent, facilitator agent and processor agent. First, we describe the characteristics of these agents before we go ahead to describe the interaction protocol in the social system.

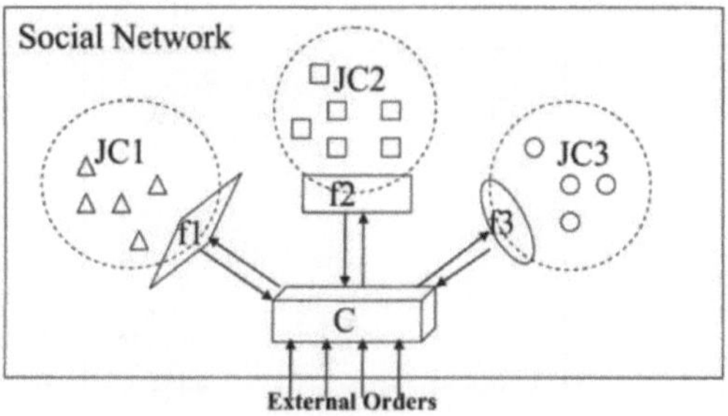


**Fig. 2.** Federated agent architecture –C is controller agent, f1, f2 and f3 are facilitator agents and JC1, JC2 and JC3 are task processing centers containing processor agents

### 3.1 Agent Description

**Controller Agent.** This agent behaves in such a ways as to maximize the overall utilization of production facilities. The controller agent is the only agent that is aware of the external world. It achieves its goal by accepting as many orders as possible subject to production feasibility constraints. The controller agent has a schedule bank where it keeps schedules of all orders that have been accepted. Fig. 3 shows a schematic diagram of the controller agent. The internal I/O interface is for communicating with facilitator agents while the external I/O interface is for communicating with the external world.

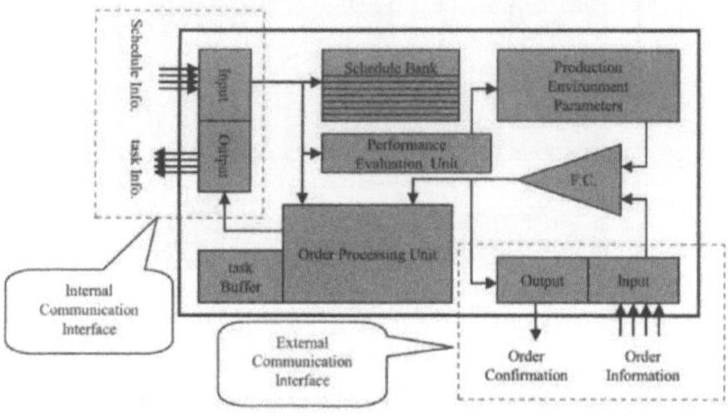


**Fig. 3.** Pictorial representation of the controller agent

**Facilitator Agent.** A facilitator agent provides a transaction interface between the controller agent and processor agents in the task processing center it represents. In order to perform its task, a facilitator agent has complete information about the status of processor agents in its domain with regards to processing capacities and availability. As an auctioneer, the facilitator agent is responsible for coordinating auction activities among its processor agents. The facilitator agent has an availability schedule which it updates for every tasked that is processed. This schedule keeps track of the number of machines that are available at every time bucket.

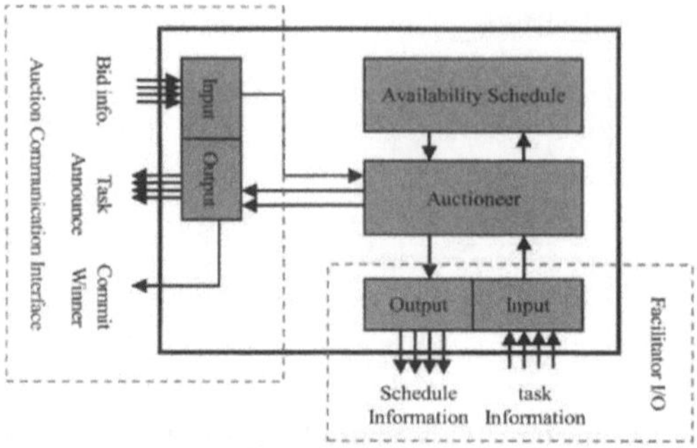


**Fig. 4.** Pictorial representation of the facilitator agent

**Processor Agent.** The actual processing of tasks is carried out by processor agents. Processor agents in the same processing center have same processing capability. A processor agent has a competitive behaviour because of its desire to make as much money as possible by winning as many bids as it can from the auction of tasks in its primary environment. Processor agents are only aware of themselves and their facilitator agent. This keeps all of them from speculative bidding resulting in a reduction in communication overhead at each task processing center. Every processor as its own scheduler which keeps track of task processing commitments it already has.

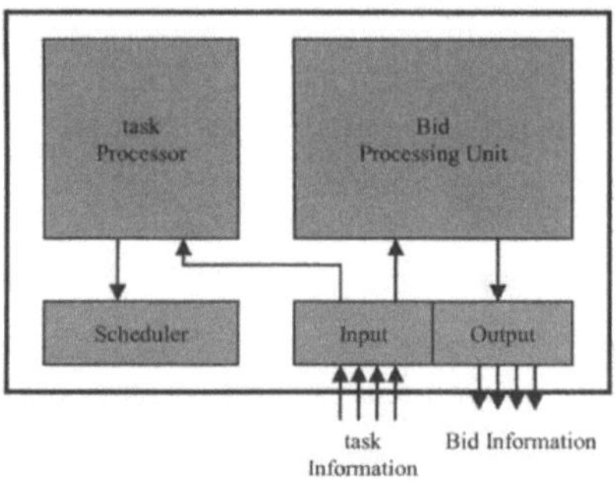


**Fig. 5.** Pictorial representation of the processor agent

### 3.2 Scheduling Algorithm

Fig. 6 shows a flowchart of the scheduling procedure. The interaction protocol of the social system is based on local competitive behaviour among processor agents in a processing center and a global cooperative behaviour among the facilitator agents in the social network.

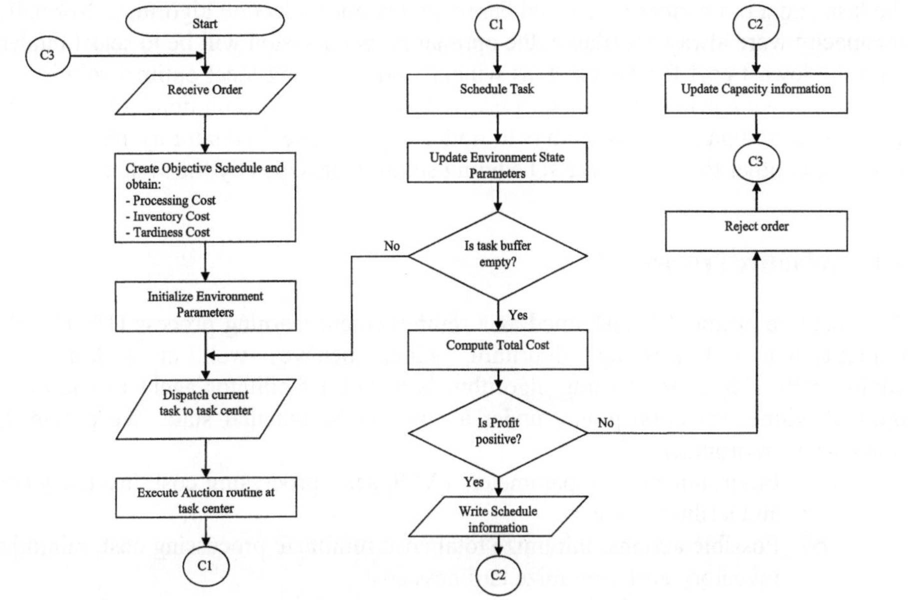


**Fig. 6.** Flowchart of scheduling process

## 3.3 Auction Routine

This part of the algorithm represents the competitive interaction that takes place among processor agents in a processing center. Fig. 7 shows a timeline diagram.

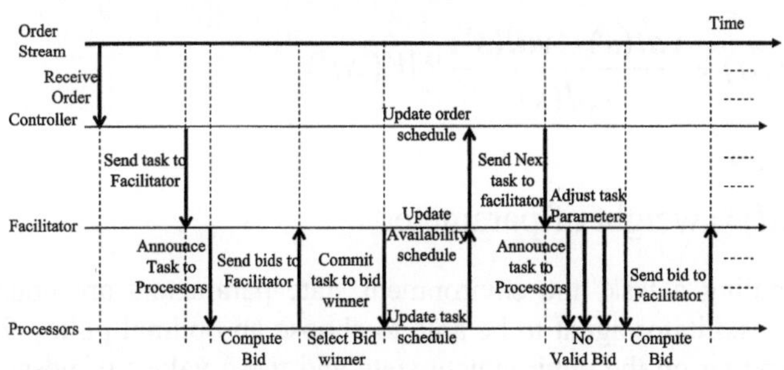


**Fig. 7.** Timeline of task dispatch and auction process

When an order comes in, the controller agent computes an objective schedule using the most cost efficient machine routes with considering capacity constraints. It is from this schedule, it obtains expected release dates and due dates for each task in the order. It then loads these parameters into its task buffers and dispatches the tasks sequentially. When a facilitator agent receives a task processing request, it proceeds to announce the task among processing agents in its domain. Processor agents bid for the task and a bid winner is selected based on a winner selection algorithm. Normally, if capacity were always available, the optimal global decision will be to select bidders with the lowest cost for each task as winners, but choosing local optimal solutions at every processing center will not necessarily result into a good solution in the long run. One way to mitigate this problem is to find a way to make facilitator agents cooperate with one another to improve the scheduling solution provided by the system.

### 3.4 Adaptive Protocol

The adaptive protocol is fashioned as a reinforcement learning process [14, 15]. We implemented a Q-learning algorithm, which behaves well in a Markovian environment [16]. The learning algorithm is to aid a facilitator agent to choose an optimal winner selection policy under a given environmental state. We define the Markovian environment as:

- Environment state parameters ($N^3$ Space): processing cost, inventory cost and tardiness cost.
- Possible actions: minimize total cost, minimize processing cost, minimize inventory cost, minimize tardiness cost
- State transfer function:

$$T(s,s') = \{(i(s)-i(s'), t(s)-t(s'), p(s)-p(s')\}$$
$$p(.) = \text{processing cost}; i(.) = \text{inventory cost}; t(.) = \text{tardiness cost} \tag{3}$$
$$val(s') - val(s) = \begin{cases} 0; \text{if } val(s) \le val(s') \\ 1; \quad \text{otherwise} \end{cases}$$

- Reward Function:

$$r(s,a) = r(s,a) + \frac{val(s) - val(s')}{val(s)} * W(val)$$
$$for\ val = i, t, p. \tag{4}$$
$$where\ W(val) = \text{weight of parameter}$$

After scheduling a task, the environment state parameters are updated so as to allow the next facilitator agent to be able to choose an optimal policy for bid winner selection depending on the environment state and the q values of possible actions in that state. This forms a kind of cooperation dynamics among the facilitator agents, which enable the system to adapt to stochastic order patterns. The q-learning algorithm creates a continuous learning environment and a delayed reward mechanism that updates Q values of actions taking in a given state depending on how close a scheduled task is to its objective schedule parameters as initially computed by the controller agent. In order words, the facilitator agents cooperate to make their schedule as close as possible to the objective schedule of the system, subject to capacity and availability constraints.

## 4 Experiments and Results

In order to verify the performance of our model, we ran simulations using a flexible job shop with 5 task processing centers and a minimum of two machines at each center. The machines in a processing center have the same processing capabilities but different speeds and processing costs. There are also 7 different types of task sequences that can be processed by the system. We fed the production system with a stream of 100 randomly generated order volumes and task sequences. Simulation was done in two parts; we first ran simulations using a strictly competitive strategy at each center, i.e., bid winners were selected by choosing the bidder with the lowest cost criterion at each processing center. Next, we introduced the cooperation mechanism which allows the system to be able to select from a list of cost criteria thereby given it an adaptive capability. We present the results of our simulations using cumulative profit, cumulative processing cost, and cumulative inventory cost as performance indices.

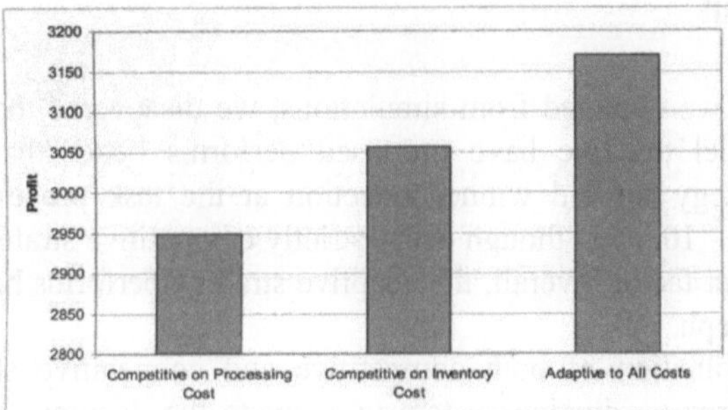


**Fig. 8.** Cumulative profit generated from order stream for different strategies

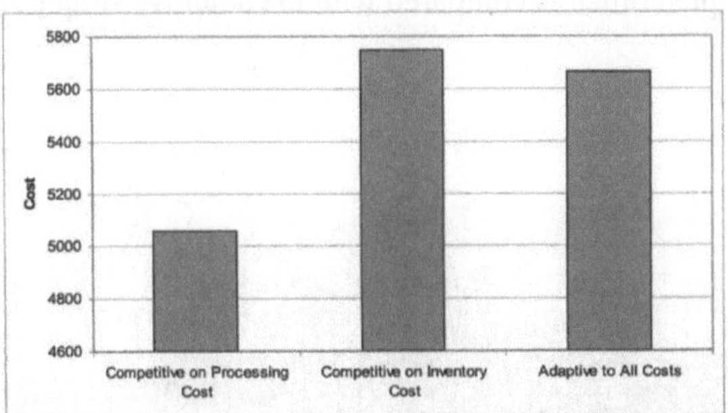


**Fig. 9.** Cumulative processing cost incurred from order stream for different strategies

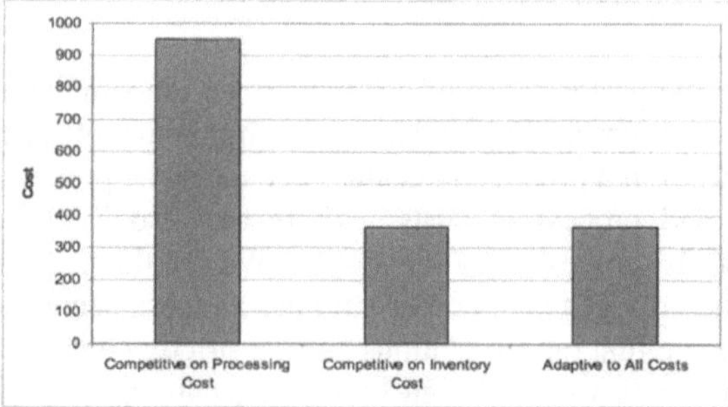


**Fig. 10.** Cumulative inventory cost incurred from processing order stream for different strategies

# 5 Discussion

From the results we obtained from simulations, we discovered that the cooperative-competition model that we have proposed performs better than using a strictly competitive strategy for bid winner selection at the task processing centers. As shown in Figs. 8 – 10, even though every strictly competitive strategy performs better in a particular cost factor, overall, the adaptive strategy performs better as revealed in the total profit graph.

Using a combination of both competitive and cooperative strategies (adaptive strategy) allows the production system to adapt to the dynamic order stream since, there is no prior knowledge of what the nature of the order would be. We show Figs. 11 – 12 to illustrate how the two competitive strategies fair in profit generation for all instances of accepted order as compared with the adaptive strategy.

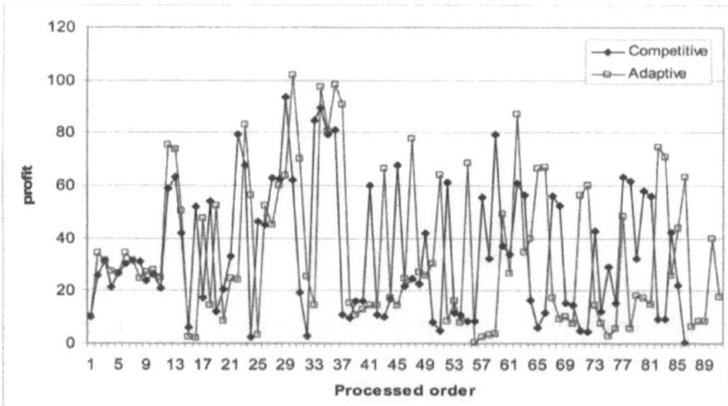


Fig. 11. Profit generated at accepted order instances (processing cost minimization strategy)

In Fig. 11, the mean profit generated from the competitive strategy is 34.3 with 85 processed orders while the adaptive strategy has a mean profit of 34.8 with 91 processed orders. It is expected that the greater the number of orders in the simulation, the greater the difference in overall profit will be since the performance of the adaptive strategy is expected to improve over time.

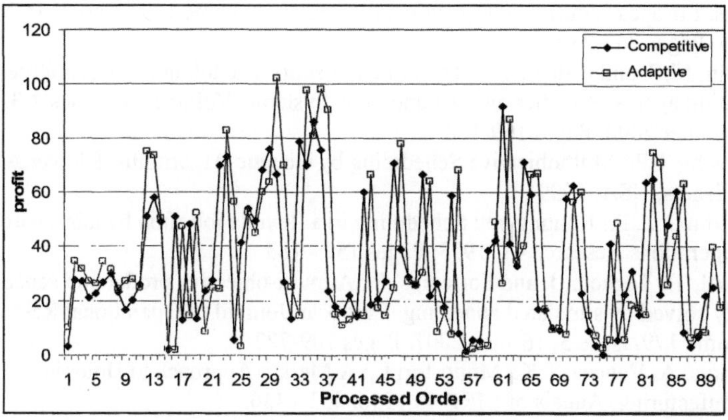


**Fig. 12.** Profit generated at accepted order instances (inventory cost minimization strategy)

In the case of Fig. 12, the mean profit is 34 with 90 processed orders for the competitive strategy based on inventory cost minimization as compared with the 34.8 and 91 processed orders of the adaptive strategy. The results show that the adaptive strategy performs better both in terms of net revenue generated and the number of orders processed.

# 6 Conclusion

Using the scheduling problem defined in this work, we have been able to show how an adaptive social network approach can be used to provide improved solutions to complex systems existing within a dynamic environment. The proposed model is based on a local competition and global cooperation strategy which will allow the system to adaptively respond to changes in the environment with a focus of predefined objectives. The same approach can be extended to other scheduling objectives like the total completion time minimization, weighted tardiness minimization etc. Also, the state parameter space defined in the Markovian environment for Q-learning can be extended to include some other parameters of concern and the possible actions in the environment can also be increased. This will improve the robustness of the system.

As stated in the research overview, our effort is geared towards more flexible supply chains, therefore, an area of interest in future research activities will be to integrate this model with a consumer layer in which orders to the system will be placed by rational consumers also seeking to maximize their utility objectives. A model like this will call for some form of negotiation mechanism between the production system and the consumer layer.

## References


[1] Allahverdi, A. and Al-Anzi, F.S.: A branch-and-bound algorithm for three-machine flowshop scheduling problem to minimize total completion time with separate setup times, European Journal of Operational Research, Volume 169, Issue 3, 16 March 2006, Pages 767-780

[2] Aydin, M.E. and Öztemel, E.: Dynamic job-shop scheduling using reinforcement learning agents, Robotics and Autonomous Systems, Volume 33, Issues 2-3, 30 November 2000, Pages 169-178

[3] Baguchi, T.P.: Multiobjective Scheduling by Genetic Algorithms, Kluwer Academic Publishers, USA, 1999

[4] Brandimarte, P.: Routing and Scheduling in a flexible job shop by tabu search, Annals of Operations Research 41, 1993, Pages 157 - 183

[5] Loukil, T., Teghem, J. and Fortemps, P.: A multi-objective production scheduling case study solved by simulated annealing, European Journal of Operational Research, Volume 179, Issue 3, 16 June 2007, Pages 709-722

[6] Markus, A, Vancza, T.K., Monostori L.: A Market Approach to Holonic Manufacturing. Annals of CIRP 45 1996. 433 - 436.

[7] Moursli, O. and Pochet, Y.: A branch-and-bound algorithm for the hybrid flowshop, International Journal of Production Economics, Volume 64, Issues 1-3, 1 March 2000, Pages 113-125

[8] Pinedo, M.: Scheduling, Theory Algorithms and Systems (2nd Edition). Prentice Hall, USA. 2002

[9] Sandholm, T.: Distributed Rational Decision Making. Multiagent Systems: A Modern Approach to Distributed Artificial Intelligence (Ed. By Weiss, G.). The MIT Press, USA. 1999. 201 - 258.

[10] Sandholm, T.: An Algorithm for Optimal Winner Determination. in Combinatorial Auctions. IJCAI 1999, 542–547.

[11] Shen, W., Norrie, D.H., Barthes, J.A.: Multi-agent system for concurrent intelligent design and manufacturing. Taylor & Francis Inc. USA. 2001

[12] Vancza, J., Markus, A.: An Agent Model for Incentive-Based Production Scheduling. Computers in Industry, Elsevier, 43 2000. 173 - 187.

[13] Walsh, W.E., Wellman, M.P.: A Market Protocol for Decentralized Task Allocation. Proceedings of the Third International Conference on Multi-Agent Systems, IEEE Computer Society 1998. 325 - 332

[14] Wang, Y and Usher, J.M.: Application of reinforcement learning for agent-based production scheduling, Engineering Applications of Artificial Intelligence, Volume 18, Issue 1, February 2005, Pages 73-82

[15] Weiss, G. (Ed.): A Modern Approach to Distributed Artificial Intelligence, The MIT Press, USA. 1999

[16] Zotteri, G. and Verganti, R.: Multi-level approaches to demand management in complex environments: an analytical model, International Journal of Production Economics, Volume 71, Issues 1-3, 6 May 2001, Pages 221-233

# A Simulation Analysis of Shop-around Behavior in a Commercial District as an Intelligent Agent Approach -A Case Study of Osu District of Nagoya City-

**Takumi Yoshida**[1], **Toshiyuki Kaneda**[1]

[1]Graduate School of Engineering NAGOYA Institute of Technology

**Abstract.** In our on-going 'shop-around' pedestrian agent project, the model is designed and implemented that performs planned actions as solutions of the scheduling problem under his/her own constraints, and has a reinforcement learning algorithm that updates the preference on the visited shops. In this paper, based on the data obtained from the questioning survey of Osu District in 2004, we examined the parameters and tried simulations under the conditions of this actual district. The validity was checked through the comparative analysis between the simulation and survey result. We also found further improvement works. In addition, by attempting an application to policy-case simulation, the research demonstrated the potentials of this simulation model.

**Keywords:** Agent-Based Social Simulation, Pedestrians' Shop-Around Behavior, Planned Action, Improvised Action, Osu District, Nagoya

## 1. Research Background and Objectives

In large modern cities, the behavior patterns of visitors to commercial districts have become increasingly diversified. For this reason, when the composition of a bustling commercial district is considered, it presents an excellent opportunity to analyze pedestrian micro behavior using a bottom-up approach. In such an analysis, pedestrians' shop-around behavior within the commercial district is the key factor to focus attention upon. Because, it is evident that each pedestrian's shop-around behavior consists of multiple units of activity; at first planned action in accordance with a preference pattern of the visitor, and later improvised action, such as the search for alternative facilities or information gathering. Furthermore, the pattern of behaviors built up is closely related to the effect of the positioning and accumulation of facilities in a commercial district. Accordingly, development of a simulation model of pedestrians' shop-around behavior in a commercial district can be a useful tool for analyzing the composition of a commercial district.

The following research on pedestrian behavior models are all worthy of mention: an absorbing Markov chain model by Sakamoto (1984), prediction of change employing a Markov chain by Saito and Ishibashi (1992), and a fusion of a Huff model and a Markov chain model by Yokoi, et al. (2000). However, these models do not give an explicit expression of planned action and improvised action, both of which are key characteristics of pedestrian behavior. In addition, with regard to diversification in the selection patterns of facilities by visitors, a modeling method that is suitable for a greater variety of expressions should be developed. To cover this point, it would seem a valuable model is Agent-Based Social Simulation (ABSS) involving an autonomous individual with intelligence; however, a model utilizing this method has not been established yet. Therefore, there is a need for a simulation model that addresses these issues.

Taking the above into account the authors designed and developed a simulator using a pedestrian behavior agent model incorporating planned behavior and immediate response behavior (Yoshida, Kaneda, 2007). In this paper, based on the data obtained from the pedestrian behavior survey of the Osu District, Nagoya City (Oiwa et al.), we established the parameters to enable this model to carry out a simulation. and we examined the validity of the simulation results through comparative analysis of the simulation and survey result, and considered points that needed improvement.

## 2. General Concepts of a Pedestrian Behavior Agent Model

### *2.1 Examination of the Characteristics of Pedestrian Behavior*

First of all, planned action and improvised action, which are key characteristics of pedestrian behavior, were examined. In the research, pedestrian behavior is classified into 4 levels. Level 1 is Planned Action: before visiting a commercial district the visitor schedules their destination and the facilities to visit within a certain period of time. The remaining three levels are all categorized under the heading improvised action: Level 2 is Alternative Visit: when the visitor trips to facilities, if fails to achieve their errand then tries to visit alternative facilities. Level 3 is Erratic Visit: a visitor unexpectedly drops into other facility other than the planned or alternative ones. And Level 4 is Detour Action: the visitor deviates from the shortest route. Table.1 shows the hierarchical characteristics of the above-mentioned pedestrian behavior. The research addressed behavior model architecture for an agent that conducts Planned Action and Alternative Visit.

**Table 1.** Functional Levels of Pedestrians' Shop-Around Behavior

| Level of action | Significance | Motivation |
|---|---|---|
| Detour Action(DA) | Walking out of the shortest route | Route preference, Information gathering |
| Erratic Visit(EV) | Visit a facility expect AV & PA | Facility preference, Information gathering |
| Alternative Visit(AV) | Visit an alternative facility expect PA | Try to visit another facility when the errand failed |
| Planned Action(PA) | Visit planned facilities | Efficient errand achievement under the constraints |

## *2.2 Assumptions Introduced for Establishing a Model*

To establish the model, the following assumptions were introduced:

**(1) A city model has only one unipolar commercial district, and there is no other prominent commercial district in any other district.**

**(2) Direct interaction between agents does not occur.**

The model can express the following interactions: interactions inside the home of each agent, and group behavior with acquaintances and friends in a commercial district; such interactions, however, are not included in the model.

**(3) All agents already know all facilities and routes.**

Essentially, it is preferable that an agent increases and updates their knowledge of facilities and routes through learning by themselves; however, at the present stage, to simplify the model, this process was omitted.

**(4) Erratic Visit and Detour Action are not examined.**

For the establishment of the model, expression of planned action and improvised action were given priority and therefore, erratic and detour behavior, which occur spontaneously, were omitted. As our further research is to plan to include these two assumptions.

## *2.3 Agent Behavior Model*

When a unipolar commercial district is assumed, the agent behavior model consists of the following three models: (1) Planning Model at home; (2) Shop-Around Model in the commercial district; and (3) Travel Model between their home and the commercial district. In Model (1), a variety of errands and time-budget are given to each agent; based on this given information and their own knowledge an agent generates the date and time to visit the commercial district, and a preliminary plan for their behavior in the commercial district. In Model (2), in accordance with the plan generated in (1), an agent who visits a commercial district walks to each of the facilities for the purpose of fulfilling their errands. If the agent fails to achieve their errand, improvised action, in which the plan is changed as required, then occurs. Moreover, based on the results of their behavior within the district, the agent updates their own knowledge base and makes use of it for a future visit to the commercial district. Model (3) connects the above (1) and (2), and expresses a round trip between the agent's home and the commercial district. Fig. 1 shows the concepts for an agent behavior model as mentioned above.

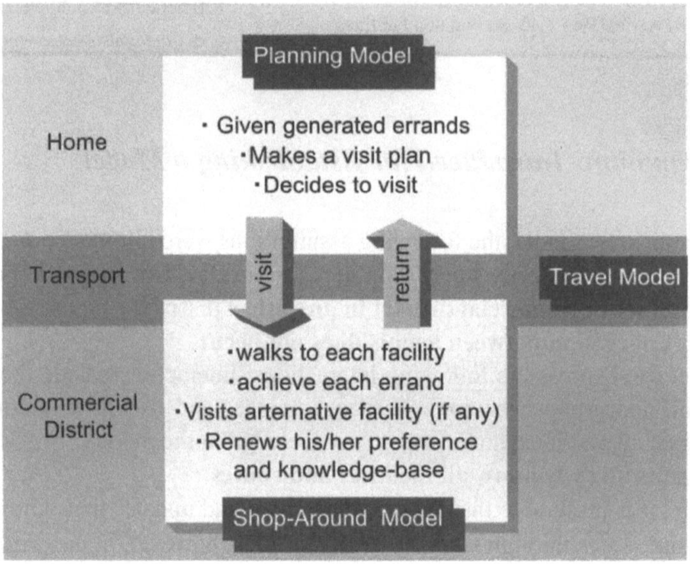


**Fig. 1.** Concept of Agent Behavior Model

## *2.4 Hierarchical Model of a Commercial District*

A spatial model of a commercial district is expressed as three levels of hierarchical structure made up from the following five categories: (1) Whole District Model that expresses the whole commercial district; (2) an Area Model that expresses each area of the commercial district; (3) Routes across areas that connect with other areas; (4) facilities that exists in each area; and (5) routes that connect the facilities in each area. District Model (1) holds the highest ranking within the entire model and includes Models (2) and (3). In the same way, Area Model (2) includes Models (4) and (5). Fig. 2 shows the basic concepts for a spatial model of the commercial district mentioned above.

# 3 Verification of the Model with Simulation: Case Study of Osu District

## *3.1 Application to Osu District*

Based on existing research data, in order to evaluate the validity of the simulation model, simulation was conducted using a case study of Osu District, Naka-ku, Nagoya City (Fig. 2).

When Osu District was modeled, the whole area of Osu District was applied as Whole District Model, and each shopping street was an Area Model. This resulted in 37 areas(includes 12 crossroads), 7 facility types, and a total of 685 facilities (Fig. 3). For a probability of an agent visiting the commercial district, $\alpha=2.13$ was assumed in Expression 2, and an errand type selection matrix was set in Table 2. The initial value of a preference variable for all facilities was set in accordance with a normal distribution, using $\mu=5$ and a standard deviation of $\rho=2$. With regard to the attribute that expresses an agent's place of residence, a short distance refers to places inside Nagoya City (it takes a half hour), a middle distance refers to places outside of Nagoya City and inside Aichi Prefecture (it takes one hour), and a long distance refers to places outside of Aichi Prefecture (it takes one and a half hours).

With regard to routes, connect the centers of each area, and each area connect each facility. In this simulation, the distance of the route in an area was assumed as all $0$; therefore, there is no difference in distances between facilities in the same area.

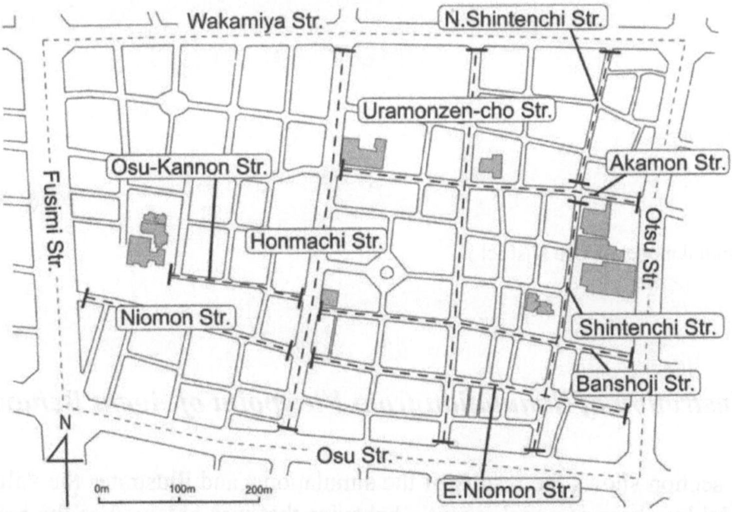


**Fig. 2.** The map of Osu District, Naka-ku, Nagoya City

**Table 2.** Distribution of errand type each attribute of agent

| | | (%) | | | | | | |
|---|---|---|---|---|---|---|---|---|
| Attribute of Agent | | Deli & Grocer's | General Goods & Furniture | Appli-ances | Clothing | Park & Temples | Second-Hand Store | Others |
| Sex | Male | 21.7 | 10.0 | 38.6 | 7.2 | 5.6 | 10.8 | 6.1 |
| | Female | 25.8 | 13.5 | 8.7 | 21.2 | 8.9 | 14.4 | 7.5 |
| Age | Less Than 30 | 21.5 | 12.8 | 22.8 | 23.4 | 5.8 | 8.5 | 5.2 |
| | 30 To 49 | 25.3 | 11.6 | 26.9 | 11.4 | 6.3 | 13.8 | 4.8 |
| | 50 or More | 24.6 | 11.8 | 18.7 | 11.0 | 11.0 | 15.4 | 7.4 |
| Trans port | Railway | 23.4 | 11.4 | 22.9 | 15.7 | 5.7 | 13.5 | 7.4 |
| | Car | 25.0 | 10.4 | 27.7 | 14.6 | 5.1 | 11.5 | 5.8 |

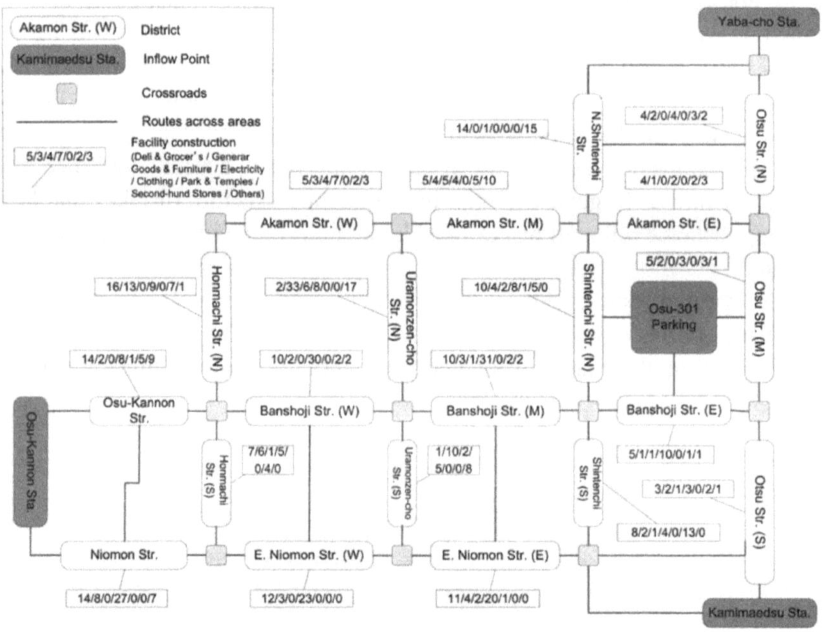


**Fig. 3.** Spatial model of Osu District

## *3.2 Illustration of Simulation from Viewpoint of Agent Behavior*

This section shows the results of the simulations, and illustrates the validity of the model by observing each agent's behavior that was obtained as the results of the execution of the simulation of the previously described model using the case study of Osu District.

As the sample, Fig.4 shows the behavior locus when on the 70$^{th}$ day an agent No.65 visited the commercial district. This agent is a housewife in their 30s or 40s and resident in the long distance zone; the agent planned to visit Facilities (1): Appliances, (2), (3): Deli & Grocer's, (4): Second-Hand Store and (6): Deli & Grocer's and traveled to the commercial district by car. When checking the behavior results, after the agent parked her car in "Osu-301 Parking Lot", she visited (1) and in this facility failed to achieve her errand; improvised action then chose (5), as an alternative facility. This alternative reorganized of the plan resulted in (5) being inserted between (4) and (6). The agent then visited (2), (3) and (4), and in Facility (4) she again failed to achieve her errand and as an alternative facility, (7) was chosen. After this, she visited (5) and achieved the errand originally planned for (1); moreover, the agent visited (6) and (7) and left for home from the "Osu-301 Parking Lot", her original appearance point. From the results, it was found that the model expresses the following situation: according to the success or failure of an errand, the agent generates improvised action and while rearranging the behavior plan as needed, the agent walks around the district. From those behavior results, it was confirmed that planned action and improvised action, which are key features of random behavior, are expressed in the model and it can be said that the objective mentioned at the beginning was achieved.

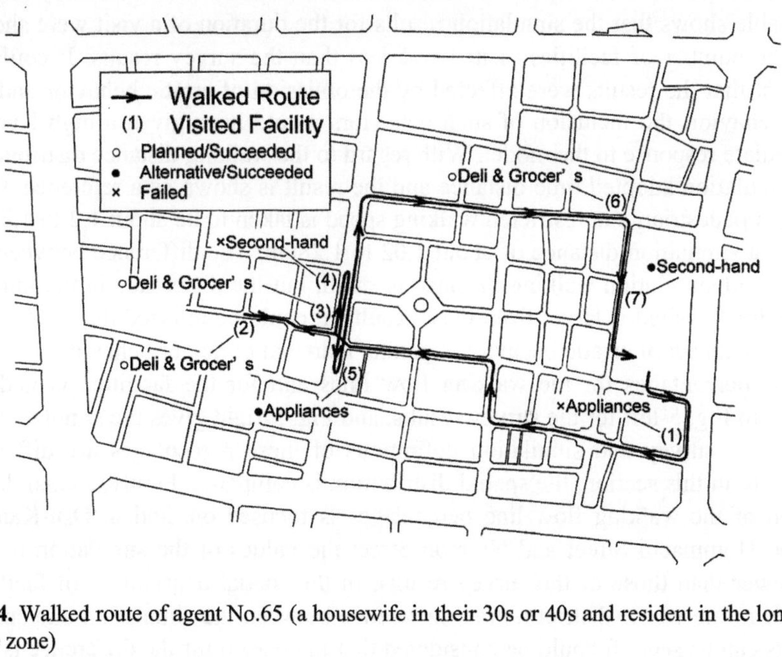


**Fig. 4.** Walked route of agent No.65 (a housewife in their 30s or 40s and resident in the long distance zone)

# 4 Comparative Analysis of the Simulation Results and the Survey Result

## *4.1 Outline of Simulation Setting*

This section presents the analysis of the results of the agents' behavior we obtained from the simulation and by comparing them with the survey results of the actual district, checks the operation, and verifies the validity of the model.

When the simulation was conducted, we established a condition of 1,500 agents living in the city and conducted a 10-day trial run; we then ran the simulation for 30 days, and calculated the average daily values of pedestrian behavior in the district as shown in the figure.

## *4.2 Comparative Analysis of Behavior in the District*

Table 3 shows the comparison of the simulation results with the survey results. The table shows that the simulation results for the duration of a visit were shorter and the number of facilities visited was less than the survey results. It could be thought that the results were affected by the omission of erratic behavior and detour behavior; the inclusion of such behaviors would have given a high level of immediate response to the model. With regard to the walking distance distribution, the simulation adopted time distance and the result is shown as a reference. Generally a pedestrian's unrestricted walking speed is taken to be about 1.2 to 1.5m/s, giving a simulation distance of about 1.02 to 1.28km. The difference between the actual and simulation walking distance is small, but the difference in the number of facilities visited is large; from these results, it could be inferred that in this district, the impact of erratic behavior is greater than that of detour behavior.

The percentages for the walking flow lines and for the facilities visited are shown in Fig. 5-left for the survey results, and Fig. 5-right gives the simulation results. The survey and simulation definitions of these percentages are different; therefore in this section, the spatial distribution is compared. Firstly, spatial distribution of the walking flow line percentages is focused on, and in Osu-Kannon Street, Hommachi Street and Niomon Street the values of the simulation results are lower than those of the survey results. In the spatial distribution of facilities visited in the whole district a great difference between the simulation and survey results can be seen. It could be considered that the reason for the difference is that in the simulation, any difference in each facility, such as gross floor area or a difference in the targeted group, is not expressed.

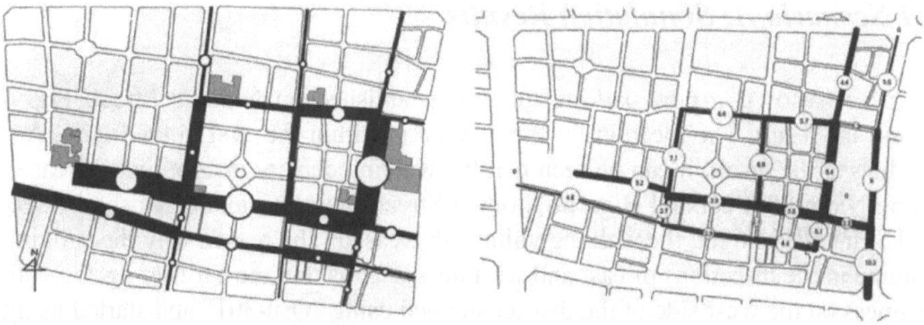

**Fig. 5.** Walking flow lines and facilities visit (left: by the questioning survey in 2003 (Oiwa et al. 2005), right: by the simulation results).

**Table 3.** Comparison between simulation result and survey result

| | Simulation Results | Survey Results |
|---|---|---|
| Stay Time in Osu District | 122min | 148min |
| Walk Time / Walk Length | 14.2min | 1.15km |
| Number of visited Facilities | 2.6 places | 5.0 places |
| Number of visited Facilities (scheduled) | 1.4 places | (Uninvestigation) |

# 5. An Attempt to Use the Simulation for Policy-making

As an example of applying the above-mentioned simulation model, we estimated how changes in the district affected agent behavior and verified the possibility of the simulator as a means of contributing to the revitalization of a shopping district. In the policy-making simulation the following two scenarios were examined.

· **Scenario A: Increase of Parking Lots**

As shown in Fig. 3, in the base case, a single parking lot is located at "Osu-301 Parking Lot", which fronts on Otsu Street, Banshoji Street and Shintenchi Street. We then provided a second parking lot on Niomon Street, verified how agent behavior changed due to this increase in parking spaces, and estimated the effect of the establishment of a new parking lot on the district as a whole.

· **Scenario B: Decrease in the Number of Facilities**

By radically decreasing the number of facilities on a certain street from the level of the base case, a so-called "shutter town" situation was created. We verified the behavior of agents in response to such a situation and examined the impact of such conditions.

### ***5.1 Scenario A: Simulation Results***

In scenario A, a new parking lot was established on Niomon Street. Fig. 6 shows the result. The location for the new lot is within the oval in the figure. According to the result, it can be seen that the walking density distribution increased along Niomon Street and Higashi-Niomon Street; however, along Banshoji Street and Shintenchi Street, the walking volume decreased. The reason why the walking volume in the directions of east and west decreased is that the car user agents with business on the west side of the district stopped using "Osu-301" and started using the newly established parking lot. With regard to the number of facilities visited, on the whole it can be seen that the number increased within the Osu District; it could be inferred that the establishment of a new parking lot encouraged revitalization of the district.

From the above, although average walking distance tended to decrease, the number of facilities visited increased; therefore, it could be thought that establishment of a new parking lot will contribute to the revitalization of a shopping district. However, as can be seen in the present Osu District, the walking volume in the east and west directions is the distinctive behavior of visitors, wandering from one shopping street to another shopping street, but in the simulation the decline in walking volume is an indicator of the atrophy of this typical behavior; therefore, on a long-term basis it is possible that the establishment of a new parking lot may damage the integrity of the district and have an overall negative effect.

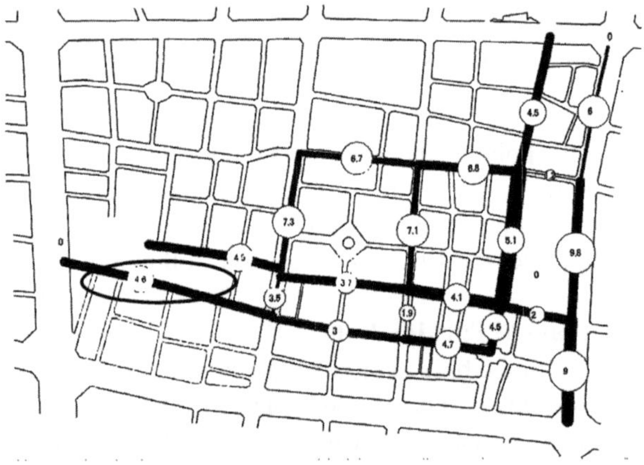


**Fig. 6.** Simulation results of scenario A -increase of parking lots-

### ***5.2 Scenario B: Simulation Results***

In Scenario B, the number of facilities in the northern part of the Uramonzen-cho Shopping Street was decreased by 77% from 66 stores to 15 stores. Fig. 7

shows the result (the decreased area is within the oval in the figure). It can be seen that in addition to Uramonzencho Street, where a decrease in walking density distribution was expected, both Banshoji Street and Higashi-Niomon Street located in the southern part of the Uramonzencho Shopping Street, suffered a relatively large decrease in walking volume too. In addition, on most of the shopping streets a slight decrease in walking volume can be seen; the exception to this trend was on Otsu Street where the walking volume increased slightly. From these results, it appears that a significant decrease in the number of facilities in Uramonzencho lowered the pedestrian circulation throughout the whole Osu District and would affect the crowded and busy feeling of the district apart from the station and the area around the parking lot. However, the result did show that the number of facilities visited in the whole Osu District hardly decreased.

In the model, a unipolar city was assumed; if in an actual city the same situation occurs, we must consider that the visitors would flow out to other commercial districts and commercial complex facilities. It can be said this point is beyond the scope of the model and is an issue that should be improved in the future.

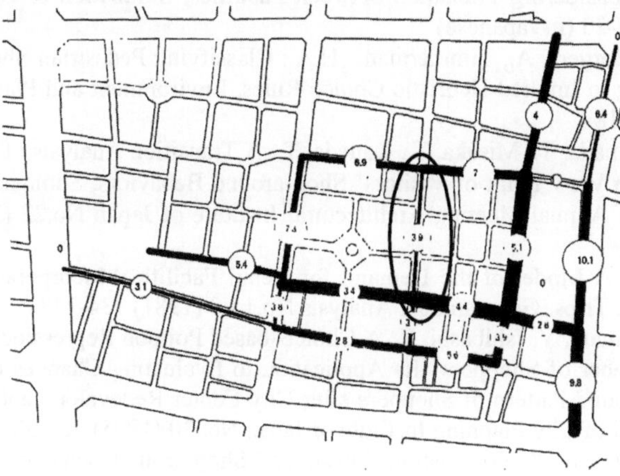


**Fig. 7.** Simulation results of scenario B - decrease in the number of facilities -

## 6 Conclusion

In this simulation when a pedestrian agent model that expressed planned behavior and one category of immediate response behavior – alternative behavior – was applied to the Osu District, Nagoya City, the research demonstrated the specific behavior of pedestrian agents, and compared the behavior of agents with actual survey results, and thus verified the validity of the simulation model. In addition, by attempting an application to policy-making simulation, the research demonstrated the usefulness of the simulation model. The following points were shown to be issues to be focused upon in the future: it is particularly necessary for an agent to have functions of erratic behavior and detour behavior – essentially a higher degree of responsiveness – and for the city model to be improved by the expansion of a unipolar model and the expression of facilities within the model. Moreover, it is necessary to make more suitable and established analysis technique for the verification of the function of the model and the simulator.

In the future, through the adjustment of parameters, the expansion of functions based on the above points, and by analyzing the evaluation of a commercial district we will continue to develop simulators, so as to be able make further contributions to the design and maintenance of commercial districts.

## REFERENCES


1. Borgers, A.,Timmermans, H.A.: A Model of Pedestrian Route Choice and Demand for Retail Facilities within Inner-city Shopping Areas, Geographical Analysis No.18 (1986) 115-128
2. Kaneda, T., Yokoi, S., Takahashi, S.: Development of Pedestrian Shop-around Behavior Model Considering Transition between Facilities, Simulation & Gaming volume 11 (2001) 17-23 (in Japanese)
3. Kurose, S., Borgers, A., Timmermans, H.A.: Classifying Pedestrian Shopping Behavior according to Implied Heuristic Choice Rules, Environment and Planning B No.28 (2001) 405-418
4. Oiwa, Y.,Yamada, T., Misaka, T., Kaneda, T.: A Transition Analysis of Shopping District from the View Point of Visitors' Shop-around Behaviors, Summaries of Technical Papers of Annual Meeting Architectural Institute of Japan No.22 (2005) 469-474 (in Japanese)
5. O'kelly, M.: A Model of the Demand for Retail Facilities, Incorporating Multistop, Multipurpose Trips, Geographical Analysis No.13-2 (1981) 134-148
6. Saito, S., Kumata, Y, Ishibashi, K. A Choice–based Poisson Regression Model to Frecast the Number of Shoppers –Its Application to Evaluating Changes of the Number and Shop-around Pattern of Shoppers after City Center Redevelopment at Kitakyushu City-, Journal of City Planning Institute of Japan No. 30 (1995) 523-528 (in Japanese)
7. Saito, S., Ishibashi, K. Forecasting Consumers' Shop-around Behaviors in an agglomerated within a City Center Retail Environment after its Redevelopments using Markov Chain Model with Covariates, Journal of City Planning Institute of Japan No. 27 (1992) 439-444 (in Japanese)
8. Sakamoto, T. An Absorbing Markov Chain Model for Estimating Consumers' Shop-around Effect on Shopping Districts, Journal of City Planning Institute of Japan No. 19 (1984) 289-294 (in Japanese)
9. Takadama, K.: Multiagent Learning –Exploring Potentials Embedded in Interaction among Agents-, Corona Publishing Co., Ltd. (2003) (in Japanese)
10. Yoshida, T., Kaneda, T.: An Architecture and Development Framework for Pedestrians' Shop-Around Behavior Model inside commercial district by using Agent-Based Approach, Computers in Urban Planning And Urban Management 2007 (2007)

# Interacting Advertising and Production Strategies - A Model Approach on Customers' Communication Networks

Jürgen Wöckl

**Abstract** In this paper we describe a simulation approach to explore different advertising and production strategies in a heterogeneous consumer market. The main focus is to model the dynamics of interacting marketing and production strategies. Such models are needed to find the right tradeoff between two general main targets: adopting the product to customer needs and/or communicate the products' features to the market. One essential key factor for a successful product launch is to set up the right product profile, which fulfills the market needs or respectively the needs of the targeted segment. The development process of new innovative products is quite complex and cost-intensive and due a potentially strong competition in most high-tech markets, the companies are forced to launch new products regularly to fulfill the steadily increasing needs of the customers. The model approach presented in this study can be used to determine the optimal product release cycles and to define suitable advertising claims to succeed in highly competitive markets. One considered advertising channel effects all consumers at the same time representing traditional large-area advertising instruments like broadcasting or print media, and a second represents the dispersion of post-purchase information in the customers' social circle - so-called word-of-mouth advertising. Here a model of an artificial consumer market has been used to provide an experimental environment for the simulation and optimization task - modelling typical stylized facts of software business. So the stylized facts are modeled using a hybrid approach of combining a continuous process described by an ordinary differential equation with discrete update processes of cellular automata and further network structures like 'random' and 'scale-free'-networks. The stability of the model is shown by comparing different market scenarios with different communication structures. Additionally the gap between the products features and the advertising claim has been varied

Jürgen Wöckl
Institute for Production Management, Vienna University of Economics and Business Administration, Austria e-mail: juergen.woeckl@wu-wien.ac.at

to figure out the influence of the resulting dissatisfaction to sales and profit of the companies. It is shown that some exaggeration generates a higher outcome, but due to the word-of-mouth effects too much exaggeration destroys the market at all.

# 1 Introduction

In this paper we describe a simulation approach to explore different advertising and production strategies in heterogeneous consumer markets. The main focus is to model the dynamics of the interaction between marketing and production strategies. Such models are needed to find the right tradeoff between two general main targets for success: addopting the product to the customer needs and/or advertise the products' features to the market. In a former study the optimal balance of exaggeration and attribute related advertising claims in the case of software business has been focused (see [11]). In this study the interaction processes between production and marketing strategies are addressed, due to the optimal company strategy is not just a question of marketing - although marketing is necessary and highly relevant. Especially in certain high-tech industries the products' features are an at least as important factor of success. So one essential keyfactor for a successful product launch is the right product, which fulfill the market needs or at least the needs of the targeted segment. But the developing process of new innovative products is quite expansive and due the strong competition - especially in most high-tech markets, the companies are to launch new products regularly to fulfill the steadily increasing needs of the customers.
Parallelly to the developement process the companies have to provide information about the products' features to the market. Therefore the companies also have to develope optimal marketing strategies, to distribute the relevant informations and to bring the product into the focus of the customers. A basic knowledge about the product and its' features is the mostly the first step of the choice/buying process, especially for the high-tech affine potentially first customers of the product - also called 'early adopters'.
A major issue for companies is the short time to market for a new product. Especially in the focused high-tech markets the competition between some major companies is quite high and due to a high advertising level and dense flow of information via the trade press, the customers' preferences are evolving quite fast. This fast adoption of the customers' preferences - at least of the preferences of early adopters, forces the companies to adopt their products to the new technology. Here beside the product development or refinement process a tricky marketing strategy might be useful, to compete with the competitor. So a strategy might be to refine existing products with some new features, and use marketing strategies to assure customers to buy. Here we mentioned first early adopters will buy and they may redistribute their

experience and satisfaction level to other potential customers. So the companies may shorten the time to market time by providing a more prototypic product, but risking an increasing dissatisfaction level of the early adopters. If such a strategy is optimal for a company, depends on the connections of the customer to other customer. To model this aspect, two concepts of networking have been focused in this study: 'random' networks and 'scale-free' networks. Depending on the type of communication between the customer, the optimal companies' policy has to be adopted. Here we are focusing on models comparing these communication networks, in terms of the resulting optimal balance between product development and advertising strategies.

This study models a high-tech market focused on innovative and new products, but also influenced / driven by global and local advertising. Here advertising is assumed as global communication process and as a local word-of-mouth communication to exchange post-purchase experience among the customers. So it deals with a fusion of two methodological approaches used to describe two main effects of advertising impact - the effect of large-area advertising instruments modeled using ordinary differential equations (ODEs) and the effect of post-purchase word-of-mouth effects modeled using cellular automata (CA). Both effects influences the customers' perception regarding the product. To succeed on the market, additionally the production processes have to be adopted to fulfill the customer needs. The communication processes have been implemented using certain connections between the cells of the cellular automata and are represented by a 'hexagonal' neighborhood - a classic cellular automata neighborhood function, a 'random' and a 'scale-free' network. The influence of this stylized fact on the optimal production/advertising strategies will be presented in the following sections. The concepts of the artificial consumer market and former studies ([4], [8], [11]) have been adopted to this issue. In the next section, first the ideas the original artificial consumer market (ACM) and later on the model adoptions to enable the fusion of the ODE and the CA approach will be presented.

## 2 Modeling the Artificial Consumer Market

Structurally the artificial consumer market is made up of a constant number of consumers represented by the cells of a cellular automata. The position of the cells in the population lattice of the cellular automata represents a local position of the consumer. Additionally to the local position each consumer has a set of internal states. Especially each consumer has an individual aspiration point of attributes which the preferred product should possess. Generally different sized segments can be assumed each representing a group of consumers with similar aspiration points and different products are established in the market competing for a market share in this artificial market. In this environment it's assumed that each firm just provides one product and therefore the profit of the firm equals the price times sales of its product reduced by the budget spent for advertising. This assumption can be made without any limitation to the model as firms providing two products can be handled as two separated firms but finally aggregating their profit.

The choice process of a artificial consumer depends on the consumer's knowledge about the products offered at the market and the attitude he gained by comparing his aspiration point to the perceived attributes of the products - induced by their advertising claims. Additionally the variables price and budget are influencing the evolvement of the attitude of a consumer regarding to a product (see equ. 1). Initially ($t_0 = 0$) no consumer knows anything about the products and the firms on the market, and their choices cannot be made rationally - in the sense of choosing a product best fullfilling their aspirations. Primary through the advertising the consumers get information about the products and their attributes and so they are able to choose the best fitting product. This evolvement of the attitude is formulated with an ODE (equ. 1). After the purchase of the supposed best fitting product the consumer gains additional information about the gap between his pre-purchase brand knowledge induced just by advertising and his own post-purchase satisfaction. Additionally in each period each consumer gets post-purchase information of his neighbors - even about products not bought by himself - and provides his own post-purchase information to his next neighbors - (word-of-mouth advertising). This word-of-mouth process is modeled using a BGK-LBM cellular automata (Bhatnagar, Gross, Krook - Lattice Boltzmann Model). So each consumer forms his own brand information based on continuous advertising and its own purchase experiences. The success of each firm/product depends on the price, the real and advertised attributes and the advertising budget invested. Generally these variables has to be optimized by all firms in the market.

In the following the functionality and the associated model equations used in this study merging the traditional ACM with the cellular automata approach is presented, short ACM-CA. The advertising impact of wide-area advertising is modeled using differential equations as mentioned above - and the local effects of post-purchase word-of-mouth effects are described by the cellular automata. Due to the CA automata is a discrete system with a discrete state space of quantities and time and the ODEs represent a continuous system, the resulting model has to deal with the resulting hybrid system.

## 2.1 Definition of the continuous system using ODEs

Following the continuous ACM ([8],[10]) the differential equation modelling the evolvement of the attitudes $att_{ijk}$ of a consumer $i$ regarding to the attribute $k$ of product $j$ is defined as:

$$\frac{d\,att_{ijk}(t)}{dt} = \frac{claim_{jk}}{\lambda_i\, price_j^*}\left[aif(budget_j)\,(1 - att_{ijk}(t))\right] \qquad (1)$$

In this equation $claim_{jk}$ indicates the advertising claim for wide-area advertising communicated by the firms $j$ regarding the attributes $k$ of their product and the parameter $\lambda_i$ describes the price sensibility of the consumer $i$. Additionally $aif(budget_j)$ indicates the advertising impact function depending to the advertising budget of product $j$, $price_j^*$ refer the relative price of product $j$ and the utilities $uti_{ij}$ are measured by the weighted Euclidean distance between the consumers' attitude and aspiration point:

$$aif(budget_j) = e^{\alpha - \frac{\beta}{budget_j}}$$

$$price_j^* = \frac{price_j}{\frac{1}{J}\sum_{j=1}^{J}(price_j)}$$

$$uti_{ij} = \frac{max(distance_{ij})}{distance_{ij}}.$$

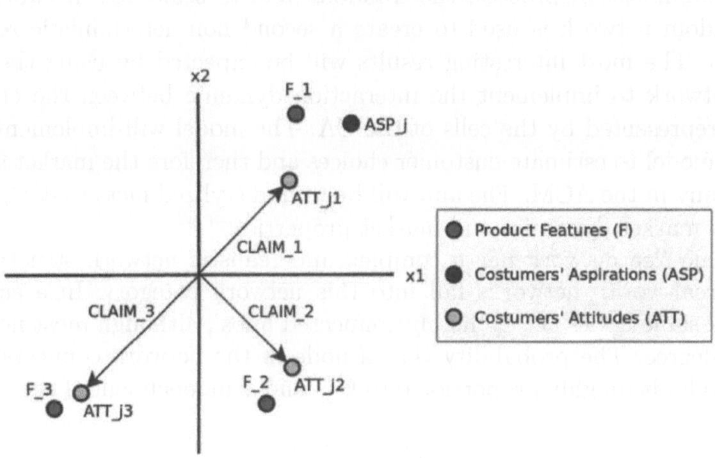


**Fig. 1** Market overview

## *2.2 Design of the Cellular Automata / Communication Network*

The design of the cellular automata (CA) environment is done object-oriented and has been implemented in Python. The lattice has a dimension of 40x50 cells and the neighborhood of the cells has been defined using different neighborhood functions. First a classical cellular neighborhood function has been used to create a stable and deterministic solution, used as a first reference

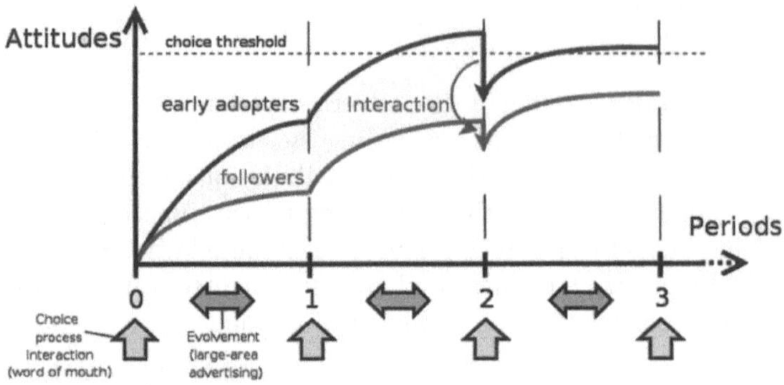


**Fig. 2** Evolvement and decision processes

solution. Additionally two kind of networks have been used to model typical communication process, the 'random' and a 'scale-free' network. Here the random network is used to create a second non-deterministic reference solution. The most interesting results will be expected by using the 'scale-free' network to implement the interaction dynamic between the customer agents represented by the cells of the CA. The model will implement a predictive model to estimate customer choices and therefore the market share of a company in the ACM. The aim will be to find stylized facts that represents realistic market dynamics and market properties.

A scale-free network has a complex and realistic network structure and many 'real-world' networks fall into this network category. In a scale-free network some nodes act as 'highly connected hubs', although most nodes are of low degree. The probability that a node in the network connects with k other nodes is roughly proportional to $k^{-\gamma}$ and is independent of the system's size:

$$P(k) \sim k^{-\gamma} \tag{2}$$

This function fits quite good for many observed network interaction data, where the coefficient $\gamma$ varies from 2 to 3.

In each generation of the CA the differential equation is solved numerically updating the attitude of the consumer evolving by the impact of the wide-area advertising. The update rule of the CA describes the information flow between consumers sharing their post-purchase satisfaction. At each discrete update-periode of the CA each consumer chooses the best-fitting product, which enables a correction of his pre-purchase attitude by comparing it with the perceived real attributes/features $f_{ijk}$ by:

$$\Delta att_{ijk} = att_{ijk}^{t+1} - att_{ijk}^{t} = \frac{f_{ijk} - att_{ijk}^{t}}{2}$$

$$\Rightarrow att_{ijk}^{t+1} = \frac{f_{ijk} + att_{ijk}^{t}}{2}.$$

The firms' target is to optimize their profit, especially the strategic decision between implementing and promoting real features (production-side) or using claims optimized to the segments (marketing-side).

## 3 Design of the study

The experimental setting assumes different levels of overstating the product features - which is necessary due to a gap in the production. The lattice of the CA defines the entire market and each cell represents an individual consumer $i$. Concerning the need for a manageable and significant experimental design in this study just one firm/product has been assumed to exist in the market and just one dimension of the k-dim. attribute/feature space has been observed. The consumers' attitude evolvement in time is described by the ordinary differential equation (ODE) solved in each generation of the cellular automata, where the cells' neighborhood have been assigned randomly using the 'scale-free network' (see equ. 2). Initially the price of the product is set constant for the whole market and therefore without loss of generality it has been assumed as 1 ($price_1 = 1$). The following parameters are set as follows: budget ($budget_1 = 1000$), level of the aspirations ($asp_{i1} = 1$), claim for the observed attribute ($claim_{11} = 1$) and real feature ($f_{i11} = 1$). To set different levels of exaggeration the advertising claim has been varied between 0.8 and 2.2 - means from a understatement to a exaggeration concerning the product feature.

A distribution of the aspiration patterns of the consumers have been assumed to generate a heterogeneous initial setting consumer market. In detail this stylized fact has been modeled by splitting up the market into two segments with different price sensibilities. Additionally this study distinguish three cases of neighborhood functions of the cellular automata to model different communication networks: First the 'hexagonal' deterministic function, second a totally random network and third 'scale-free' network, which assigns the 'early adopters' randomly to 'followers'. This linkage ensures a know-how transfer about the products' features from the 'early adopters' to the 'followers'. Due to the random neighborhood connections the CA looses the locality diffusion property, but overall the information interaction is ensured.

In detail the parameter $\lambda_i$ of the consumer has set to one for the first segment ($\lambda_i = 1$), but with the probability of $p = 0.03$ the consumer is assumed to be less price sensible and so a member of the second segment ($\lambda_i = 0.5$). So 3% of the population are assumed to be 'early adopters' willing to pay double of the price of the common consumer. This assumptions, the resulting random initial state of the aspiration patterns and the stochastic neighborhood function for 'scale-free networks' cause a stochastic component in the model influencing the evolvement of the attitude of a specific consumer. The price sensibility of each segment is assumed as normal distributed with a variance of 0.05 - so $N(1, 0.05)$ for the main segment and $N(0.5, 0.05)$ for the segment of 'early adopters'. The comparison between the three cases shows the stability of the model and makes the resulting stylized facts comprehensible. Further it has been assumed that each consumer may buy one product in each periode, but there is no force - the consumer can decide to buy or not to buy. The choice process is modeled using a threshold function. If the attitude reaches a certain level of the aspiration level $\xi$, the consumer decides to buy the product. The choice process is modeled as:

$$choose\ product\ j,\ if\ att_{ijk} \geq \xi\ asp_{ik}$$

which means that in this setting the consumer buy if their attitude cross the threshold of $\xi = 0.8$ (because of the assumption $asp_{i1} = 1$). Finally, the initial conditions, the choice rules and the resulting different evolvement speed of the two customer groups regarding the different price sensibility ($\lambda_i$) cause the 'early adopters' to buy the products earlier than the common consumer. The consumer which buy the product are able to obtain an individual evaluation comparing the advertised attributes with the real product features. Here it is assumed that the advertising has been exaggerated and that the consumers are dissatisfied ($f_{i11} = 1 > att_{i11}$). The update rules of the cellular automata generate a local dispersion of the post-purchase experience among the neighborhood communicating the dissatisfaction and lowering the attitudes of the adjacent consumers.

## Results

In this section the results are presented demonstrating the designated stylized facts. Especially some plots of the states of the cellular automata and the evolvement of the attitudes are provided to give some insights into the model dynamics. As mentioned above some consumer are less price sensible than others and so their attitude rises faster. This triggers an early choice decision. This first purchaser are also called 'early adopters'. Due to the features of the products are not fulfilling the consumers' aspirations the dissatisfaction rises after the purchase. The word-of-mouth advertising modeled by the cellular automata communicates the post-purchase experience to the neighbors also lowering their attitudes regarding the product.

The following figures show the states of the cellular automata and the attitudes for two different communication networks after 100 calculated generations: the deterministic 'hexagonal' neighborhood function (figure 3) and a scale-free network function (figure 4), assigning a random number of neighbors to the 'early adopters'.

The rise of the attitudes depends on the price sensibility of the customer. Due to the model settings assume two different sensibility levels, there are just two different attitude states. Our assumtions provoke that the 'early adopters' are dissatisfied after their first purchase and communicate the dissatisfaction to their neighborhood. The cell states of the cellular automata show the dispersion of the dissatisfaction to the adjacent customers lowering their advertising driven attitudes.

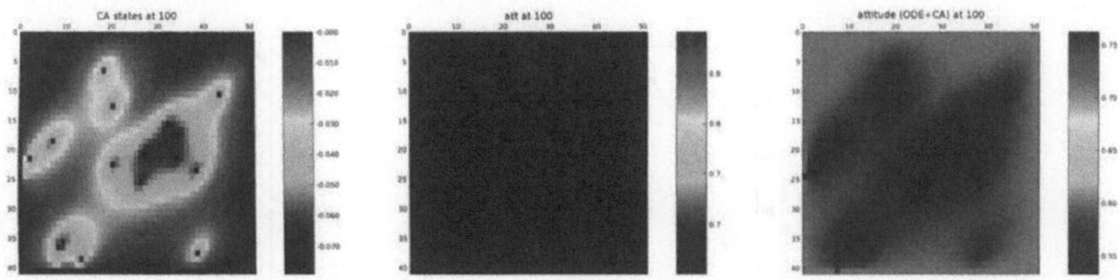


**Fig. 3** Advertising driven attitudes, CA states and total attitudes at time 100; using deterministic 'hexagonal' neighborhood function

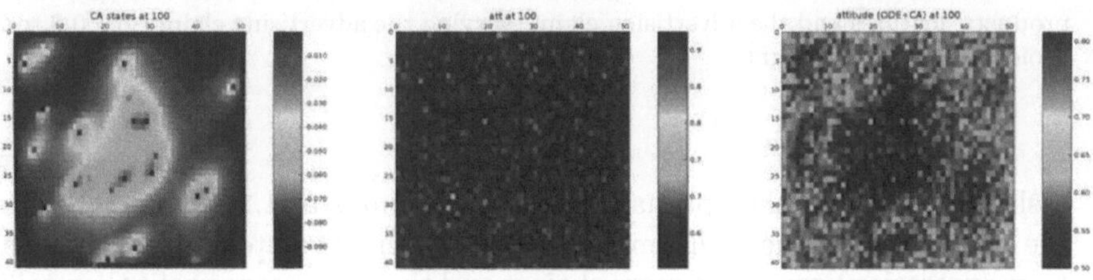


**Fig. 4** Advertising driven attitudes, CA states and total attitudes at time 100; using a scale-free network function

The results show the interaction of the two advertising strategies using a 'scale-free' network communication. The formation of local areas of dissatisfaction triggered by the word-of-mouth communication among the consumer can be seen clearly. The word-of-mouth process has also a relevant impact in the case of high-tech business - triggering the disposition to buy among the consumer in their social circle. Figure (5 - left) shows the evolvement of the decision process of time. Figure (5 - right) shows the result of the optimization

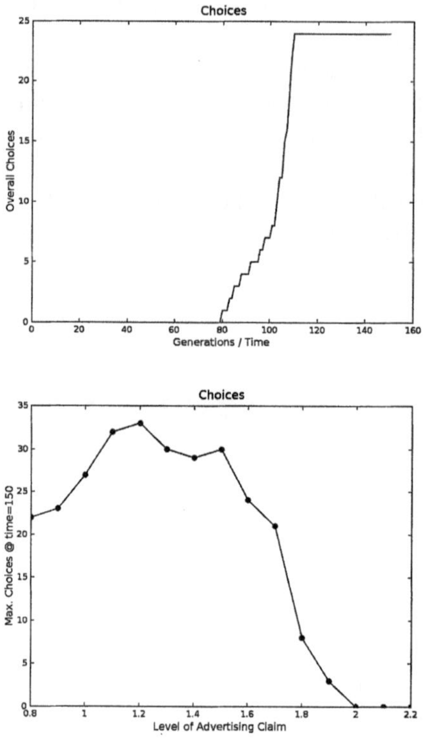


**Fig. 5** Evolvement of the choice process of the consumer market using a 'scale-free' communication network (left) and the objective function of the optimization of the gap between products' features and the advertising claim (varying the advertising claim from 0.8 to 2.2; choices at time 150) (right)

task. It shows that the optimal advertising claim is at 1.2 - thus higher than the true feature value. This implies that here an overstated advertising claim (=exaggeration) generates more choices and therefore more profit than telling the true. But there is a limit as exaggerating too much results in no choices at all. Especially if the dissatisfaction among the 'early adopters' caused by the exaggerating is too high, the early nevertheless start buying in early periods, but the high degree of dissatisfaction brandy by the word-of-mouth process, blocks all future choices of the entire market. This result is also quite realistic in real-world high-tech markets. The optimal choice of the marketing claim is important in such markets driven by advertising. Especially if the product attributes are not easily appraisable by the customer, the advertising claim is the first trigger of choice and thus important for the marketing strategy of such products. Therefore an exaggeration in advertising gains more sales and more profit - but the word-of-mouth process among the customers is a dangerous limitation - able to destroy the reputation of the product.

# References

[1] Albert, R., Barabási, A.L.: Statistical mechanics of complex networks. Reviews of Modern Physics **74**, 47–97 (2002)

[2] Bakos, Y., Brynjolfsson, E.: Bundling information goods: Pricing, profits, and efficiency. Management Science **45**(12), 1613–1630 (1999)

[3] Barabási, A.L., Bonabeau, E.: Scale-free networks. Scientific American pp. 50–59 (2003)

[4] Buchta, C., Mazanec, J.: SIMSEG/ACM A Simulation Environment for Artificial Consumer Markets. Working Paper 79, SFB - Adaptive Modelling: www.wu-wien.ac.at/am (2001)

[5] Cusumano, M.: The business of software. Free Press, New York (2004)

[6] Feichtinger, G., Hartl, R.F., Sethi, S.P.: Dynamic optimal control models in advertising: Recent developments. Management Science **40**(2), 195–226 (1994)

[7] Lawson, M.V.: Finite automata. CRC Press (2004)

[8] Schuster, U., Wöckl, J.: Optimal Defensive Strategies under Varying Consumer Distributional Patterns and Market Maturity. Journal of Economics and Management **1**(2), 187 – 206 (2005)

[9] Shapiro, C., Varian, H.R.: Information Rules: A Strategic Guide to the Network Economy. Harvard Business School Publishing (1998)

[10] Wöckl, J., Schuster, U.: Derivation of stationary optimal defensive strategies using a continuous market model. In: AMS Annual Conference, pp. 305–311 (2004)

[11] Wöckl, J., Taudes, A.: A hybrid approach to modelling advertising effects - an application to optimal lying in software business. In: The First World Congress on Social Simulation, vol. I, pp. 175–182. Kyoto, Japan (2006)

[12] Wolf-Gladrow, D.A.: Lattice gas cellular automata and lattice Boltzmann models: an introduction. Springer (2001)

# Agent-Based Approaches to Social Systems

# A Method to Translate Customers' Actions in Store into the Answers of Questionnaire for Conjoint Analysis

**Hiroshi Sato[1], Masao Kubo[1], Akira Namatame[1]**

[1] Dept. of Computer Science, National Derense Academy of Japan,

1-10-20 Hashirimizu, Yokosuka, Kanagawa 238-8686, Japan

{hsato, masaok, nama}@nda.ac.jp

**Abstract.** This paper proposes a new method for analyzing people's hidden intention that is made from the combining of old statistical method and new IT tools. We extend conjoint analysis so as to it can deal with something other than questionnaires. Conjoint analysis is a statistical technique to reveal customers' invisible preference using series of questions regarding tradeoffs in products. The proposed method interprets customers' actions in store, such as flow line or sojourn time, as the series of questions. We demonstrate the effectiveness of this method through the experiments done in a convenience store. The results of the experiments show that the preference of the customers is clearly changed between before and after meal.

**Keywords:** Service, Science, Conjoint Analysis, Multi-attitude Compositional Model, Random Utility Maximization Model, Multinominal Logit Model, Behavior Analysis

## 1 Introduction

To know customers' preference is the most important, but at the same time, the most difficult thing in service industry. There are two directions regarding customer survey: (1) direct approach: asking customers, and (2) indirect approach: observing customers. Questionnaire [1] is the typical example of the former and behavior analysis [2] is the typical example of the latter. Both have pros and cons.

Questionnaire is simple and traditional way, but it usually needs incentives such as money to have people answer the questionnaire and what is worse that people may not answer correctly to the questionnaire. Behavior analysis is based on the observation of the agents (animal or human). It can obtain more precise result because the action of agents represents what they think. But this method needs special instruments for recording or analyzing.

Conjoint analysis is a variation of questionnaire [3] [4]. Through the series of questions regarding tradeoffs of products, it can reveal the relative strength of attributes of customers for the products. This method can prevent examinee to deceive examiner because the question is indirect and it is difficult to estimate what the examiner want to know.

Conjoint analysis usually uses the cards (conjoint cards) and it required the examinees to sort or choice these cards. The examples of the products that have tradeoff of attributes such as price, size, or performance are printed in the cards. The results of sorting or choice can be used to estimate relative importance of the attribute to the examinees.

When we think about buying behavior, the actions of customers in a store such as turning the corner to some direction or stopping at the front of some shelf can be thought as choice or sort. POS system is used for long time to analyze customers but it can store only purchase data. Given the recent and rapid development of IT tools, it is relatively easy to store, retrieve and analyze these customers' actions.

We then propose the extension of conjoint analysis which can be carried out without conjoint card. In this method, each actual product in the store represents is considered to be each conjoint card, and the stored records of customers' actions are considered to be the choice or sort of the conjoint cards.

The rest of this paper consists of the following sections. In Section 2, conjoint analysis – the theoretical background of this study – is reviewed. In Section 3, we propose a new method that extends conjoint analysis. The effectiveness of the method is demonstrated in Section 4. Section 5 concludes the paper and discusses future work.

## 2 Conjoint Analysis

### *2.1 Basic Concept*

Conjoint analysis is the statistical technique used in marketing research to determine the relative strength of the attributes of the product or service. It originated in mathematical psychology [3] [4].

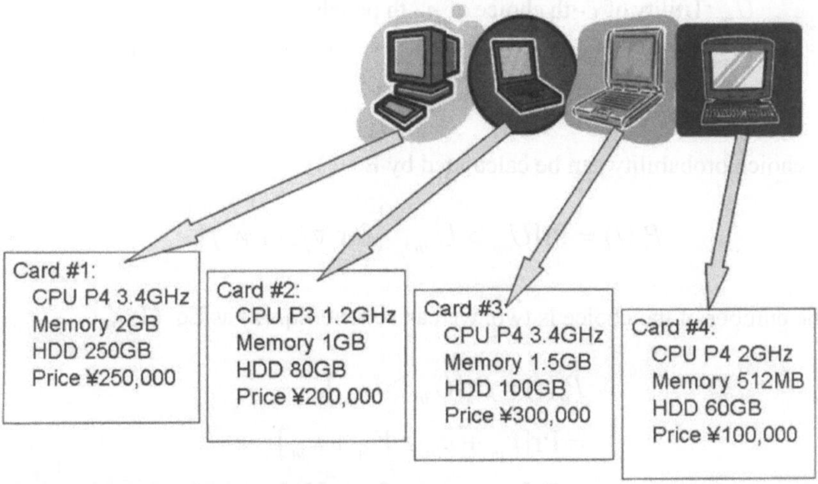


**Fig. 1** An example of conjoint card: The combination of attributes of PC makes conjoint cards. Examinees should asked to choice or sort these cards by their preferences.

The following is the typical procedure of conjoint analysis: A product or service is described in terms of a number of attributes. Figure 1 shows an example of Personal computers. PCs may have attributes of size, mobility, memory, hard drive disk, CPU, and so on. Each attributes can be broken into a number of levels. Examinees see a set of cards created from an appropriate combination of the levels of the products and are asked to choice, sort, or rate bye their preferences. Using the regression analysis to their sorting data, the implicit utilities for the levels can be calculated.

## *2.2 Theoretical Background*

Conjoint analysis assumes that agents have utilities and they behave based so as to get higher utilities. We use Random Utility Maximization (RUM) Model [3]. In RUM, the utility is formulated in the shape of linear functions (Eq. (1)).

$$U_{in} = \beta_1 x_{1in} + \beta_2 x_{2in} + \cdots + \beta_k x_{kin} + \varepsilon_{in} = V_{in} + \varepsilon_{in}, \tag{1}$$

where,

$U_{in}$ : Utility of $i$ - th choice of $n$ - th people,
$x_{kin}$ : $k$ - th explaining variable of $i$ - th choise of $n$ - th people,
$\beta_k$ : $k$ - th parameter.

The choice probability can be calculated by Eq. (2).

$$P_n(i) = \Pr[U_{in} > U_{jn}, \quad \text{for } \forall j, \;\; i \neq j] \tag{2}$$

If the number of the choice is two, we can rewrite Eq. (2) as Eq. (3).

$$\begin{aligned} P_n(i) &= \Pr[U_{in} \geq U_{jn}] \\ &= \Pr[V_{in} + \varepsilon_{in} \geq V_{jn} + \varepsilon_{jn}] \\ &= \Pr[\varepsilon_{jn} - \varepsilon_{in} \leq V_{in} - V_{jn}] \\ &= \Pr[\varepsilon_n \leq V_{in} - V_{jn}] \\ &= \mathrm{CDF}_\varepsilon(V_{in} - V_{jn}) \end{aligned} \tag{3}$$

**Binary Logit Model**

If the CDF is logistic distribution, Eq. (3) will be Eq. (4).

$$P_n(i) = \frac{1}{1 + \exp(-\mu(V_{in} - V_{jn}))} = \frac{\exp(\mu V_{in})}{\exp(\mu V_{in}) + \exp(\mu V_{jn})} \tag{4}$$

where, $\mu$ is a scale parameter.

**Multi-nominal Logit Model**

When the number of choice will be more than three, Eq. (4) is modified to Eq. (5).

$$P_n(i) = \frac{\exp(\mu V_{in})}{\sum_{j=1}^{J} \exp(\mu V_{jn})}, \quad i = 1, \ldots, J \qquad (5)$$

Parameter estimation is done by Maximum Likelihood Estimation (MLE). Log likelihood is often used for actual calculation (Eq. (6)).

$$L = \prod_{n=1}^{N} \prod_{i=1}^{J} P_n(i)^{d_{in}}, \text{ or } \ln L = \prod_{n=1}^{N} \prod_{i=1}^{J} d_{in} P_n(i) \qquad (6)$$

where,

$$d_{in} = 1 \text{ (if n - th person selects i - th choice)}, \\ = 0 \text{ (if not)}$$

## 3 Extension of Conjoint Analysis: Conjoint Analysis without Conjoint Cards

In this section, we propose the extension of conjoint analysis. Usually, we use the cards to carry out conjoint analysis. As we mentioned in Section 2, conjoint analysis consists of the following two parts: (1) making conjoint cards, and (2) asking examinees to sort or choice the cards. We can substitute each part using IT tools. We discuss the possibility of extension of these two parts in order.

**Table 1** Attributes and levels of the foods and drinks sold in convenience store.

| Attributes | Level |
|---|---|
| Category | Food, Snack, Drink |
| Temperature | Hot, Normal, Drink |
| Life time | Short, Long |
| Price | High,, Normal, Low |

### 3.1 Making Conjoint Card from Store Shelf

We have to remember the each conjoint card represents a possible product that has some combination of attributes. Fortunately, there are many kinds of possible products on a shelf in usual convenience store. Then we can approximately think the products on the shelf as the display of conjoint cards and the action taken at the front of the shelf can be translated into the action against conjoint cards.

In particular, if the products in a convenience store are treated, foods and drinks will be the good object of conjoint analysis. Table 1 shows the example of possible attributes and levels of foods and drinks. Fig. 2 and Fig. 3 show the shelf layout in a convenience store and the translation of products on the shelf to conjoint card.

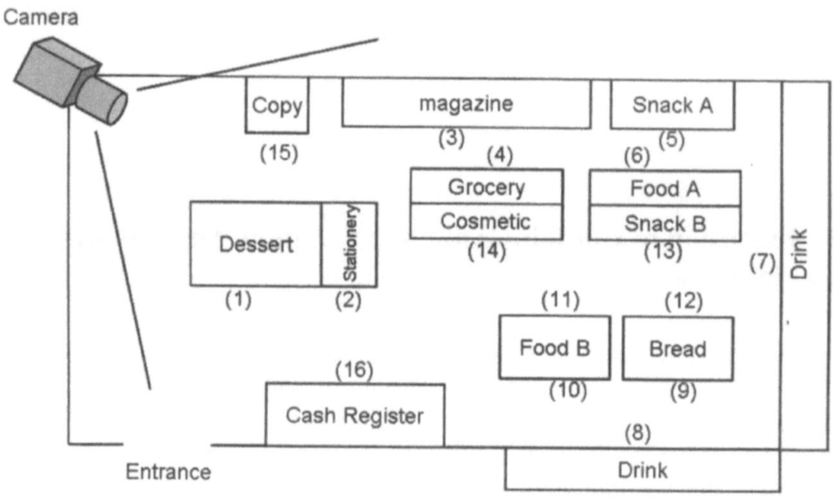


**Fig. 2** The layout of a convenience store: Each shelf in the store is classified by the products.

### 3.2 Reading Mind from Customers' Action Instead of Asking Questions

The above idea is maybe considered at the time in the past, but there is no means to capture the customers' action at the time. However, recent development of IT tools allows us to monitor and record customers every action. For example, RFID or very smart image processing methods can be used for this purpose.

Card #1 (Food, Hot, Long, Low) --> Position (6)
Card #2 (Food, Hot, Shot, High) --> Position (11)
Card #3 (Food, Normal, Short, Normal) --> Position (12)
Card #4 (Snack, Cold, Short, Normal) --> Position (1)
Card #5 (Snack, Normal, Long, Low) --> Position (5)
Card #6 (Snack, Normal, Long, High) --> Position (13)
Card #7 (Drink, Cold, Long, Normal) --> Position (7)
Card #8 (Drink, Cold, Short, Normal) --> Position (8)

**Fig. 3** An example of making conjoint card: the product on the shelf can be translated into the conjoint cards. "Position" represents the shelf number in Fig. 2.

We select spending time at the front of each shelf in the store as the substitution of customers' choice for conjoint analysis. Longer spending time can represents they are interested in the products and vice versa. POS data is unsuitable for this purpose because it only tells us whether it sold or not. We need more detailed data such can represent how they want or how they wander to buy some products.

In this study, we use the video-based IT tools because it is easy to check the log (just watch the video). We adopt Vitracom's SiteView [5] for the analysis of video image. It can count, detect and track objects crossing the counting line in both direction at very high density. Fig. 4 shows the screenshot of counting people.

## 4 Experiments

In order to examine the effectiveness of our method, experiments were done in two situations at the convenience store located in our campus. Table 2 shows the condition of the experiments. Fig. 5 and Fig. 6 are scenes of these two situations.

**Table 2** Condition of the experiments.

| Date | March 16th, 2007 |
|---|---|
| Location | Convenience store in our campus (National Defense Academy of Japan) |
| Time | 16:00-17:00 (Situation 1)<br>18:00-19:00 (Situation 2) |
| Objects | Food and drink |

**Fig. 4** Analyzing customers' action: Vitracom Siteview is used to analysis of customers' behavior. Siteview is video-based counting device. It can count objects crossing the counting line and can detect and track the objects using optimization algorithms.

**Fig. 5** Typical scene of Situation 1 – Before meal: 16:00-17:00. In this situation, the store is sparsely populated (2 ~ 3 people in the store).

**Fig. 6** Typical scene of Situation 2 – After meal: 18:00-19:00. In this situation, the store is densely populated (10 ~ 15 people in the store).

## *4.1 The Results of Experiments*

SPSS conjoint [6] is used to carry out these investigations. Fig. 7 shows the relative importance of attributes of the products in both experiments. Importance among attributes are almost same in situation 1 (before meal), but the category is the most important attributes in situation 2 (after meal).

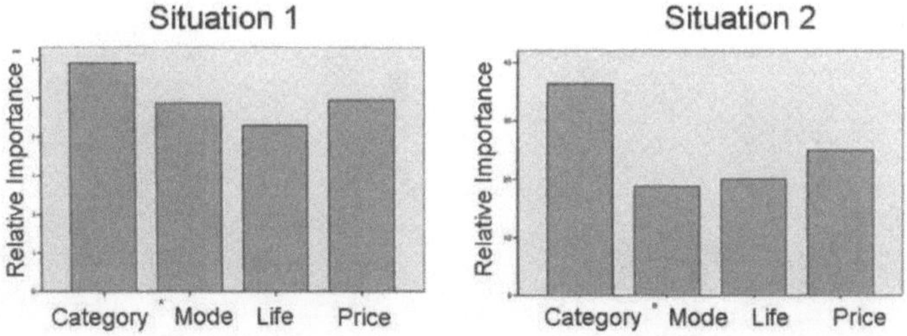


**Fig. 7** The difference of relative importance to the utility function.

Fig. 8 (a)-(d) shows the effects of each attributes on the utility functions of the customers. These results clearly show the change of the parameters of the utility functions.

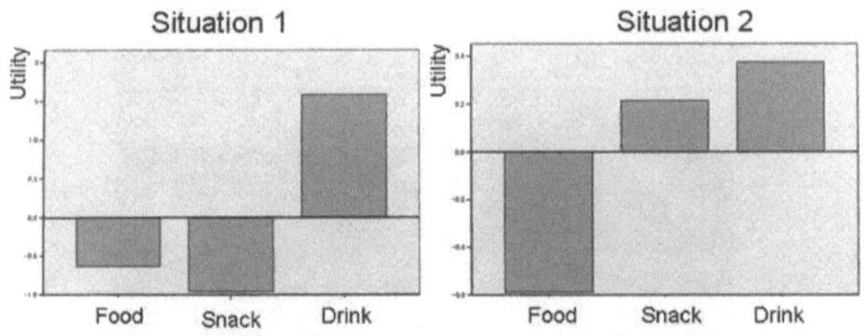


(a) Category of the Food

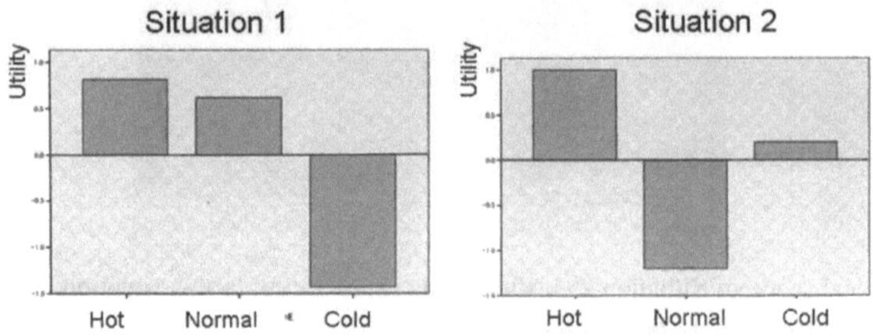


(b) Temperature of the food

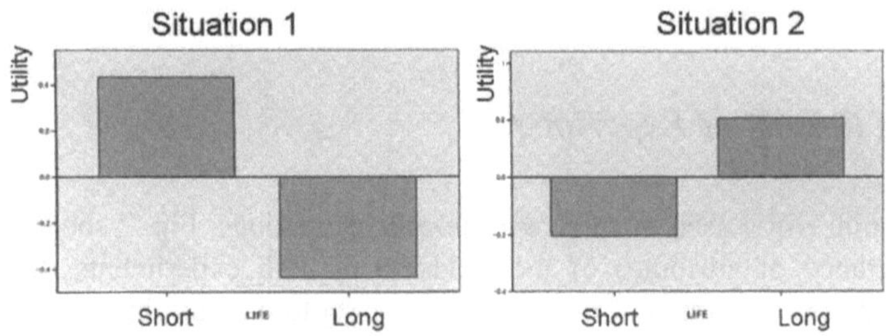


(c) Lifetime of the food

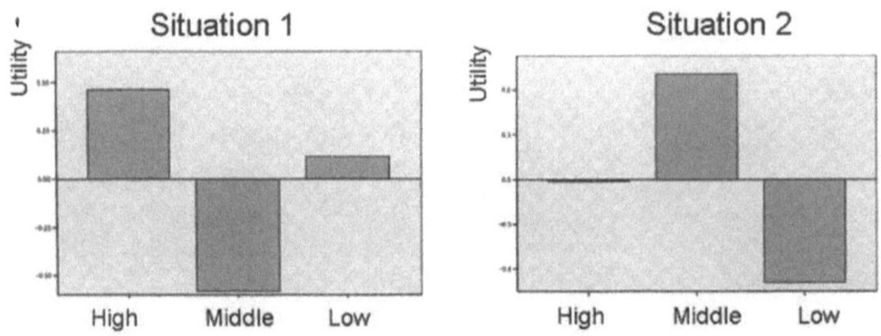


(d) Price of the food

**Fig. 8** The effects of each attributes on the utility function of customers

## 4.2 *Validity of the methods*

We now compare this result to the real sales. Because the data we can obtain is daily base, we arranged the analysis in daily base. Figure 9 is the results of conjoint analysis.

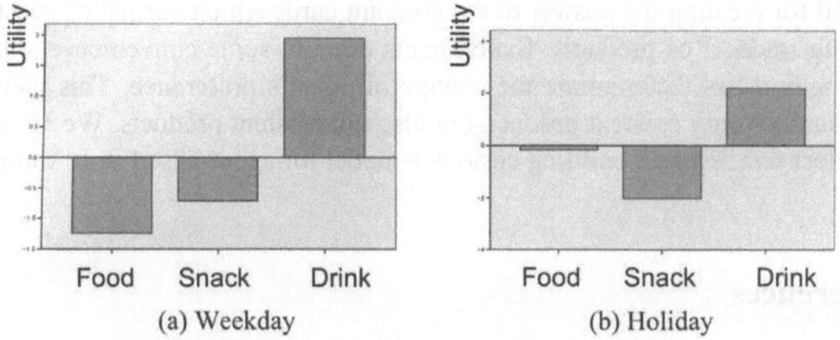


**Fig. 9** The utilities of the products classified its category.

The interesting finding is that the order of the utilities is coincident with the order of the real sales figures: In weekday, the preference of customer is drink > snack > food. On the other hand, the preference is drink > food > snack in holiday. This fact means that people tend to buy the product which they wonder whether they buy or not. This result suggest that the store we investigated has a appropriate layout and display.

## 4.3 *Simulating the efficiency of the possible products*

When we get the estimation of the parameters of utility function for customers, we can simulate the rating of nonexistent products by calculating the utility function. The following two products show the opposite utilities between the situations.

- (Drink, Normal, Short, Low) gets high rating in situation 1 (= 5.61), but low rating in situation 2 (= 1.24).
- (Snack, Hot, Short, Middle) gets low rating in situation 1 (= 2.22), but high rating in situation 2 (= 4.00).

## 5 Conclusion

Knowing customers' preference is the most important but difficult thing in marketing. We propose new investigating method which combines questionnaire and behavior analysis. In this method, customers are modeled as agents that maximize their utilities. The parameter of utility function of the agent is estimated with their actions in store such as flow line and sojourn time. More precisely, agents' action is used for creating the answer to the conjoint cards which consist of questions regarding tradeoff of products. Experiments done in some convenience store show this method can differentiate the change of agent's preference. This method can simulate not only existent product, but also nonexistent products. We are planning to reflect this result in building customer-model for agent-based store simulator.

## References

1. P. Hague, Questionnaire Design, Kogan Page (1993)
2. R. W. Malott, M. E. Malott, E. A. Trojan, Elementary Principles of Behavior, Prentice Hall College Div. (1999)
3. P. Green, V. Srinivasan, Conjoint analysis in consumer research: Issues and outlook, Journal of Consumer Research, vol. 5, (1978), 103-123
4. A. Gustafsson, A. Herrmann, F. Huber (ed.), Conjoint Measurement: Methods and Applications, Springer-Verlag (2006)
5. Vitracom Siteview Web Page: http://www.vitracom.de
6. SPSS Web Page: http://www.spss.com

# Agent-Based Simulation of Learning Social Norms in Traffic Signal Systems

Kokolo Ikeda[1], Ikuo Morisugi[2], and Hajime Kita[1]

**Abstract** Formation and maintenance of social norms are important to keep multi-agent systems. Though the mechanisms how such norms are formed and maintained or collapsed are not clear, agent-based simulation is a promising approach for analysis. In this paper a compact framework of simulation, with learning agents in traffic signal system, is presented. Through some preliminary experiments, for example to analyze how the traffic volume effects the norm formation, the potential of the framework is shown.

## 1 Introduction

"Social Norm" is one of the most important factor of multi-agent systems such as our society, for both improving individual comfort and keeping the whole system efficient and stable. For example, it is a well-authorized norm in railway stations that the people who are riding should wait until the people who are getting off. If this norm is broken, for example someone try to ride first, the total cost of boarding and exiting will be fairly increased. On the other hand, considering that the social agents behave seeking their own goals, it is not clear how such social norms are formed, maintained, and collapsed.

Roughly speaking, there are two prominent types of pressures on social norm formation, that is, "bottom-up" pressures and "top-down" pressures. Bottom-up pressures are pressures arising naturally by the mutual interactions of the agents, such as the eventual loss by the greedy actions, or being crushed around doors in the aforesaid example. However, there is no guarantee that such pressures lead the policies of agents to the socially desirable direction.

1. Kyoto University, Yoshida-Nihonmatsu, Sakyo, Kyoto 606-8501, JAPAN,
e-mail: kokolo@media.kyoto-u.ac.jp kita@media.kyoto-u.ac.jp ·
2. NTT WEST Corporation, Baba 3-15, Chuou, Osaka 540-8511, JAPAN

By contraries, top-down pressures are pressures given by certain organization ruling the society in order to control it efficient and stable. In many cases, top-down pressures are given in the form of laws, the members of the society are forced to keep them, and someone violating them is punished if it is detected. One of the most serious disadvantages of top-down pressures is the expensive cost to plan them, to inform them, to monitor the behavior of the agents, and to detect and punish the violators. Considering the above advantages and disadvantages of the both types, it is important and practical to lead the bottom-up formation of norms to the desirable direction by supporting with the small and reasonable top-down pressures. Our research goal is to analyze the relationship of top-down and bottom-up aspects of pressures by agent-based simulations

Such purpose has been widely studied by a lot of researchers. For example, [Hobbes 2004] and [Durkheim 1985] clarified the problem of controlling society, and argued top-down pressures are necessary for cooperating together and stabling the society. On the other hand, [Axelrod 1986], [Shoham and Tenneholtz 1993] and [Shoham and Tenneholtz 1997] discussed the bottom-up pressures. Axelrod approached the problem by a computer simulation, and Shoham and Tenneholtz approached by a stochastic game-theoretic framework. By them and other researchers, the knowledge about norm formations has been accumulated.

In a large part of previous researches, the target problem were highly abstracted such as using game theoretical framework of payoff matrix, to make study simple. However, with the model highly simplified compared with the real problem, it may miss the important factors. With the developments of fast and inexpensive computers, models much more complex and closer to the real problem than payoff matrix have being employed. For example, [Shibata, et. al 2003] showed how the norm of stations, the order of getting on and off is formed, with the agent-based simulations. Such realistic simulations probably give us more informative knowledge than simple simulations. In this paper, a framework of realistic simulation of learning agents in traffic signal systems is presented, for being utilized to accumulate the knowledge specific to such real problems. Traffic signal system is of course important real social system. At the same time, this target is favorable to study because it is compact, *i.e.* relatively independent on the other systems and problems, scalable, and well-defined.

In Section 2, the framework of simulation is shown. In Section 3, the plain implementation of it with a Genetic Algorithm is presented. In Section 4, some preliminary experiments are produced to show the potential of the framework and the implementation. Section 5 gives conclusion of this paper.

## 2 Problem Description

In this paper, we discuss norm formation through the learning of the drivers in traffic signal systems. The drivers learn their policies to go through the crossing as fast as possible, and at the same time as safe as possible. For this purposes, the drivers adjust their policies to the signals and other drivers, through the interactions.

Considering the norm formation through the interactions among the multiple and various driver agents, we employ the framework of Genetic Algorithm (GA) for the simulation of learning, because multiple solutions are kept and evolved in GA. In this simulation, many policies of drivers are kept as solutions and evaluated in the traffic, the inferior policies are removed and new policies are introduced by GA operators.

While in the real world we have some established norms for crossing, we assume that there is no established norm such as "go if blue, stop if red" in prior, and assume that there are many policies before learning. If the policies after learning are converged, we regard the norm is formed. The purpose of this paper is to know whether the desirable norm is formed or not, to clarify the mechanism behind it, and find efficient and effective way to control it.

## 3 Plain Implementation of the Proposed Framework with Genetic Algorithm

In this section, the simulation framework of learning agents in traffic signal systems is presented. At first, the whole simulation consists of two main layers, "Learning Simulator" and "Traffic Simulator". The agents with their own policies are thrown into the traffic simulator, the chances of testing their policies are given there, and by using the evaluation values the agents revise their policies in the learning simulator, by GA operators.

### *3.1 Traffic Simulator*

As the traffic simulator, we employ a specifically described cellular-automata-type simulator, though there are many candidates which are more realistic. The components, terms, and procedures are defined as follows (See Fig. 1 and Table 1).

- **clock** : The simulator uses a discrete time step.
- **map** : The simulation space is 12×12 two dimensional cells. A cell is empty or occupied by *a* car. The central 2×2 cells are defined as the **crossing**.

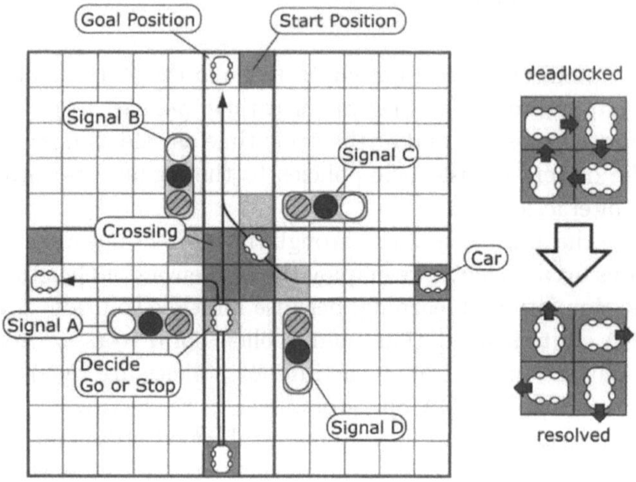


**Fig. 1** An example of traffic simulator, used for the experiments

- **signal** : There are several signals, usually around crossing, representing a color. We use an finite set if colors, such as $\{blue, red\}$, and the color changes depending only on the clock (not on the existence of cars). In this setting, signals to the opposite directions show the same color.
- **car** : The car is the main agent of the traffic simulator.
  - **route** : The route of car $car_i$ is defined as $Route_i = \{(x_{i,j}, y_{i,j})\}_j$ where $j = step_i \in \{0, 1, ..., m_i\}$ when the car appears.
  - **policy** : Each car $car_i$ has its policy $P_i$. The policy is parameterized by the matrix $P_{i(c,t)}$ where $c$ is the color of the signal, and $t \in \{straight, turn - left, turn - right\}$ is the turn(direction) of the car derived from its route. The car goes ahead if the position is not at one-cell short of crossing (signal position). At the signal position, the car goes at the probability $P_{i(c,t)}$.
- **waiting list of cars** : The list of cars $List_{\text{waiting}}$ are created at the initialization phase. When the car appears, is not defined in prior.
- **list of existing cars** : The set of cars in the cells are listed in $List_{\text{on}}$. The cars move step by step following the order of $List_{\text{on}}$.
- The procedure of updating the state is as follows.
  1. **deadlock detection** : When $car1$ blocks the next position of $car2$ and $car2$ blocks the next position of $car1$, they cannot reach the goal forever, because the car moves step by step. We call such situation "deadlock" (See Fig. 1 also). Deadlocks are easily detected by tracking the next position of each car one after another. The complexity is $O(\mid List_{\text{on}} \mid)$.
  2. For each the car $car_i \in List_{\text{on}}$, in order,

a. **goal detection** : If the car $car_i$ is at the last position of $Route_i$, the car is considered to reach its goal, and $car_i$ is drawn from $List_{\mathrm{on}}$.
b. **decision making** : If not, the policy $P_i$ decides whether the car $car_i$ goes or stops, depending on the signal color. If the car stops, return to step 2.
c. **crush** : If a car $car_j$ has moved at this timestep, blocks the next position of car $car_i$ in **crossing**, the two cars are considered to be lightly-crushed. In this case, the latter car $car_i$ doesn't move in this timestep.
d. **move** : If the next position of the car $car_i$ is empty, the car moves to there.

3. **deadlock resolution** : By the definition, the cars deadlocks couldn't move in the above procedures. Then, the deadlock is resolved by a special procedure. In this procedure, all cars deadlocks are moved to their next positions at once.
4. **car appearance** : According to a probabilistic distribution, several cars appears on the map. The cars are selected from $List_{\mathrm{waiting}}$ and added to the tail of $List_{\mathrm{on}}$. In this paper, there are four cells where the car appear, and the probability is $N_{\mathrm{traffic}}/(4\times 3600)$.

- Normally, the simulator is terminated if both $List_{\mathrm{on}}$ and $List_{\mathrm{waiting}}$ are empty.
- **evaluation** : The average timesteps $timestep_i$ to reach the goal, the average number of crushes $crush_i$, and the average number of deadlocks $deadlock_i$ are given for each policy $P_i$. Finally, the evaluation value $e_i$ of $P_i$, to be maximized, is calculated as

$$e_i = -timestep_i - \rho_{\mathrm{crush}} \times crush_i - \rho_{\mathrm{dead}} \times deadlock_i.$$

**Table 1** Parameters and Notations

| Symbol | Explanation | Value set $E_1$ | Value set $E_2$ |
|---|---|---|---|
| $N_{\mathrm{color}}$ | Number of colors of the signals | 2 | 3 |
| $\rho_{\mathrm{crush}}$ | Penalty if a car is crushed | 1000 | 1000 |
| $\rho_{\mathrm{dead}}$ | Penalty if a car is deadlocked | 1000 | 1000 |
| $N_{\mathrm{traffic}}$ | Traffic volume in 3600 timesteps | 100-5000 | 600-5000 |
| $N_{\mathrm{agent}}$ | Number of agents in the learning simulator | 30 | 100 |
| $N_{\mathrm{alter}}$ | Number of new policies created by crossover | 3 | 5 |
| $N_{\mathrm{evals}}$ | Number of runs for one policy in the traffic simulator | 200 | 200 |
| $N_{\mathrm{gnrs}}$ | Number of generations (learning period) | 1000 | 2000 |
| $\alpha_{\mathrm{BLX}}$ | Crossover parameter for BLX | 0.5 | 0.5 |
| $P_{i(c,t)}$ | Policy of $i$-th agent, representing the probability of going when the car's turn(direction) is $t$ and the signal color is $c$. | (2-dims) | (9-dims) |

## 3.2 Learning Simulator

The learning simulator works by the following procedures (See Fig. 2 also).

1. **Initialization** : $N_{\text{agent}}$ agents are created. Normally, the parameters of policies $P_{i(c,t)}$ are randomly selected from $[\emptyset_p, 1]$, where $\emptyset_p = 0.0001$ is the lower bound of probability.
2. **Evaluation** : All agents are evaluated by the traffic simulator. $N_{\text{evals}}$ cars are created for each policy $P_i$, $N_{\text{agent}} \times N_{\text{evals}}$ cars are added to $List_{\text{waiting}}$. The average of evaluated values are used for selection.
3. **Selection of unfavorable policies** : $N_{\text{alter}}$ policies with the least evaluation value are selected. For each such policy $P_i$,
   a. **Selection of parents** : The worse-evaluated agent is going to refer the better-evaluated agents. So, two policies $P_{k1}$ and $P_{k2}$ are randomly selected from the policies other than these $N_{\text{alter}}$ policies.
   b. **Crossover** : The new policy for the agent with $P_i$ is created by the crossover operator BLX-$\alpha$ [Eshelman and Schaffer 1993], *i.e.* $P_{i(c,t)} := P_{k1(c,t)} + \beta_{(c,t)}(P_{k2(c,t)} - P_{k1(c,t)})$ where $\beta_{(c,t)}$ is randomly selected from $[-\alpha_{\text{BLX}}, 1 + \alpha_{\text{BLX}}]$. Each $P_{i(c,t)}$ is bounded to $[\emptyset_p, 1]$.
4. **Next Generation** : Repeat $N_{\text{gnrs}}$ times from step 2.

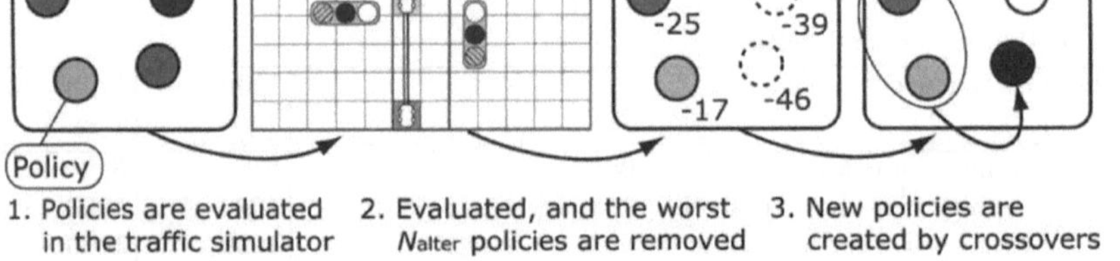


**Fig. 2** Overview of learning procedures

# 4 Experiments

In this section, we show some experiments on the framework and the implementations, to show their potential. At first, the simplest case where the car never turn, *i.e.* the goal cell is the front of start cell, and the signal shows only blue and red.

## *4.1 The Simplest Case : No Turn and Two Colors*

### 4.1.1 Settings and Purpose

The parameters used for this subsection are summarized in Value set $E_1$ of Table 1. Signals repeat a pattern as shown in Fig. 3, in this setting asymmetry is intentionally introduced. The purpose of the simplest example is to examine the essential behavior of this system. The advantages of using this setting are that the distribution of policies can be easily plotted, and that the learning results are easily categorized to some groups.

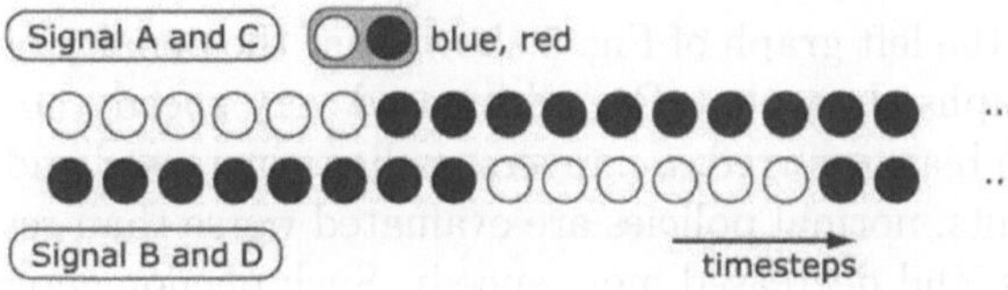


**Fig. 3** Signal patterns using 2 colors

In particular, the policy is represented as $(P_{blue}, P_{red})$ where $P_{blue}, P_{red} \in [\emptyset_p, 1]$ are the probabilities of going when the signal color is blue and red respectively. We categorize these 2-dimensional policies to four types.

- **normal** : $P_{blue} > 0.1$ and $P_{red} \leq 0.1$. This type can reduce the risk of crush well.
- **reverse** :$P_{blue} \leq 0.1$ and $P_{red} > 0.1$. When all signals are red, cars with such policies may crush.
- **rude** : $P_{blue} > 0.1$ and $P_{red} > 0.1$. This type can reach the goal fast, but the risk of crush can not be reduced.
- **other** : Ohter cases, *i.e.* $P_{blue} \leq 0.1$ and $P_{red} \leq 0.1$.

### 4.1.2 Experiments

At first, $N_{\text{traffic}}$ is fixed to 300, and 100 trials are performed for the setting. In all the trials, policies are converged to a category, normal, reverse and rude.

Fig. 4 shows the transition of the evaluation values and the distribution of policies, of a typical trial converged to a normal category. The left graph shows that the penalty of crush decreased and finally almost zero. The right graphs show that $P_{red}$ decreased speedy at early stage of GA, and $P_{blue}$ increased slowly after $P_{red}$ is almost zero. This is a typical successful trial.

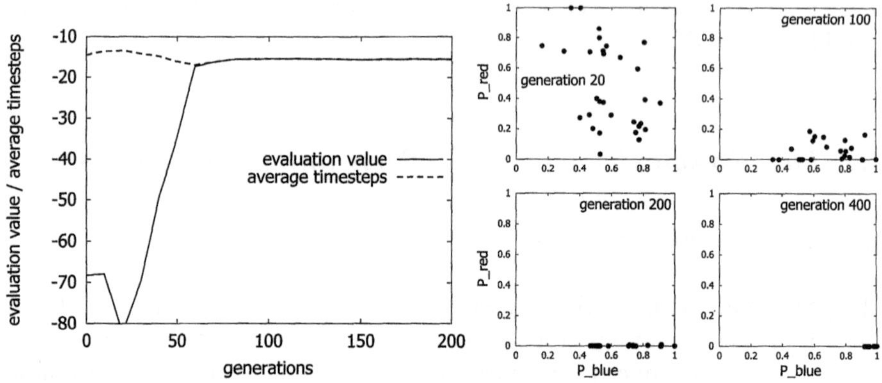


**Fig. 4** Transition of the evaluation values and average timesteps of a successful trial (left) and transition of the distribution of policies (right), 20, 50, 60, 100 generations.

Fig. 5 shows the transitions of a typical trial converged to reverse category. Compared with Fig. 4, the left graph of Fig. 5 shows that the penalty of crush remains. The right graphs show that $P_{blue}$ decreased very speedy before generation 20. Once such bias occurred, *i.e.* reverse policies increased and become dominant in the agents, normal policies are evaluated worse than reverse policies by interactions, and decreased more speedy. Such phenomenon is called "lock-in phenomenon" and this is very important problem.

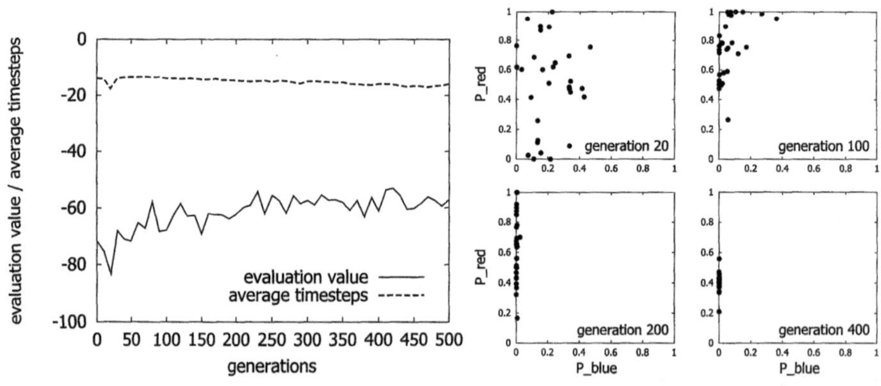


**Fig. 5** Transition of the evaluation values and average timesteps of a failed trial (left), and transition of the distribution of policies (right), 20, 100, 200, 400 generations.

At last, Fig. 6 shows the transitions of a typical trial converged to rude category. The left graph shows that the penalty of crush never improved like as reverse case. The right graphs show that both $P_{blue}$ and $P_{red}$ increased through the optimization.

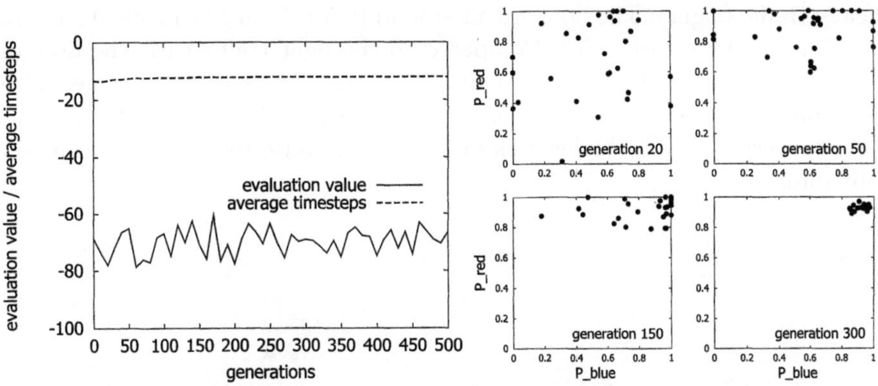


**Fig. 6** Transition of the evaluation values and average timesteps of a failed trial (left), and transition of the distribution of policies (right), 20, 50, 150, 300 generations.

### 4.1.3 The effect of traffic parameter

In this section, influence of traffic density is examined changing $N_{\text{traffic}} = 100$ to $N_{\text{traffic}} = 5000$.

Fig. 7 shows the number of trials the policies are converged to the categories. For example, if $N_{\text{traffic}} = 100$, the policy is converged to rude category in 84 trials (/100 trials) and to normal category in 16 trials.

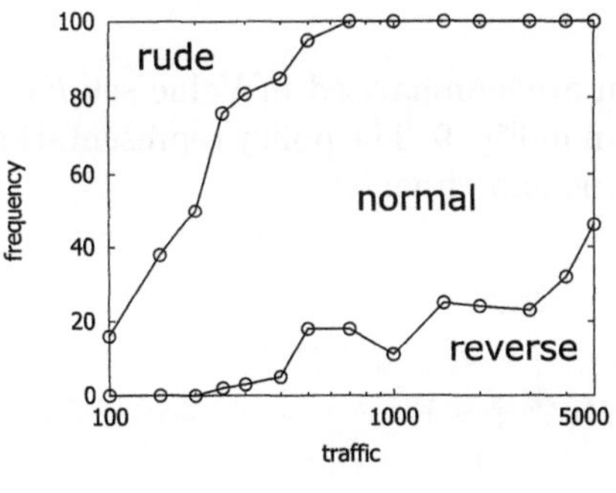


**Fig. 7** Dependency of policy convergence on traffic density.

From the graph, we find two trends:

1. The lighter the traffic is, the more frequently the rude policy is attained.
2. The heavier the traffic is, the more frequently the reverse policy is attained.

The reason of the first trend is clear. In the light traffic case, the crush occurs rarely, so the risk of crush is smaller than the heavy traffic case, then the rude policies are evaluated better than the normal/reverse policies in many trials.

We consider the second trend is caused by the speed of evolution. Fig. 8 shows the averages of policies at generation 20, at light traffic (left) and heavy traffic (right). The dots are all around $(0.5, 0.5)$ in generation 0 and are dividing to three categories. Compared to the light traffic case, the dots of the heavy traffic case are well dividing to normal or reverse directions. This division means that the speed of evolution is fast, and may cause the premature convergence. The higher risk of crush may cause the higher alternation pressure.

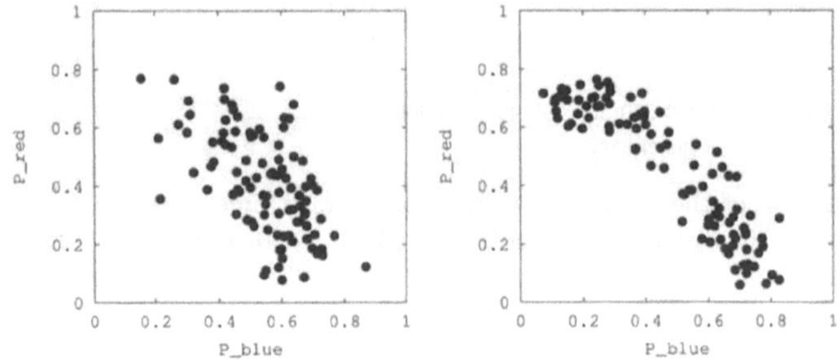


**Fig. 8** A dot is the average of $P_{blue}$ and $P_{red}$ in generation 20 of a trial. 100 trials, $N_{\mathrm{traffic}} = 500$(left) and $N_{\mathrm{traffic}} = 5000$(right).

## *4.2 Case of Three Directions and Three Colors*

The parameters used for this subsection are summarized in Value set $E_2$ in Table 1. Signals repeat a pattern as shown in Fig. 9. The policy representation is more complex than that of the previous experiments:

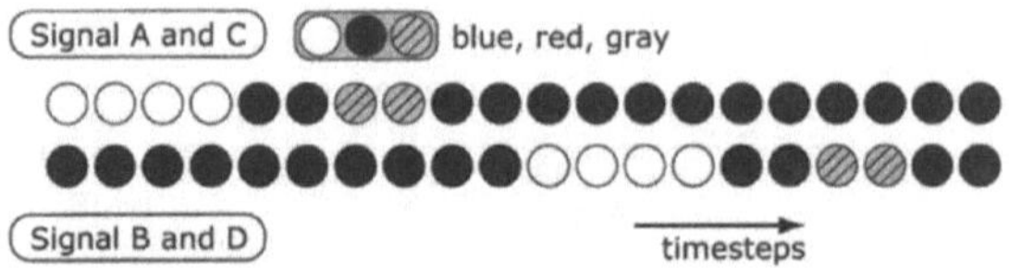


For example, $P_{good1} = ((1, \emptyset_p, \emptyset_p), (1, \emptyset_p, 1), (\emptyset_p, \emptyset_p, 1))$ is known as the good policy to reduce the risk of crush. At the same time, $P_{good2} = ((\emptyset_p, \emptyset_p, 1), (1, \emptyset_p, 1), (1, \emptyset_p, \emptyset_p))$ is far from $P_{good1}$ but can avoid crush also.

Cases from the light-traffic $N_{\mathrm{traffic}} = 600$ to the heavy-traffic $N_{\mathrm{traffic}} = 5000$ are tested. Table 2 shows the number of trials the policies are converged to the categories. It is natural that the lighter traffic, the rude policies are learned (even if rude one is not favorable). In heavy traffic, not-rude but not-good policies, such as $P = ((\emptyset_p, 1.0, \emptyset_p), (0.2, 1.0, 0.7), (0.7, \emptyset_p, \emptyset_p))$ are often learned. The reason is fast evolution and premature convergence we guess.

| $N_{\text{traffic}}$ | rude | $P_{good1}$ | $P_{good2}$ | other |
|---|---|---|---|---|
| 600 | 98/100 | 0/100 | 2/100 | 0/100 |
| 800 | 48/100 | 31/100 | 21/100 | 0/100 |
| 1000 | 5/100 | 65/100 | 30/100 | 0/100 |
| 1500 | 1/100 | 71/100 | 28/100 | 1/100 |
| 3000 | 0/100 | 54/100 | 30/100 | 16/100 |
| 5000 | 0/100 | 6/100 | 18/100 | 76/100 |

**Table 2** Convergence of policies in three directions/three signal case

Fig. 10 (left-top) shows the transitions of the evaluation values in three trials (not so typical, there are various patterns) when $N_{\text{traffic}} = 800$. Case-A and case-C finally reach the $P_{good1}$ and $P_{good2}$, and in case-B the evaluation value is not improved.

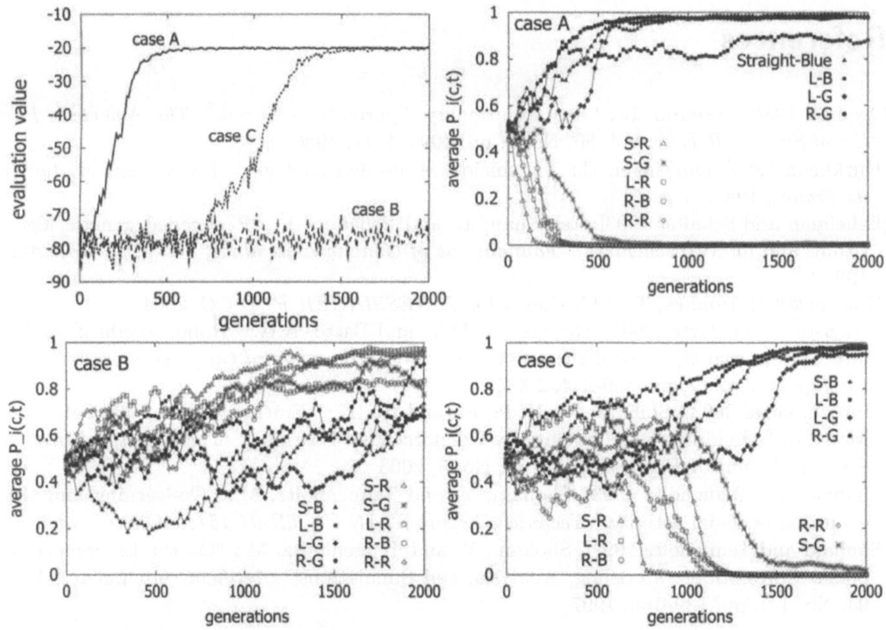


**Fig. 10** Transitions of the evaluation values (left-top) and transitions of the possibility averages of case A, B and C.

Fig. 10 also shows the transitions of the averages of $P_{blue,straight}$, ..., of case A, B and C. In the case A, the averages shifted to favorable direction from early stage of search. It should be noted that the shift occurred not simultaneously but one after another. In the case B, the averages shifted to rude direction gradually. In the case C, the early stage is similar to case-B. This would have caused pull-in, but in this trial good norm was achieved finally. This is very interesting phenomenon, there may exist a key to lead the bottom-up formation of norms to the desirable direction by supporting with the reasonable top-down pressures.

## 5 Conclusion

In this paper, we presented a compact and enough realistic framework of simulation with learning agents in traffic signal system. Through the implementation of the framework with Genetic Algorithm and some preliminary experiments, the potential of the framework was shown. The results from only simple settings were very rich in diversity, and many approaches to analyze the result will be required and fruitful.

## References

[Axelrod 1986] Axelrod, R.: "An Evolutionary Approach to Norms," *The American Political Science Review*, Vol. 80, No. 4, pp. 1095–1111, 1986.

[Durkheim 1985] Durkheim, E.: "Le suicide: etude de sociologie," *Presses universitaires de France*, 1985.

[Eshelman and Schaffer 1993] Eshelman, L. and Schaffer, J.: "Real-coded genetic algorithms and interval schemata," *Foundations of Genetic Algorithms*, Vol. 2, pp. 187–202, 1993.

[Hobbes 2004] Hobbes, T.: "The Leviathan," *KESSINGER PUB CO*, 2004.

[Rosenstein and Barto, 2001] Rosenstein, M.T. and Barto, A.G.: "Robot weightlifting by direct policy search," *Proceedings of the 17th International Joint Conference on Artificial Intelligence*, vol. 2, pp. 839-844, 2001.

[Shibata, et. al 2003] Shibata, K., Ueda, M. and Ito, K.: "Emergence and Differentiation Model of Individuality and Socility by Reinforcement Learning," *The Society of Instrument and Control Engineers*, Vol. 39, No. 5, 2003

[Shoham and Tenneholtz 1993] Shoham, Y. and Tennenholtz, M.: "Co-learning and the evolution of social activity," *Technical Report STAN-CS-TR-94-1511*, 1993

[Shoham and Tenneholtz 1997] Shoham, Y. and Tennenholtz, M.: "On the Emergence of Social Conventions: Modeling, Analysis, and Simulations," *Artificial Intelligence*, Vol. 94, No. 1-2, pp. 139–166, 1997

# Discovery of Family Tradition with Inverse Simulation

Setsuya Kurahashi

**Abstract** In this study, we investigate what would happen in a Chinese historical family line. We analyzed a particular family line, which had so many successful candidates, who passed the very tough examinations of Chinese government officials over 500 years long. First, we studied the genealogical records 'Zokufu' in China. Second, based on the study, we implemented an agent-based model with the family line network as an adjacency matrix, the personal profile data as an attribution matrix. Third, using "inverse simulation" technique, we optimized the agent-based model in order to fit the simulation profiles to the real profile data. From the intensive experiments, we have found that both grandfather and mother have a profound impact within a family to 1) transmit cultural capital to children, and 2) maintain the system of the norm system of the family. We conclude that advanced agent-based models are able to contribute to discover new knowledge in the fields of historical sciences.

## 1 Introduction

It is more than 30 years since Pierre Bourdieu introduced the structure of reproduction in relation to cultural capital and education. He introduced the system of the norm (Habitus) within a family which reproduces cultural capital and plays a critical role in the selection of social stratification [Bourdieu(1979)]. Furthermore, he referred to the civil service examination [1] which was used as the selection system for government officials in former days in China and also indicated the role played

Setsuya Kurahashi
Graduate School of Systems Management, University of Tsukuba, Otsuka 3-29-1, Bunkyo, Tokyo, Japan, e-mail: kurahashi@gssm.otsuka.tsukuba.ac.jp

[1] The term "Civil Service Examination" in the historical science field means the very tough examination for government officials of higher classes in China. The examination system runs for about 1,300 years.

by cultural capital in the selection mechanism of examinations, when he pointed out the importance of examinations in the French education system [Bourdieu(1970)].

However, in modern societies, the traditional concept of the family is forced to change in various areas because of the changes in the social system and local society through the advance of globalization. Under these circumstances, it is getting more important to know the function of the family as the fundamental element of the social system [Gayle et al(2002)Gayle, Berridge, and Davies].

There have been many changes in sociological approaches to family study. Now, there are approaches based on historical demography which interprets the time-series changes of family composition / population size, approaches based on network logic which borrows from the analysis method of the social network, and also traditional approaches based on the comparative institution, historical sociology and exchanges.

As mentioned above, the method of approach changes gradually to computable sociology in the field of family study. In this study, we construct an ABM based on the viewpoint of historical demography and the social network, and analyze the family system of a particular Chinese family line over a period of about 500 years. We then clarify the system of the norm which is maintained by the family, through a simulation of time-series changes of the attribution of family members, and then by inverse simulation.

## 2 Related work

Study of the family has been promoted from various angles: sociology, historical science, anthropology and biology. We begin with the sociology of the family.

### *2.1 Theory of Cultural Capital*

In "La Distinction [Bourdieu(1979)]" Pierre Bourdieu defined cultural capital as the coming together of tangible and intangible property related to culture in the broad sense of the term. He classified it into the following three categories;

a) The variety of knowledge, accomplishment, expertise, liking and sensitivity that each individual has accumulated through his/her family circumstances and school education. This is where cultural capital is embodied in the individual (embodied cultural capital).
b) Cultural property which can be held in the form of materials such as books/ pictures/ tools/ machines. These cultural goods can be transmitted physically as an exercise of economic capital, and symbolically as cultural capital (objectified cultural capital).

c) Qualifications and titles from the education system and examinations. This is institutional recognition of the cultural capital held by an individual, most often understood as academic credentials (institutionalized cultural capital).

According to Pierre Bourdieu, education institutions do not monopolize the production of cultural capital, itself, although they have the ability to change inheritance cultural capital to educational status capital by monopolizing the issuance of certificates. This shows that such accomplishment may have an advantage for seemingly fair examinations of the implicit norms of the ruling class, and that culture and Habitus are actually invisible standards of selection. Habitus means the system of norms as the system of all tendencies, which produces a manner of action and sense, specific to certain classes / groups. The concept cultural capital is fundamentally linked to the concepts of fields and habitus. Cultural capital can be derived from an individual's habitus. This cultural capital is handed on to the next generation within the family.

## 2.2 *Theory of Social Capital*

The importance of social reliance is recognized by the study results of social capital and social system, described by Patnum [Patnum(1993)][Patnum(2000)]. The norm of reciprocation is maintained by imitation, socialization and forcing by sanction. Reciprocation is classified into the following two categories; "Balanced Reciprocation", which is to exchange reciprocally with a specified person. "Generalized Reciprocation", which benefits in a one-sided way and lacks balance at a particular time, but it is to be returned in the future. The norm of generalized reciprocation is a very productive component of social capital and is related to the close network of social exchanges. It can be said that family lines not only maintain within a family the system of the norm which reproduces the norm of reciprocation, but they also make use of marriage which implies the system of the norm between families that have produced many excellent descendents. This study intends to examine the above assumption.

## 2.3 *Analysis of the Family System*

The sociology of the family commences from a fundamental sociological concept and assumption. In other words, the sociology of the family consists of the concept of reward, benefit, and exchange, and the assumption that regards human being as a rational existence. The sociology of the family and its analysis is specified from the notion that the family is one of the social systems. This analysis is an approach to the family from the point of a system based on the fundamental concept that the system is the body which organizes and regulates the norms/ rules / customs of the human being,

Where the data is arranged in time series order, it is an historical comparison. Where the data is arranged by country or by area, it is an international comparison. "Clan, Caste and Club" by Hsu is regarded as a typical comparison analysis [Hsu(1963)]. Hsu analyzes the influence of relationships within the family, based on the framework of the inheritance system as family typology.

Study of the traditional family shows clearly that the line maintains its continuity and formality through various strategies such as traditional practices within the family [Fujii(1997)][Shimizu(1999)][Hiroshima(1998)].

Historical demography has shown the change of population and the social conditions of the past through the empirical approach of drawing on parish records in Europe on historical trends of families/households. The footsteps of individuals over the past hundreds of years can be reconstructed from such data as the historical records of individual religious affiliation/ official family registry in Japan which recorded the details of names, ages, relationships to head of household, births, deaths, marriages and moving destinations of members of each house [Hayami(1997)]. The data from genealogical records is similar to the above records.

## 2.4 Agent Approach

In addition to the above mentioned sociological approaches, studies of family and history have also been carried out by social simulation employing agent technology. The appearance of mating was reproduced in SugarScape, Epstein by the interactions between the birth rate of the agent and the population density [Epstein and Axtell(1996)]. The mating of agents was phrased as the network of family line and the relationship of the agents by blood was shown. Timothy A. Kohler and other scholars applied agent simulation to archeology in the Village Project and brought out the connections between change of vegetation, migration and change of population [Kohler et al(2000)Kohler, Kresl, Van, Carr, and Wilshusen]. Furthermore, Cathy A.Small analyzed the influence of the marriage system in Polynesian society by employing a simulation model [Small(2000)]. As mentioned above, use of the agent model is common in sociology and anthropology.

These studies have shed light on social structures and historical matters. Most of them are done from the analytical point of view. However, the agent approach focuses attention on the active aspect of history and has reproduced the change of matters by employing a simulation model. However, these simulation models are built based on historical facts which are already known and have been analyzed. This study tries to bring out undiscovered historical facts and structures by employing the above agent models and the inverse simulation method which estimates such acts and variables that may meet the historical facts.

## 3 History of Civil Service Examinations

In the past in China, there were examinations for the recruitment of government officials which were institutionalized as the civil service entrance examinations in the Tang Era. The golden age of the examinations was during the Sung Era, when politicians who passed the examinations displayed great abilities and reached the heights of politics. The examinations comprised a provincial examination, a metropolitan examination and a palace examination, with an entrance examination for school in each prefecture as a preliminary step. Successful candidates who passed the final step of the palace examination were called "chin-shih". As there was no way to make money other than passing the examinations over the hundreds of years from the start of the examination in the 6th or 7th centuries, many people tried hard to pass the examination. As a result, the competition heated up and the environment surrounding individuals became more important than the ability bestowed upon individuals in order to beat the competition. If an individual had the same abilities as others, then being rich became advantageous, it was better to be from an educated family and to live in an urban area with more advanced culture rather than being poor, from unlearned parents and living in the country. This phenomenon factored in the progress of the skewed distribution of culture and wealth. When European civilization surged into China at the ending of the Ching dynasty, it became more important to have an education in such things as natural science, experimentation and industrial art. The civil service examination system was finally brought to an end in 1904 by the Ching Government.

As mentioned above, it was a qualifying examination system for high-ranking officials which had been implemented for more than one thousand three hundred years. Those who successfully passed the examinations were qualified and earned the title of chin-shih. There were two categories of the examinations: the department of literature and that of military service. There was a big difference between the two categories. Only the chin-shih from the department of literature was respected and the other chin-shih were hardly respected at all. There were the following titles awarded as qualifications by the department of literature:

- chu-jen for those who passed the local examination.
- sheng-yuan for the status of student at a prefecture school, as a preliminary step.
- kung-sheng and chien-sheng for those who were recommended as central students.

High officials were recognized as legitimate career-track bureaucrats who were accepted from chin-shih, chu-jen and kung-sheng. Additionally, there was another category of chuan-na, where money was paid for obtaining such titles such chien-sheng.

According to Benjamin A. Elman [Elman(1991)], where the state emphasized the examination system for the production of loyal bureaucrats to the state, the candidates regarded the system as the most authoritative method for achieving their own personal success through successfully passing the said examinations. However, it took a huge investment in time, effort and training in order to achieve such success.

Such candidates set their family, clan and lineage as the strategic targets of the social reproduction of their community. In the Ming and Ching eras, the school education system accepted only those candidates who already had a good command of the official language and were literate in classical Chinese. It was the responsibility of each house to obtain and maintain these elite positions as the "House of the Bureaucrat" at the initial stage of educating a son and preparing for his entering government service. Further more, Elman points out that it is possible for seemingly ordinary candidates to have achieved academic success because they had bureaucrats among their close relatives or affinities to the same lineage. Under these circumstances, the examination system proved successful in that it created elite families as areas of cultural reproduction, and it guaranteed that the right background gained superiority for a successful future social and political career and that the candidates came from a family which had the tradition of learning classics and spoke the official language.

However, there was a tendency that certain family lines could produce more successful candidates even from among these elite families, and there was a big difference between each family line. In China, records of family trees had been made from old times and kept as genealogical records: "Zokufu". Zokufu refers to records relating to family tree and lineage. It is a paternal record from the primogenitor and includes name, birth year, year of death, antemortem achievement, wife's name, number of children, place of residence and other information for each family member.

In this study, we used the Zokufu of the Y Family in the Ming and Ching Eras. The Zokufu mainly consists two parts: the "sekei" which shows generally the family tree, and the "sehyo" which records the details of the profile of each member. Each example is shown in Fig.1.

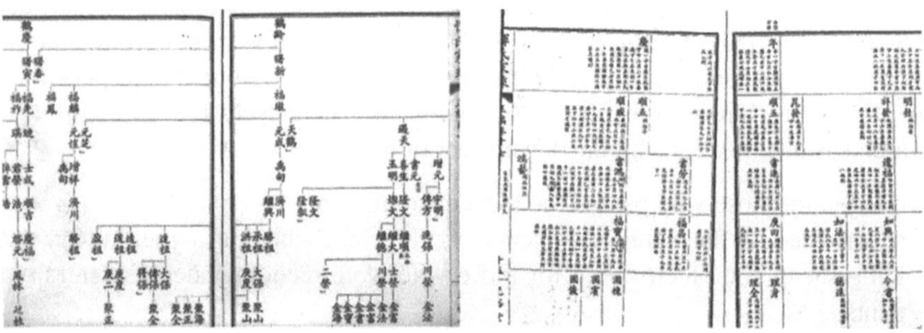

**Fig. 1** Family Tree(left) and Personal Profile(right)

Changzhou, Jiangsu, the home of the Y Family, is located in the Gangnam region, which produced the highest number of successful examination candidates, who ranked 1st or 2nd throughout the country in the Ming and Ching eras. It was clear that most of these candidates were from certain families, and there were twenty seven families which kept producing chin-shih and chu-jen for the period of more than five generations during the Ming and Ching eras. Among these families, the Y Family was one of the typical cases, it produced twenty two successful candidates for the period of more than twelve generations. By analysis based on agent simulation, we began to know why so many such successful candidates were produced from the same family, by employing the Zokufu of the Y Family.

## 4 Agent Simulation of Zokufu

We prepared the types of data as available for the simulation from the "sekei" and "sehyo" Zokufu data. The sekei data show the relationship between father and son, and we were able to prepare the adjacency matrix from this data. The Zokufu of the Y Family contains data for a total of 1237 persons. It makes the adjacency matrix of $1237 \times 1237$ which shows the relationship between parent and child, and "0, 1" represents this relationship. In the same manner, we prepared the attribution matrix of each person from the sehyo data. The attribution involves chin-shih, chu-jen, kung-sheng, sheng-yuan, chien-sheng, chuan-na, merchant, painter, poet, qualified status of the examinations of wife's family home / daughter's married family and others. Each of these elements is represented by 0, 1 Fig.2. We can reproduce the family tree with attribution from the above two matrixes and implement the simulation based on this family tree. Each member of the family tree is grouped by birth year and tallied as successful candidate by cohort.

Outline of the Agent Simulation as follows:

- Each agent can transmit cultural capital from parent to child, from grandfather to grandson, from great-grandfather to great-grandson, by face to face along the family tree shown by the adjacency matrix.
- There are two categories of cultural capital: knowledge cultural capital and art cultural capital.
- Where there is a successful examination candidate on the mother's side of the family, his cultural capital is transmitted from parent to child in the same manner.
- Children have by birth the character of knowledge and art.
- The degree of a child's cultural capital depends on the synergetic effect of the character of the child and the cultural capital transmitted by others. However, only knowledge cultural capital affects success in the examinations and art cultural capital does not directly affect the rate of success in the examinations.

The agents can take the above mentioned actions. At the same time, they have parameters that decide each pattern of action. The parameters are in common with all the agents, which are as follows:

- Who is the transmitter? (father, grandfather, great-grandfather)
- Degree of effect on individual cultural capital (rate of transmission from father and others).
- Degree of effect by education (the increasing rate of the effects by cultural capital and by education of character).

| | 1 | 2 | 3 | 4 | 5 | 6 | 7 | 8 | 9 | 10 | 11 | 12 | 13 | 14 | 15 | 16 |
|---|---|---|---|---|---|---|---|---|---|---|---|---|---|---|---|---|
| 1 | 0 | 1 | 1 | 1 | 1 | 0 | 0 | 0 | 0 | 0 | 0 | 0 | 0 | 0 | 0 | 0 |
| 2 | | 0 | 0 | 0 | 0 | 1 | 1 | 0 | 0 | 0 | 0 | 0 | 0 | 0 | 0 | 0 |
| 3 | | | 0 | 0 | 0 | 0 | 0 | 1 | 0 | 0 | 0 | 0 | 0 | 0 | 0 | 0 |
| 4 | | | | 0 | 0 | 0 | 0 | 0 | 0 | 0 | 0 | 0 | 0 | 0 | 0 | 0 |
| 5 | | | | | 0 | 0 | 0 | 0 | 1 | 1 | 1 | 1 | 0 | 0 | 0 | 0 |
| 6 | | | | | | 0 | 0 | 0 | 0 | 0 | 0 | 0 | 1 | 1 | 1 | 1 |
| 7 | | | | | | | 0 | 0 | 0 | 0 | 0 | 0 | 0 | 0 | 0 | 0 |
| 8 | | | | | | | | 0 | 0 | 0 | 0 | 0 | 0 | 0 | 0 | 0 |
| 9 | | | | | | | | | 0 | 0 | 0 | 0 | 0 | 0 | 0 | 0 |
| 10 | | | | | | | | | | 0 | 0 | 0 | 0 | 0 | 0 | 0 |
| 11 | | | | | | | | | | | 0 | 0 | 0 | 0 | 0 | 0 |
| 12 | | | | | | | | | | | | 0 | 0 | 0 | 0 | 0 |
| 13 | | | | | | | | | | | | | 0 | 0 | 0 | 0 |
| 14 | | | | | | | | | | | | | | 0 | 0 | 0 |
| 15 | | | | | | | | | | | | | | | 0 | 0 |
| 16 | | | | | | | | | | | | | | | | 0 |

| | Shins | Kyoi | Kous | Saii | Taig | Enou | Gunk | Riin | Saik | Sais | Dghts |
|---|---|---|---|---|---|---|---|---|---|---|---|
| 1 | 0 | 0 | 0 | 0 | 0 | 0 | 0 | 0 | 0 | 0 | 0 |
| 2 | 0 | 0 | 0 | 1 | 0 | 0 | 0 | 0 | 0 | 0 | 0 |
| 3 | 0 | 0 | 0 | 0 | 0 | 0 | 0 | 0 | 0 | 0 | 0 |
| 4 | 0 | 0 | 0 | 0 | 0 | 0 | 0 | 0 | 0 | 0 | 0 |
| 5 | 0 | 0 | 0 | 0 | 0 | 0 | 0 | 0 | 0 | 0 | 0 |
| 6 | 1 | 0 | 0 | 0 | 0 | 0 | 0 | 0 | 0 | 0 | 0 |
| 7 | 0 | 0 | 0 | 1 | 0 | 0 | 0 | 0 | 0 | 0 | 0 |
| 8 | 0 | 0 | 0 | 0 | 0 | 0 | 0 | 0 | 0 | 0 | 0 |
| 9 | 0 | 0 | 0 | 0 | 0 | 0 | 0 | 0 | 0 | 0 | 0 |
| 10 | 0 | 0 | 0 | 0 | 0 | 0 | 0 | 0 | 0 | 0 | 0 |
| 11 | 0 | 0 | 0 | 0 | 0 | 0 | 0 | 0 | 0 | 0 | 0 |
| 12 | 0 | 0 | 0 | 0 | 0 | 0 | 0 | 0 | 0 | 0 | 0 |
| 13 | 0 | 0 | 0 | 1 | 0 | 0 | 0 | 0 | 1 | 0 | 0 |
| 14 | 0 | 0 | 0 | 1 | 0 | 0 | 0 | 0 | 0 | 0 | 0 |
| 15 | 0 | 0 | 0 | 0 | 1 | 0 | 0 | 0 | 0 | 1 | 0 |
| 16 | 0 | 0 | 0 | 0 | 0 | 0 | 0 | 0 | 0 | 0 | 0 |
| 17 | 0 | 0 | 0 | 0 | 0 | 0 | 0 | 0 | 0 | 0 | 0 |
| 18 | 0 | 0 | 0 | 0 | 1 | 0 | 0 | 0 | 1 | 0 | 0 |
| 19 | 0 | 0 | 0 | 0 | 0 | 0 | 0 | 0 | 0 | 1 | 0 |
| 20 | 0 | 0 | 1 | 0 | 0 | 0 | 0 | 0 | 0 | 0 | 0 |

**Fig. 2** Adjacency matrix(left), Attribution matrix(right)

- Mode of transmission of cultural capital (how knowledge cultural capital and art cultural capital are transmitted).
- Degree of effect of the mother's side of the family (transmission rate of cultural Capital.

# 5 Inverse Simulation

The agent simulator model and code are shown in Fig.3, Fig.4. As shown in this chart, cultural capital is transmitted to children by the system of norms and the parameter which characterizes the norms that is transmitted along with the family line. The agent simulations are plurals implemented at the same time by way of the rule. The profile information of all the agents, which appears as a result, is compared with the actual profile information based on the attribution data prepared by the sehyo. These profile data are employed after tallying by cohort. The objective function sets the error of mean square of this simulator profile information and actual data profile information. The objective function is as follows:

$$min : CohortFitness = \sum_{i=1}^{n} \sum_{j=1}^{m} (c_{ij} - sc_{ij})^2, \tag{1}$$

where n : the number of cohort, m : the number of caltural capital, $c_{ij}$ : caltural capital degree, $sc_{ij}$ : simulated caltural capital degree. We select the better models by way of tournament with the value of the objective function by each generation and produce agent models that have the next generation parameter after the process of crossover and mutation. As a result, we can obtain an agent model that indicates results similar to the actual profile information. By analyzing the parameters of this model, we can estimate the strategies of the family lines which produced many successful examination candidates.

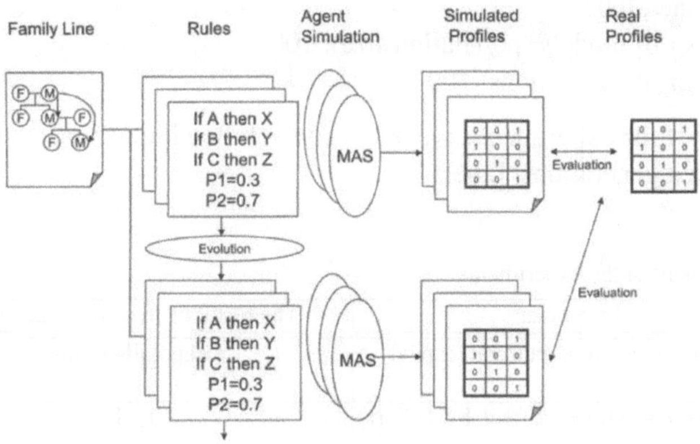


**Fig. 3** Inverse simulation model

```
Inverse-Simulation(realData)
  set parameters and rules of each society to random
  for each society in the world
    Create-Society(parameters, rules)
  end for
  while generation < maxGeneration
    for each society in the world
      Simulate-Society(parameters, rules)
      fitness <- fitness-function(realData)
    end for
    SelectRecombinate-Society(fitness)
  end while
  optimumParameters-and-rules <-
            parameters and rules of society[maxFitness]
  return optimumParameters-and-rules
end
```

**Fig. 4** Inverse simulation code

# 6 Experimental Results and Consideration

At the beginning, we implemented simulation experiments for which parameters were arbitrarily provided. However, we knew it was difficult to find a family that could keep producing the high level of cultural capital to be able to have successful examination candidates. This is why there are too many parameters, variables, action patterns to be set for discovering suitable combinations. Therefore, as a second step, we implemented inverse simulations under the following conditions:

- The term of a simulation:1450 - 1750
- Selection by tournament
- Selection rate: 0.8
- Crossover rate: 0.8

- Mutation rate: 0.05
- The number of models (population size): 100
- The number of generations: 100

As the result of the experiments, we obtained the following parameters after a lapse of one hundred generations(Tabel. 1).

**Table 1** The result of the experiments

| Parameter | The result of Inverse Simulation |
|---|---|
| People who transmit cultural capital to child | Grandfather only |
| Effect by education : Effect by cultural capital | 1 : 1 |
| Effect by mother | 100%, if maternal grandfather is a successful candidate. |
| Crossover rate between knowledge and art cultural capital | 20% |
| Transmission method of cultural capital | Both cultural capitals of knowledge and art are transmitted to the child from parents. |

From the above parameters, the following five patterns are prepared for people who transmit cultural capital to a child. Among them, only No.4 (from grandfather) is selected. This may show that the influence of the grandfather is stronger than that of the father. 1) from father, 2) from father and grandfather, 3) from father, grandfather and great-grand father, 4) from grandfather only, 5) from great-grand father only. Based on these results, we implemented a statistical analysis by family line data and attribution data, and examined the effect of the data. The results are as follows:

- where grandfather is a successful examination candidate, father is a non-successful candidate and grandson is a successful candidate, and where grandfather is a non-successful candidate, father is a non-successful candidate and grandson is a successful candidate, the odds ratio is 3.4 between the above two cases. The effect on grandson is 99% screened and shows a significant difference where only grandfather is a successful candidate.
- where father on mother's side of the family is a successful examination candidate and son is a successful candidate, and where father on mother's side of the family is a non-successful candidate and son is a successful candidate, the odds ratio is14.2 between the above two cases. The effect on son is 99% screened and shows a significant difference where only father on mother's side of the family is a successful candidate.

It is contrary to what we expected that the result of the experiments shows that the effect on the child from the grandfather is stronger than from the father in transmission of cultural capital. Furthermore, we find that the transmission of cultural capital is made from the mother's side of the family to children of the married family. The result of such screenings has been verified by statistical analysis based on these findings of agent simulations. This shows that the doubling-up of three generations has had a big effect on education under the big family system in China. It shows that such a norm system can help to rebuild the family fortunes when there is marriage with girls from "good families". It is very interesting result that the effect on children by grandfather is stronger, than by father and that the effect by mother is important, considering the characteristics of the family line which kept producing the successful examination candidates. It may be possible that these facts are the custom which had been transmitted as the norms through generations. Further, the following are found as the characteristics of this particular family;

- Knowledge cultural capital and art cultural capital of parents are transmitted equally to child.
- 20% of each cultural capital affects the other.

The above matters are assumed by the family as the transmission method of cultural capital. It can be said that the following are some of the factors which resulted in keeping producing the successful examination candidates:

- Transmission of artistic ability is made to descendants, as well as of knowledge.
- It is more important that such education is provided within the family to make the most of this ability, if the child has strong artistic abilities, even if knowledge is inferior.

The result of transmission function of caltural capital is the following,

$$cl^c_k = r(cl^p_k \cdot pc^p_k) + (1-r)(cl^p_a \cdot pc^p_a), \tag{2}$$

$$cl^c_a = r(cl^p_a \cdot pc^p_a) + (1-r)(cl^p_k \cdot pc^p_k), \tag{3}$$

where $cl^i_j$ : $i$'s caltural capital about $j$, $pc^i_j$ : $i$'s personal characteristics about $j$, $i$ : $c$=child, $p$=parent, $j$ : $k$=knowledge, $a$=art, $r$ : crossing rate of cultural capital. This may imply that there is a strong bond in the exchanges between artists and intellectuals, and in the relationships between brothers and sisters, which can be seen in the present society.

## 7 Summary and Challenges

In this study, by employing agent technology, we analyzed the family line for a family which produced many successful examination candidates for a period of five hundred years. We implemented an inverse simulation based on a multi-agent model which expresses the family line network and the personal profile data as an adjacency matrix and as an attribution matrix, respectively, and sets the real profile data as an objective function. As a result, we found that grandfather and mother had a strong effect on the child for the transmission of cultural capital within the family.

This was verified by statistical analysis. With this model, we showed the possibilities that an agent based model could contribute to new discoveries of facts in the fields of historical science and sociology. We also showed the possibility of historical simulation by ABM.

As challenges for the future, we need to research the changes of family motto which can be obtained by inverse simulation through analyzing by generation and by branch of family, not using the method employed in this study, to all the family lines without exception. In addition, we plan to construct a model that can simulate the influence of the daughter's married family and of artists such as painters, and then to refine the models.

## References

[Bourdieu(1970)] Bourdieu P (1970) Reproduction in Education, Society and Culture. Sage Pubns

[Bourdieu(1979)] Bourdieu P (1979) Distinction: A Social Critique of the Judgment of Taste. Harvard University Press

[Elman(1991)] Elman BA (1991) Political, social, and cultural reproduction via civil service examinations in late imperial china. The Journal of Asian Studies 50(1):7–28

[Epstein and Axtell(1996)] Epstein JM, Axtell R (1996) Growing Artificial Societies. The MIT Press

[Fujii(1997)] Fujii K (1997) Historical Sociology of The Family and Kinship (in Japanese). Tosui Shobou

[Gayle et al(2002)Gayle, Berridge, and Davies] Gayle V, Berridge D, Davies R (2002) Young people's entry into higher education: quantifying influential factors. Oxford Review of Education 28(1):5–20

[Hayami(1997)] Hayami A (1997) The world of Historical Demography (in Japanese). Iwanami Shoten

[Hiroshima(1998)] Hiroshima K (1998) Eurasian Project on Population and Family History, Working Paper Series No.4. International Research Center for Japanese Studies

[Hsu(1963)] Hsu FLK (1963) Clan, Caste, and Club. Van Nostrand

[Kohler et al(2000)Kohler, Kresl, Van, Carr, and Wilshusen] Kohler TA, Kresl J, Van C, Carr WE, Wilshusen RH (2000) Be there then: A modeling approach to settlement determinants and spatial efficiency among late ancestral pueblo populations of the mesa verge region, u.s. southwest. In: Kohler TA, Gumerman GJ (eds) Dynamics in Human and Primate Societies, Oxford University Press, pp 145–178

[Patnum(1993)] Patnum R (1993) Making Democracy Work: Civic Traditions in Modern Italy. Princeton University Press

[Patnum(2000)] Patnum R (2000) Bowling Alone: The Collapse and Revival of American Community. Simon & Schuster

[Shimizu(1999)] Shimizu H (1999) The approaches of family change (in japanese). In: Nonoyama H, Watanabe H (eds) Beginning The Sociology of The Family, Bunka Shobou Hakubun Sha, pp 42–68

[Small(2000)] Small CA (2000) The political impact of marriage in a virtual polynesian society. In: Kohler TA, Gumerman GJ (eds) Dynamics in Human and Primate Societies, Oxford University Press, pp 225–249

# Analysis of Focal Information of Individuals: Gaming Approach to C2C Market

**Hitoshi Yamamoto[1], Kazunari Ishida[2] and Toshizumi Ohta[3]**

[1] Faculty of Business Administration, Rissho University, Shinagawa-ku, Tokyo 141-8602, Japan, hitoshi@ris.ac.jp

[2] Faculty of Policy Studies, University of Shimane, 1-236 Tangi, Hachioji City, Tokyo 192-8577, Japan., k-ishida@u-shimane.ac.jp

[3] The Graduate School of Information Systems, University of Electro-Communications, Choufu-shi, Tokyo 182-8585, Japan, ohta@is.uec.ac.jp

**Abstract:** To analyze the effect of reputation management systems for promoting cooperative behavior in a C2C market, we developed a virtual C2C market system and experimented with participants to analyze transaction and information behaviors. According to the result of our experiment, we found that over 80% of participants behaved cooperatively. However, some participants accumulated high reputation in the early round of the experiment, and then exploited cooperative participants with the high reputation and defective action. The result indicates existence of vulnerability of reputation management system. Based on analysis of information behavior, we also found that cooperative participants often referred the number of defects and duration of ID unchanged. The result indicates cooperative participants prefer risk adverse to choose trustful others to make deal.

## 1 Introduction

The recent growth of C2C (consumer-to-consumer) market is one of the remarkable phenomena of the Internet, a medium which reduces the limitations of our lives in terms of distance, time, and opportunity. We can sell or buy virtually any goods to others on the Internet, something never possible before. eBay and Yahoo! Japan Auction are two successful examples of such marketplaces.

However, market growth also results in an increase of risk on transactions, namely the failure of buyers to pay for goods and the failure of sellers to send goods. In an online market, participants can easily change identities by changing handles, or on-screen user IDs, and enter or exit the market. As a result, they may take advantage of the opportunity to accept goods without payment or to accept payment without sending goods, actions which we define as defective behavior.

To discourage defective behavior, we need management systems to promote cooperative behavior among participants on an online C2C market. A reputation management system (RMS) is one effective means employed by many online marketplaces such as eBay and Yahoo! Japan Auction. An RMS provides a mechanism for participants to evaluate each other and to share their evaluations.

Several studies have revealed that an RMS allows participants to behave cooperatively as they attempt to maximize their own profit (Kollock, 1999) (Yamamoto et. al. 2004). Online C2C marketplaces are commonly employing their own systems (Dellarocas, 2003). A typical RMS calculates a reputation score for each participant by summing and averaging ratings given by those who have made transactions with the participant. An RMS also provides other information such as individual comments and transaction history.

One question concerning reputation management systems is exactly which information is effective for selecting trustworthy transaction partners. Does the effectiveness of information depend on the role of the participant as a buyer or seller, or as a cooperative or defective participant? To find the answers, we designed an experiment using a virtual online C2C market in which we observed transaction behavior in terms of cooperative and defective behavior. To determine which information is most often referenced by each type of participant, we also observed information behavior, i.e. frequently referred information. Based on our observations, we can propose an RMS design that effectively promotes cooperation.

## 2 Reputation Management Systems and Evolution of Cooperative Behavior

There are three general approaches to analyzing the effectiveness of a reputation management system in an online C2C market: case studies, computer simulations, and laboratory experiments.

In one example of the first approach, a case study of eBay transactions by McDonald (2002) reported that buyers with a high reputation score can sell their goods at high prices. Resnick and Zeckhouser (2001) reported that rate of positive evaluation was over 99% in all mutual evaluations on eBay.

Using the second approach of computer simulations, online C2C transactions can be modeled in terms of game theory, particularly iterative prisoners' dilemma (PD). Yamamoto et. al. (2004) analyzed C2C transactions with a computer simulation model and revealed that an RMS emphasizing positive feedback is effective in promoting cooperative behavior in an online market, whereas an RMS emphasizing negative feedback is effective for offline transactions.

In the third approach, a great deal of research based on laboratory experiments has analyzed cooperative behavior in reputation management systems also in terms of PD. Ruth and York (2004) investigated how the presentation of performance information affects stakeholders' attitudes towards firms that seek to enhance their reputation. The results indicate that consistency between source and information type determines the impact on degree of attitude change. Rice (2004) investigates PD in terms of existence and uncertainty of feedback information. The results indicate that social welfare and efficiencies of trade increase where feedback is allowed. Bolton et. al. (2004) compares trading in a market with feedback to a market without feedback. The results indicate that feedback mechanisms induce a substantial improvement in transaction efficiency.

However, it should be noted that these studies have not discussed what kind of feedback information is important in the decision-making process of selecting buyers or sellers, and the identification of such information is necessary in order to design an effective RMS. To identify these decision-making factors in our own experiment, we designed a virtual online C2C market in which we observed transaction and information behavior in terms of cooperative or defective behavior, as well as what type of feedback information is more important to each type of participant.

## 3 Modeling C2C Online Transactions

In order to design an experimental model of a C2C online marketplace, we discuss transactions within the framework of prisoner's dilemma, in which the players are represented by buyers and sellers.

### *3.1 Prisoners' Dilemma*

A player who participates in a C2C online transaction always has an incentive to cheat others (i.e., to defect) due to anonymity and the ease of entry and exit from transactions. On one hand, a buyer may accept goods from a seller without payment. On the other hand, a seller may accept payment from a buyer without sending goods. The situation in C2C online transactions is representative of prisoner's dilemma. We can consider these strategies within a payoff matrix, as shown in Table 1 (T>R>P>S).

**Table 1.** Payoff matrix for prisoner's dilemma

| | | Action of player-2 | |
|---|---|---|---|
| | | C | D |
| Action of player-1 | C | R, R | S, T |
| | D | T, S | P, P |

In the prisoner's dilemma of a C2C online transaction, a seller can take two possible actions: cooperation, i.e., sending goods in exchange for payment, and defection, accepting payment without sending goods. Likewise, a buyer can also cooperate or defect, i.e. pay for goods or accept goods without payment.

## *3.2 Formulation of Transactions*

Each participant in our experiment was assigned a role of either buyer or seller. We denoted the item price, production cost, and utility of the item for each buyer by P, C, and V, respectively. We assume profitable conditions, i.e. V>P>C>0.

Based on the notations, we can define a payoff for a seller (T, R, P, S) as (P, P-C, 0, -C), where T, R, P, and S denote gain or loss in four possible cases of PD: defect against a cooperating opponent, mutual cooperation, mutual defect, and cooperation with a defecting opponent, respectively. In the case of defect against a cooperating opponent, a seller accepts payment from a buyer but does not ship the item to the buyer. In this case, the seller gains T (=P) because the seller gains payment (P) without loss of production cost (C). In the case of mutual cooperation, a seller sends the item to a buyer and accepts payment from the buyer. In this case, the seller gains R (=P-C), because the seller gains payment (P) and loses the production cost (C) of the item. In the case of mutual defect, the seller does not send the item to the buyer, and the buyer does not send payment to the seller. In this case, the seller does not gain or lose anything (P=0). In the case of cooperation with a defecting opponent, the seller sends the item to the buyer, but the buyer does not send payment. In this case, the seller loses S (=-C) because the seller pays the production cost (C) of the item without gaining payment (P) from the buyer.

We can likewise define a payoff for a buyer (T, R, P, S) as (V, V-P, 0, -P). In the case of defect with a cooperating opponent, the buyer accepts the item from the seller but does not send payment to the seller. In this case, the buyer gains T (=V) because the buyer enjoys utility (V) without payment to compensate for the seller's production cost. In case of mutual cooperation, the buyer sends payment to the seller and the buyer receives the item from the seller. In this case, the buyer gains R (=V-P) because the buyer enjoys the utility of the item (V) after paying the price of the item. In the case of mutual defect, the buyer does not send payment to the seller, and the seller does not send the item. In this case, the buyer does not gain or lose anything (P=0). In the case of cooperation with a defecting opponent, the buyer sends payment to the seller, but the seller does not send the item to the buyer. In this case, the buyer loses S (=-P) because the buyer pays the price of the item (P) without receiving the item from the seller.

In our experiment, we assume that the higher the price of an item, the higher its utility to the buyer, accounting for a difference in utility between a commodity and a luxury item (e.g., daily food vs. jewelry). Based on such an assumption, we define a relation between V and P as $V = \alpha P$, where $\alpha$ is called the utility coefficient, and buyable condition is $1 < \alpha$. We also assume a relationship between the price and cost of an item, i.e. the higher the price of an item, the higher the production cost due to materials or technology. Based on this assumption, we define a relation between C and P as $C = \beta P$, where $\beta$ is called the cost coefficient. Profitable condition is then $0 < \beta < 1$. The gains and losses of the buyer and the seller in all cases are summarized in Table 2.

**Table 2.** Payoff matrix in experiment in online C2C transactions

| | Buyer Cooperates | Buyer defects |
|---|---|---|
| Seller cooperates | Buyer's gain: $(\alpha-1)$P<br>Seller's gain: $(1-\beta)$P | Buyer's gain: $\alpha$ P<br>Seller's loss: $-\beta$ P |
| Seller defects | Buyer's loss: -P<br>Seller's gain: P | Buyer's gain: 0<br>Seller's gain: 0 |

# 4 Experiment Overview

In this section, we explain how the virtual transaction system for our experiment was developed. We assigned all participants to one of two groups, sellers or buyers, and each participant participated in several transactions. In our experiment, the system not only provided each participant with an initial handle, or onscreen user ID, but also allowed other handles if the participant wished to change it. It should be noted that a handle is simply a means of identification in an online market, and not a participant's real name. On the system, participants took actions, e.g. bid and award, synchronously. Although in real online C2C markets buyers and sellers take actions asynchronously, we did not design the experiment this way in order to simplify the transaction process for ease of analysis. Before we can analyze an asynchronous situation, we must first analyze the basic mechanisms in a synchronous situation.

The system permitted a buyer to make multiple bids to sellers and a seller to make multiple awards to bidders. In the last phase of a period, each participant was given the opportunity to change handles. If the participant decided to do so, the system provided a new handle randomly selected from the database. Each participant was allowed to choose between cooperation or defect in a transaction. Cooperation meant sending payment to a seller or sending an item to a buyer. Defect meant either no payment or no shipment. Each participant was able to view sorted lists of other participants' feedback information in terms of transaction history, item prices, number of cooperative transactions, the number of defective transactions, reputation score, and the total number of transactions. The transaction history displayed not only the date/time of the transaction but the handles of both buyer and seller and their respective actions. The reputation score was calculated by subtracting the number of defective actions from those of cooperative actions.

## *4.1 Transaction Process*

In our experiment, the transaction process was composed of four steps: (a) sellers deciding item prices, (b) buyers bidding on items, (c) sellers awarding items to bidders, and (d) both buyers and sellers checking the results of transactions and deciding whether to change their handles. An administrator managed the transaction process synchronously to ensure that all participants made decisions in each step (Fig. 1).

In the first step, the item pricing phase [(a) of Fig. 1], each seller decided on the price at which to sell the item within a predefined range, e.g., between 100 and 800.

We assumed that a seller could reap benefit at a certain constant rate in relation to the item price, meaning that the difference between price and production cost for a luxury item is greater than that for a commodity item.

Hence, a seller could reap greater benefit from selling a high-priced item than a low-priced item in a successful transaction. At the same time, however, the seller also faced a high risk in losing high-priced items if the transaction failed. In other words, the transaction was high risk and high return when the seller decided to sell a high-priced item, and the transaction was low risk and low return when the seller decided to sell a low-priced item.

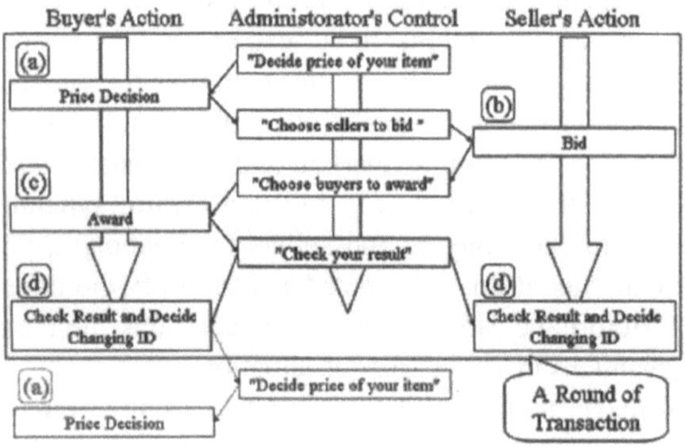


**Fig.1** Overview of transaction steps.

In the second step, the bidding phase [(b) of Fig. 1], each buyer viewed feedback information for all sellers, then selected good sellers, and decided whether to make the payment. The action was applied to all sellers selected by the buyer in each round.

In the third step, the awarding of items [(c) of Fig. 1], each seller viewed feedback information for all bidders, then selected good buyers, and decided whether to send the item. The action was also applied to all buyers selected by the seller.

At this stage, all pairs of buyer and seller who have reached a deal agreement can gain or lose points based on the payoff matrix defined in 3.2 according to their actions and the price of the item. The system records their actions in their transaction histories for future reference.

In the last step [(d) of Fig. 1], all participants checked the transaction results, then decided whether to change handles, allowing them the opportunity to avoid any negative transaction history.

# 5 Experiment Results

We carried out the experiment and observed transaction and information behavior of its participants using the system explained in Section 4. The participants were 37 students at a university in Tokyo. They were divided into two groups consisting of 18 buyers and 19 sellers. We performed 3 trial rounds so that participants could learn the system before the experiment itself, which consisted of 10 rounds. We did not inform the participants of the number of rounds in order to avoid its potential effect on decision-making. We provided rewards to the 5 highest-scoring participants as an incentive. We defined two types of behaviors among participants: transaction behavior and information behavior. The former refers to cooperative or defective behavior. The latter refers to what information a participant references during decision-making.

## *5.1 Analysis of Transaction Behavior*

Trajectories in Fig. 2 illustrate changes of the numbers of cooperative and defective actions. The figure illustrates an increase in cooperative action along with passage of time, as suggested by the high linear regression coefficient ( $\hat{\beta} = 9.85$ ).

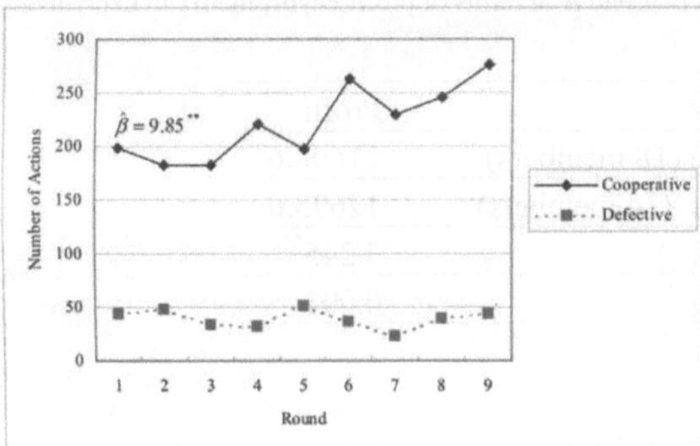


**Fig. 2.** Trajectories of cooperative and defective behavior

Fig. 3 illustrates the distribution of participants according to total profit and rate of cooperative action. We can distinguish three groups, which we will refer to as A, B, and C, in the figure based on hierarchical clustering analysis (Fig. 4). In group A, which includes 32 of 37 participants, the higher the rate of cooperative action, the higher the profit each participant made. This tendency is further supported by the result of t-test for two sub groups within group A, which divided into those who were always cooperative and those who were sometimes defective (Tab. 3). The three participants in group B obtained varying degrees of high profit with cooperative and defective actions. The final two participants in group C remained at low profit due to their consistently defective actions.

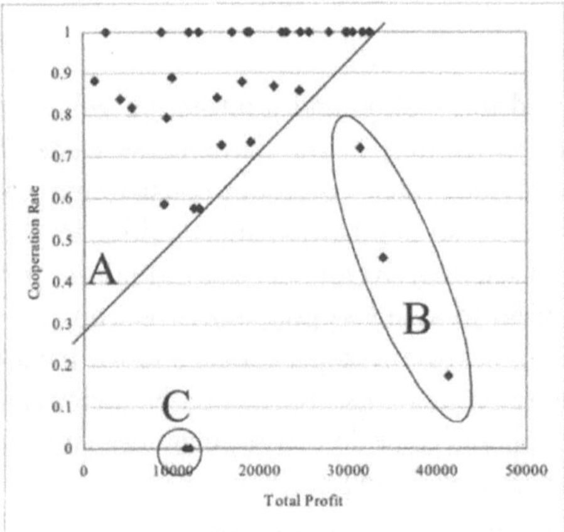


**Fig. 3.** Distribution of participants according to profit and cooperative behavior

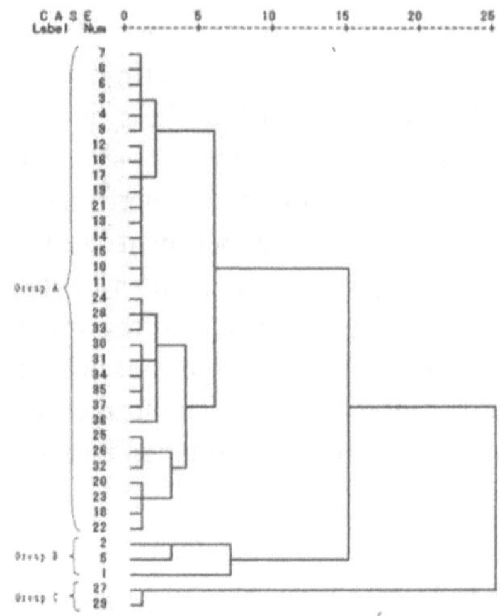

**Fig. 4.** Hierarchical clustering analysis of profit and ratio of cooperative action

**Table 3**: Result of t-test for two sub groups in group A in terms of profit Payoff matrix in experiment in online C2C transactions

| | Profit |
|---|---|
| Always cooperative (18 members) | 21698.6 |
| Sometime defective (14 members) | 12675.0 |
| t-value | 3.245** |
| p-value | 0.003 |

**$p$<.01

**Table 4**: Cumulative number of information behaviors

| | Reputation | C. | Defect | ID Duration | Detailed history | Number of Transaction |
|---|---|---|---|---|---|---|
| C. | 2.94 | 3.85 | 4.76 | 3.92 | 1.18 | 1.24 |
| Defect sometime | 3.89 | 5.81 | 2.89 | 1.06 | 2.03 | 1.54 |
| t value | 0.55 | 1.33 | 1.94+ | 3.59*** | 0.60 | 0.34 |
| p value | 0.58 | 0.18 | 0.05 | 0.00 | 0.55 | 0.73 |

+$p$<.10 *$p$<.05 **$p$<.01 ***$p$<.001 ( C. = Cooperation)

## *5.2. Analysis of Information Behavior*

During this experiment, the total number of references to information by all participants was 739. The number of references to the price of an item was only 5. The low number may suggest that finding cooperative participants is more important than finding participants who want to buy or sell high price items.

Fig 5 shows each cumulative total of information behaviors according to the criteria explained in Section 4. To analyze differences in information behavior between cooperative and defective participants, we categorized participants into two categories: always cooperative (participants who never defected against others) and sometimes defective (participants who took one or more defective actions against others) [1]. We normalized the number of information behaviors because of the difference among the numbers of references by the participants.

In addition, we ignored the information behaviors concerning price, which totaled to 11, because of the relatively low number.

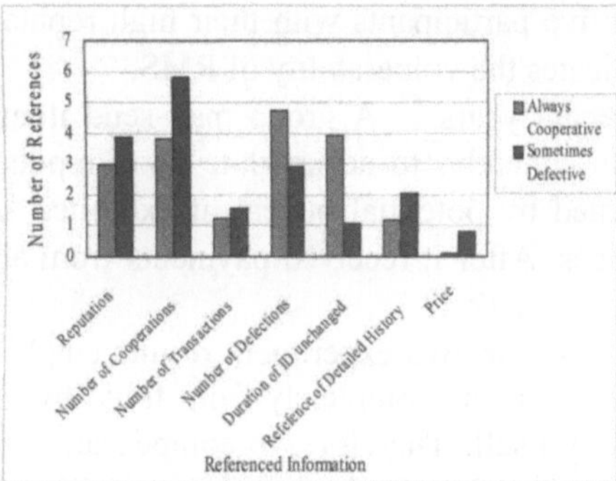


**Fig. 5.** Cumulative total of information behaviors

According to Fig. 5 and Tab. 4, cooperative participants focused on the number of defective actions taken and the duration that a user ID remained unchanged. Hence these are the most effective factors to discriminate between cooperative and defective participants. We also classified answers to the question of which strategy is the most effective for maximizing profit into three categories in table 5.

**Table 5**: Effective strategy for maximizing profit

| Strategy | Num |
|---|---|
| Always cooperative | 18 |
| Early round cooperation, high reputation obtained, and then defect in big deals | 14 |
| Always defect, while changing ID | 2 |
| Other | 2 |

# 6 Discussion

As shown in Fig. 2 and 3, a reputation management system (RMS) promotes cooperative behaviors in the market because each participant discriminates between cooperative and defective others.

## *6.1 Transaction Behavior and Fundamental Flaws of RMS*

Fig. 3 illustrates two insightful facts of a reputation management system. On one hand, group A, whose shape is a rectangle, depicts that choosing cooperative behavior gives a participant the chance to increase profit. This supports that RMS can promote cooperative behaviors by participants in the market.

However, on the other hand, in our interview after the experiment, the three participants in group B said that they had accumulated high reputation in the early rounds of the experiment, then exploited cooperative participants with their high reputation scores and defective action. The result indicates the vulnerability of RMS.

Such crimes have often occurred in recent years [4)]. A group may send items to buyers at first and respond to their claims quickly, to accumulate good reputation scores. When the group had become trusted by potential buyers, it exhibited large number of expensive items at reduced prices. After it received payments from all its buyers, the group disappeared from the online market.

The real-life examples of actual crime and our own experiment results emphasize that it is difficult to protect cooperative participants completely from fraud by malicious participants by employing an RMS by itself. Therefore, to compensate for the fundamental flaw of the ordinary RMS, it should be supplemented with another system, e.g. a legal system.

## *6.2. Information Behavior and Redesigning RMS*

Based on our analysis of information behavior, we found significant differences between purely cooperative participants who never defected against other participants, and the others. All the purely cooperative participants often referenced two types of information: the duration that the handle remained unchanged and the number of defective actions. The tendency indicates that cooperative participants prefer transactions with other risk-adverse participants.

Based on our conclusions concerning relation between transaction and information behaviors, we can formulate a strategy to detect malicious sellers. Based on transaction behavior, the detectable signs are high reputation scores and switching from a low volume of inexpensive items to a high volume of expensive that. In the post-experiment interview, participants in group B sold inexpensive items to accumulate high reputation scores in the early rounds of transactions, then defected against buyers in transactions involving expensive items. Based on information behavior, the detectable sign is that the behavior is not risk-adverse, which means that the player could defect against others as shown in Fig. 3.

Employing such signs to detect malicious sellers, online marketplaces can monitor high-risk sellers to protect buyers, and we can reduce dependency on legal systems for secure online C2C transactions. We propose a combination of the two signs to detect defective participants as one of the principles in redesigning RMS.

To make an RMS effective in protecting cooperative participants, extensive historical data must be archived to calculate reputation scores. Moreover, a participant who has not been selected by the others to make transactions might not be trustworthy, even though he or she has used the same handle for a long time. Hence, we suggest that the number of past transactions is as important as the number of defective actions and duration that the handle has remained unchanged. We also propose the combination of these three types of information as another principle in redesigning RMS.

### *6.3.Limitations of the Experiment*

In an actual market, a participant who has used a particular handle for only a short duration could be simply be new to the market. In our experiment, however, all participants entered into the market at the same time and therefore knew that there were no new users. As a result, they might conclude that any user with an ID for any duration of ID as an indicator of trustful person. Hence, the effect of the duration should be discussed again in future experiments.

Moreover, in an actual market, the rating for the quality of a transaction is evaluated by a participant based on a subjective point of view, although here we assume that the evaluation was recorded by the RMS system based on an objective point of view, i.e. cooperate or defect. We took into account the robustness of reputation information against unfair evaluation as discussed by Dellarocas (2000). In our experiment, we ignored that the evaluation method in our current system is objective (i.e., the system just records cooperative and defective actions) rather than subjective (e.g., the seller sent me a good item quickly). In future experiments, we will investigate the effects of subjective evaluation by participants.

## 7. Summary

We developed an experimental system to analyze the effects of a reputation management system (RMS) on promoting cooperative behavior by participants and to reveal information behavior in an online C2C market. According to the results, over 80% of participants behaved cooperatively due to the RMS. Moreover, based on analysis of information behavior, we also found that cooperative participants often referenced the number of defective actions and duration of handle. In the experiment, no new users entered the market. The situation was predicted by Yamamoto et. al. (2004) using computer simulation. The results conclude that negative RMS based on the number of defects are effective in promoting cooperative behavior when there are few new users. Based on the results, we proposed two principles for redesigning future RMS for secure online C2C transaction. In future research, we will investigate online C2C markets in a dynamic environment whereas participants are always changing.

## Notes

1) 18 of 37 participants were always cooperative, while the others were sometime defective.
2) The result of discriminant analysis and the distribution of discriminant points.
3) The number of effective answers was 36, while there were 37 participants.
4) News report by Asahi Online, Jan. 13, 2005 (http://www.asahi.com)

## References


1. Bolton, G. E., Katok, E. and Ockenfels, A. "How effective are online reputation mechanisms? An experimental study" Management Science 50(11) 1587-1602.2004.
2. Dellarocas C., "The Digitization of Word of Mouth: Promise and Challenges of Online Feedback Mechanisms", Management Science, Vol. 49, No. 10, pp.1407-1424, 2003.
3. Dellarocas, C., "Immunizing Online Reputation Reporting Systems against Unfair Ratings and Discriminatory Behavior", Proceedings of the 2nd ACM Conference on Electronic Commerce, pp.17-20, 2000.
4. Kollock, P., "The Production of Trust in Online Markets", Advances in Group Processes, Vol.16, pp.99-123, 1999.
5. McDonald, C., and C. Slawson, "Reputation in An Internet Auction Market", Economic Inquiry, vol.40, issue 4, pp.633-650, 2002.
6. Resnick, P. and Zeckhauser, R., "Trust among strangers in internet transactions: Empirical analysis of eBay's reputation system", Technical report, University of Michigan, 2001.
7. Rice, S.,"Online Reputations with Noisy Transactions: An Experimental Study", 2004 Workshop on Information Systems and Economics.
8. Ruth,J., and York,A.,"Framing information to enhance corporate reputation: The impact of message source, information type, and reference point",Journal of Business Research vol.57, pp.14-20,2004.
9. Yamamoto, H., K. Ishida and T. Ohta, "Modeling Reputation Management System on Online C2C Market", Computational & Mathematical Organization Theory, Vol. 10, No. 2, pp.165-178, 2004.

# Market and Economy I

# Social Network Characteristics and the Evolution of Investor Sentiment

**Nicholas S. P. Tay**

School of Business and Management, University of San Francisco, CA 94117, USA

Tel: 1(415)422 6100; E-Mail: tay@usfca.edu

**Abstract**. This paper creates a bare bone model to understand how network characteristics such as the richness of the information environment, tendency for investors to extrapolate past data and social influence affect the transmission and evolution of investor sentiment within the network. Our results replicate qualitatively the empirical characteristics of actual investor sentiment documented by [1].

**Key words:** Agent based model, social network, investor sentiment

## 1 Introduction

The ability to communicate with each other via e-mails, internet messengers, internet chat rooms and message boards has allowed us to extend our social network and reach a great number of people at lightning speed and with negligible costs and yet remain anonymous to the message recipients. In finance, these social linkages play an important role to facilitate the exchange of investment information and ideas and the spread of sentiment which ultimately move prices in the market. Not surprisingly, there have been numerous cases where unscrupulous individuals took advantage of this technology to manipulate investors' perception. Clearly market manipulations are more successful under some conditions than others. This paper seeks to understand how social network characteristics affect the transmission and evolution of sentiment[1].

With this in mind, we develop a bare bone agent based model to understand how social network characteristics such as the richness of the information environment, tendency for investors to extrapolate past data and social influence affect the transmission and evolution of investor sentiment in the network. We model the interactions of the agents between the market and a message board. We do not al-

---

[1] We are grateful for the comments from three anonymous reviewers which have helped to improve this paper.

low agents to learn and we leave out the asset pricing mechanics of the market. We do this intentionally because as with most agent based models, the model parameters can interact in complex ways to affect how the model evolves. We feel that it is prudent to begin our investigation with a simple model and gradually add to it more realism in our follow up papers as we advance our understanding of the dynamics in this model. Despite our naïve translation of the real world in the model, our results replicate qualitatively the empirical characteristics of actual investor sentiment documented by [1].

Section 2 develops the agent based model that will drive our investigations. Section 3 explains how the various sentiment related indices are computed. Section 4 describes the experiments and the control parameters for these experiments. Section 5 discusses our findings. Section 6 discusses the strengths and limitations and concludes.

# 2 The Model

## 2.1 Overview

Time is discrete and at the start of each period, agents receive exogenously generated public and private information about the prospect of a stock. The agents interpret the information received and revise their beliefs accordingly. They then decide whether or not to post their revised beliefs on a stock message board. At the end of each period, the information shared on the stock message board is aggregated to form an average opinion about the prospect of the stock. With some probability, agents will be persuaded to adopt the average opinion as their new beliefs. Agents' beliefs at the end of each period are subsequently carried forward to the beginning of the next period and hence will color their interpretations of the newly arrived information. These steps are iterated repeatedly to investigate how investor sentiment evolves under various controlled settings.

## 2.2 Exogenous Public and Private Information

The exogenously generated information are represented as having (M+P) dimensions, each of which has a value that is drawn at random from a normal distribution, N($\tilde{\mu}$,0.3). The (M+P) values represent information from (M+P) messages. For ease of exposition, we refer to them as information bits. The M bits of information are assumed to be public information and may include news, company filings, analysts' reports etc. The P bits are private information which investors may stumble upon by chance. Surprisingly it is not unusual for investors to stumble upon private information every so often. [1] has reported cases where information revealed on message boards actually foreshadowed subsequent press releases.

We set M to 100 and P to 10. The mean level, $\tilde{\mu}$, is common to all the M+P information bits at each moment in time. To model fluctuations in $\tilde{\mu}$, we let $\tilde{\mu}$ varies over time according to a random draw from N(0.1,0.3). The variable $\tilde{\mu}$ reflects the degree of bullishness or bearishness with a stock. Since $\tilde{\mu}$ has a mean of 0.1 there is a slight degree of bullishness in the information stream for the stock.[2] We allow the agents a 20 percent probability of observing P, the private information. However, even though the M bits of information are available to the public, in an environment where information is costly, investors may decide to do without such information. For instance, when there are too few stock analysts following a stock, information about this stock will not be as readily available. Consequently, investors will now have to research the information on their own making the acquisition of information more costly. We use the parameter, $\Omega$, to model how rich or poor the traditional information environment is. For a rich (poor) traditional information environment, we set $\Omega$ to 0.9 (0.5) which corresponds to a scenario where investors are able to observe without much effort 90% (50%) of the publicly available information. The observed values are represented with real numbers generated from the distributions described earlier. The unobserved values are denoted with "NA" in the data matrix.

### 2.3 *E-Information*

The agents, upon digesting the exogenous information revealed in the market, decide whether or not to post their beliefs on a stock message board. Their personal beliefs are an amalgamation of the exogenous information observed by the agents and their existing beliefs. Since each agent's belief may evolve differently, the personal information shared on the message board will likely be heterogeneous even if the agents observe the same exogenous information. Following [1], we use the term e-information to refer to the information that is embedded in messages on the stock message board.

It is evident from observing activities on message boards that the intensity of discussions and interests is not uniform across all stocks. We use the parameter, $\Phi$, to control how rich or poor the e-information environment is for a stock. Since the richness of the e-information environment is a function of the willing-

---

[2] The values generated will be converted to buy or sell intents. Positive values are more likely to result in buy intentions and so can be equated with bullishness. This will be explained in details in Section 3

ness of agents to post their information on message board, we use $\Phi$ to control the likelihood that an agent will share his information on the message board. For the sample of stocks studied by [1], there is on average about 200 messages posted per day in a rich e-information environment and less than 5 messages posted per day in a poor e-information environment. To replicate the levels of messages observed, we set $\Phi$ to 0.1 to mimic a rich e-information environment and 0.002 to mimic a poor e-information environment. Since the total number of traders is set to 2000, it works out to an average of 200 messages for the rich e-information environment and about 4 messages for the poor e-information environment.

## 2.4 Agents and Information Aggregation

There are 2000 agents in the model. At the start of each period, agents receive information from an external source. We assume that agents revise their beliefs by taking a weighted average of the content embedded in the new information and in their existing beliefs. There are evidences from [2], [3] and [4] that suggest that investors in formulating their beliefs tend to extrapolate past information. To model extrapolation of past information, we let $W$ be the weight assigned to the new information so $(1-W)$ will be the weight assigned to each agent's existing belief. In the simulations, we set $W$ to 0.1, 0.5 and 0.9 to investigate the consequence of various degree of extrapolation of past data.

Each agent's belief is a path dependent evolution of the information that the agent has been exposed to in the past and the choices that were made. As explained in the preceding section, agents, after revising their beliefs, subsequently decide whether or not to share their new beliefs on a stock message board. We can visualize the information shared on the message board as a $(M+P) \times K$ matrix of real numbers where $K$ refers to the number of posters and $(M+P)$ is the size of the information bits. Recall that $K$ is a function of $\Phi$ and the number of traders; when $\Phi$ equals to 0.1, $K$ has an average value of 200 and when $\Phi$ equals to 0.002, the average value of $K$ is 4. We aggregate the information on the message board by taking a simple average across the $K$ individuals for each information bit. The result is a $(M+P) \times 1$ vector which represents the average opinion on the message board of each of the $(M+P)$ information bits. All the agents in the model observe this average opinion and may be persuaded to adopt this viewpoint. We use the social influence parameter, $\pi$, to control the likelihood that an agent will drop his own belief about an information bit and adopt the prevailing average opinion on the message board for that information bit. In the simulations, we set $\pi$ to 0.2 and 0.8 to correspond to a 20 percent and an 80 percent chance of accepting the average message board opinion respectively. In addition, if an agent has an information bit with zero value, we assume that an agent automatically accept the average message board opinion for that information bit. The agents' beliefs that emerge at the end are carried forward to the beginning of the next period.

## 3 Measuring Sentiment

Our goal is to understand the transmission and evolution of sentiment. To measure sentiment, we follow [1] to construct a sentiment index that is defined as the number of buy messages minus the number of sell messages. In this sense, the sentiment index is a measure of the net bullish sentiment. We want to measure for each period a) the sentiment embedded in the information from the exogenous source, b) the sentiment derived from messages on a message board and c) the sentiment derived from end-of-period beliefs for all the agents. For ease of exposition, we refer to these sentiments as news sentiment, posting sentiment and market sentiment respectively. Since the data are all real numbers, we need to first transform these numbers to buy and sell signals. The data in our model generally lies within the range [-1, 1]. To transform the data to buy and sell signals, we consider values greater than 1/3 to be buy signals which we code as "1" and values smaller than -1/3 to be sell signals which we code as "-1". Values lying within [-1/3, 1/3] are considered neutral signals and are coded as "0". However, the sentiment index is not a scaled measure of sentiment and does not give an indication of the strength of the sentiment. To scale the sentiment between -1 and 1, we divide the sentiment index by the size of the information bits to derive a better measure of the extent of the bullish or bearish opinion embedded in the information set. Following, [1], we call this the Sentiment Percentage.

The equations for calculating the Sentiment Percentage are as follows:

$$\text{News Sentiment Percentage} = \frac{BUYS_1 - SELLS_1}{M + P}. \quad \mathbf{(1)}$$

$$\text{Posting Sentiment Percentage} = \frac{BUYS_2 - SELLS_2}{(M + P) \times K}. \quad \mathbf{(2)}$$

$$\text{Market Sentiment Percentage} = \frac{BUYS_3 - SELLS_3}{(M + P) \times N}. \quad \mathbf{(3)}$$

The different subscripts for the BUYS and SELLS signals refer to the specific BUYS and SELLS signals for each category. Subscript "1" refers to signals embedded in the news sentiment, Subscript "2" represents signals from posting sentiment, and Subscript "3" refers to signals embedded in market sentiment. (M+P) refers to the size of the information bits. The variable K refers to the number of posters, and the variable N refers to the total number of agents.

## 4 Experiments

In the experiments, we examine 24 different combinations of parameter values, each of which we refer to as a "case". The parameters of the model and the values examined for each of the 24 cases are tabulated in Table 1. We keep some parameters fixed throughout, including the number of agents (N), set at 2000, and the size of the externally generated information (M and P), set to 100 and 10 respectively. For each of the 24 cases, we run simulation of the model for 100 time periods and we repeat the simulation for each case 100 times to obtain results for the average behaviors. To facilitate comparisons across the 24 cases, we keep the time series for the exogenously generated information to be identical in each simulation of the 24 cases. Consequently, the simulations are independent only in the sense that the seeds of the algorithms used to generate pseudo-random variables differ across the 100 simulations for each case. This provides us with 100 separate time series of each of the various sentiment indices for each case allowing us to compute estimates of various descriptive statistics for each of the 24 cases. Besides investigating how the characteristics of the social network affect the transmission and evolution of investor sentiment in the network, we are also interested in establishing face validity for our model which we accomplish by comparing our results to the empirical characteristics of investor sentiment on actual stock message boards reported by [1].

## 5 Simulation Results

The columns of Tables 2 and 3 present summary results for the 24 cases described in Section 4.

### *5.1 Autocorrelations, Posting Activities and the Level of Sentiment Indices*

Table 2 shows a positive relationship between the level of posting activities and the average level of absolute sentiment on the message board. In constructing the absolute sentiment we have scaled the unscaled sentiment level by dividing it by the average size of the news information bits. The positive relationship suggests that herd behavior is more acute when there is a high level of activities on the message board. What we observe is consistent with the empirical characteristics for actual sentiment data found in [1].

The autocorrelation for the news sentiment level is nearly zero. This is expected by design and is done intentionally to allow us investigate if the nature of the information environment or the manner in which investors interact and revise their beliefs could induce persistence or positive autocorrelation in the posting and market sentiment levels. It is evident from the results in Table 3[3] that there is indeed persistence in the posting and market sentiment levels. The level of persistence increases with the degree of extrapolation and is most acute when there is severe extrapolation as in cases 9 to 12 and 21 to 24. This effect is less acute for cases 21 to 24 where there is a poor traditional information environment. However, social influence or the level of posting activities does not appear to have any significant effect on persistence.

The observed persistence in the sentiment level is not inherited from the news sentiment but is a consequence of extrapolation activities. Our finding is similar to [1] who has reported persistence in the posting sentiment level computed from actual data collected from various stock message boards. They remarked that this is evidence that investors change their minds slowly or in other words, investors extrapolate past data. Other studies such as [2], [3] and [4] have also reported similar findings.

## 5.2 *Correlations between Sentiment Indices*

To understand how the network characteristics affect the transmission of sentiment from an exogenous source to the message board and subsequently to the rest of the agents in the model, we examine the contemporaneous correlation between news and posting sentiment indices and between news and market sentiment indices. We expect to see a high level of correlation between the sentiment indices if the transmission is highly effective and vice versa if it is less effective[4].

On examining the contemporaneous correlations in Table 3, we observe that transmission effectiveness degrades as extrapolation of past data becomes more severe. This effect is less severe in a poor traditional information environment. Again social influence or the level of posting activities does not appear to have any significant impact on the transmission of sentiment.

Our results in section 5.1 and 5.2 suggest that when investors heavily extrapolate past data, they can impair the transmission of sentiment information and can induce persistence in sentiment levels. Further, the richness of the traditional information environment can play a secondary role in compounding the problem.

---

[3] The results in Table 3 are different from our original working paper because we corrected an error in the calculation of the averages and extended our simulations for each case from 20 to 100 runs.

[4] We consider a transmission highly effective if the transmission process maintains the integrity of the sentiment that is embedded in the original signal.

## 6 Strengths, Limitations and Conclusion

A primary weakness of our model is its lack of realism which in turn places limitations on the usefulness and relevance of our results to real markets. There are several issues with the way we model how agents aggregate information and how they adapt over time. We assume that agents use a constant weight to compute the weighted average of the new information and their existing beliefs when formulating their new beliefs at the start of each period. We also assume that agents take a simple average of the opinions posted on the message board and then decide based on a fixed probability whether they would replace their beliefs with the average opinion on the message board. Surely, real agents do not behave in such simplistic manners. Even relatively unsophisticated investors in formulating their beliefs do not use a constant weight to aggregate the new information and their existing beliefs. Instead, they are more likely to employ varying weights depending on the conditions in the market, the reputation of the sources and whether there are other collaborating evidences. Moreover, investors that participate actively on message board are likely to develop varying degree of trust for participants on message board and therefore will weigh the opinions from various contributors differently. Further, rather than using a fixed probability to determine whether they would adopt the average opinion on the message board, investors are more likely to use this information to update their beliefs in some Bayesian manner or via some conditional rule of thumbs. Our model has ignored all these aspects of learning. Nonetheless, if changes in the above parameters take place slowly over time, our analysis would still be reasonable.

As we have argued earlier in this paper, our intent is not to develop a realistic model at this stage. Our goal is to develop a base platform to help us understand how the model parameters influence the model dynamics before making the model too complex. Despite the flaws with our model, our results are consistent with the empirical characteristics of actual sentiment data documented by [1]. We find a) a high level of persistence in the sentiment level, b) strong positive correlations between investor sentiment and news sentiment, and c) sentiment is positively related to the volume of postings. Further, we observe that social network characteristics such as the richness of the information environment and the extent to which agents extrapolate past data can interact to either enhance or degrade the quality of information in the transmission process.

## References

1. Das, S., A. Martinez-Jerez, and P. Tufano. (2005). e-Information: A Clinical Study of Investor Discussion and Sentiment. *Financial Management,* 34(3), 103-137.
2. De Bondt, W., and R. Thaler. (1985). Does the Stock Market Overreact? *Journal of Finance,* 40(3), 793-805.
3. Lakonishok, J., A. Shleifer, and R. Vishny. (1994). Contrarian Investment, Extrapolation, and Risk. *Journal of Finance*, 49(5), 1541-1578.
4. Kroll, Y., H. Levy, and A. Rapoport. (1988). Experimental Test of the Mean-Variance Model for Portfolio Selection. *Organizational Behavior and Human Decision Processes*, 42(3), 388-410

**Table 1.** Parameter settings for the various cases

This table provides the parameter settings for each of the 24 cases examined in our study. The four parameters: $\Omega$, $\Phi$, W and $\pi$ measure the richness of the traditional environment, the richness of the e-information environment, the extrapolation weight and the likelihood of social influence respectively.

| CASES | 1 | 2 | 3 | 4 | 5 | 6 | 7 | 8 | 9 | 10 | 11 | 12 |
|---|---|---|---|---|---|---|---|---|---|---|---|---|
| $\Omega$ | 0.9 | 0.9 | 0.9 | 0.9 | 0.9 | 0.9 | 0.9 | 0.9 | 0.9 | 0.9 | 0.9 | 0.9 |
| $\Phi$ | 0.1 | 0.002 | 0.1 | 0.002 | 0.1 | 0.002 | 0.1 | 0.002 | 0.1 | 0.002 | 0.1 | 0.002 |
| W | 0.9 | 0.9 | 0.9 | 0.9 | 0.5 | 0.5 | 0.5 | 0.5 | 0.1 | 0.1 | 0.1 | 0.1 |
| $\pi$ | 0.8 | 0.8 | 0.2 | 0.2 | 0.8 | 0.8 | 0.2 | 0.2 | 0.8 | 0.8 | 0.2 | 0.2 |
| **CASES** | **13** | **14** | **15** | **16** | **17** | **18** | **19** | **20** | **21** | **22** | **23** | **24** |
| $\Omega$ | 0.5 | 0.5 | 0.5 | 0.5 | 0.5 | 0.5 | 0.5 | 0.5 | 0.5 | 0.5 | 0.5 | 0.5 |
| $\Phi$ | 0.1 | 0.002 | 0.1 | 0.002 | 0.1 | 0.002 | 0.1 | 0.002 | 0.1 | 0.002 | 0.1 | 0.002 |
| W | 0.9 | 0.9 | 0.9 | 0.9 | 0.5 | 0.5 | 0.5 | 0.5 | 0.1 | 0.1 | 0.1 | 0.1 |
| $\pi$ | 0.8 | 0.8 | 0.2 | 0.2 | 0.8 | 0.8 | 0.2 | 0.2 | 0.8 | 0.8 | 0.2 | 0.2 |

**Table 2.** Average Absolute Sentiment and Postings

This table reports the average of the absolute level of sentiment on the message board and the average level of messages posted for each of the 24 cases. To compute these averages, we first average over time for each simulation and then we calculate the average of this time average over the 100 simulations for each case. We denote the average of the absolute level of sentiment as $\overline{I}$ and the average level of posting activity as $\overline{Post}$

| CASES | 1 | 2 | 3 | 4 | 5 | 6 | 7 | 8 | 9 | 10 | 11 | 12 |
|---|---|---|---|---|---|---|---|---|---|---|---|---|
| $\overline{I}$ | 66.6 | 1.3 | 48.8 | 1.0 | 78.9 | 1.7 | 54.4 | 1.1 | 251.2 | 5.9 | 141.4 | 3.0 |
| $\overline{Post}$ | 201.2 | 4.0 | 202.0 | 4.1 | 202.1 | 4.0 | 202.1 | 4.1 | 201.7 | 4.1 | 202.3 | 4.1 |
| **CASES** | **13** | **14** | **15** | **16** | **17** | **18** | **19** | **20** | **21** | **22** | **23** | **24** |
| $\overline{I}$ | 35.5 | 0.7 | 13.6 | 0.3 | 51.3 | 1.0 | 18.4 | 0.4 | 195.0 | 4.0 | 69.7 | 1.5 |
| $\overline{Post}$ | 202.2 | 4.0 | 202.1 | 4.0 | 201.8 | 4.0 | 202.3 | 4.1 | 202.0 | 4.1 | 201.6 | 4.0 |

**Table 3.** Autocorrelations and Contemporaneous Correlations

We use, $\rho_j$ to denote the autocorrelation of the sentiment percentage index for j and $\rho_{i,j}$ to refer to the contemporaneous correlation between the sentiment percentage index i and j. We use the subscripts '1', '2' and '3' to denote respectively a) the news sentiment, b) the posting sentiment, and c) the market sentiment. The number reported in the top row for each case is an average over 100 simulations. Directly beneath each average value we report the standard deviation over the 100 simulations for each case.

PANEL A: RICH TRADITIONAL INFORMATION ENVIRONMENT ($\Omega$=0.9)

| | POOR E-INFORMATION ($\Phi$=0.002) | | | | | | RICH E-INFORMATION ($\Phi$=0.1) | | | | | |
|---|---|---|---|---|---|---|---|---|---|---|---|---|
| W | 0.9 | 0.9 | 0.5 | 0.5 | 0.1 | 0.1 | 0.9 | 0.9 | 0.5 | 0.5 | 0.1 | 0.1 |
| $\pi$ | 0.8 | 0.2 | 0.8 | 0.2 | 0.8 | 0.2 | 0.8 | 0.2 | 0.8 | 0.2 | 0.8 | 0.2 |
| **CASES** | **2** | **4** | **6** | **8** | **10** | **12** | **1** | **3** | **5** | **7** | **9** | **11** |
| $\rho_{1,2}$ | 0.939 | 0.939 | 0.817 | 0.801 | 0.454 | 0.419 | 0.969 | 0.966 | 0.838 | 0.815 | 0.462 | 0.367 |
| | 0.025 | 0.026 | 0.023 | 0.028 | 0.061 | 0.065 | 0.018 | 0.018 | 0.020 | 0.019 | 0.055 | 0.062 |
| $\rho_{1,3}$ | 0.958 | 0.986 | 0.837 | 0.847 | 0.484 | 0.464 | 0.977 | 0.988 | 0.852 | 0.841 | 0.498 | 0.424 |
| | 0.016 | 0.002 | 0.017 | 0.016 | 0.057 | 0.077 | 0.011 | 0.002 | 0.015 | 0.015 | 0.054 | 0.070 |
| $\rho_1$ | -0.021 | -0.021 | -0.021 | -0.021 | -0.021 | -0.021 | -0.021 | -0.021 | -0.021 | -0.021 | -0.021 | -0.021 |
| | 0.097 | 0.097 | 0.097 | 0.097 | 0.097 | 0.097 | 0.097 | 0.097 | 0.097 | 0.097 | 0.097 | 0.097 |
| $\rho_2$ | 0.058 | 0.060 | 0.362 | 0.379 | 0.690 | 0.687 | 0.060 | 0.065 | 0.391 | 0.419 | 0.749 | 0.818 |
| | 0.099 | 0.098 | 0.098 | 0.104 | 0.090 | 0.120 | 0.098 | 0.097 | 0.105 | 0.108 | 0.086 | 0.100 |
| $\rho_3$ | 0.057 | 0.069 | 0.365 | 0.421 | 0.702 | 0.816 | 0.060 | 0.068 | 0.385 | 0.431 | 0.737 | 0.848 |
| | 0.099 | 0.101 | 0.098 | 0.103 | 0.081 | 0.083 | 0.098 | 0.100 | 0.104 | 0.106 | 0.083 | 0.070 |

PANEL B: POOR TRADITIONAL INFORMATION ENVIRONMENT ($\Omega$ =0.5)

| | POOR E-INFORMATION ($\Phi$ =0.002) | | | | | | RICH E-INFORMATION ($\Phi$ =0.1) | | | | | |
|---|---|---|---|---|---|---|---|---|---|---|---|---|
| W | 0.9 | 0.9 | 0.5 | 0.5 | 0.1 | 0.1 | 0.9 | 0.9 | 0.5 | 0.5 | 0.1 | 0.1 |
| $\pi$ | 0.8 | 0.2 | 0.8 | 0.2 | 0.8 | 0.2 | 0.8 | 0.2 | 0.8 | 0.2 | 0.8 | 0.2 |
| **CASES** | **14** | **16** | **18** | **20** | **22** | **24** | **13** | **15** | **17** | **19** | **21** | **23** |
| $\rho_{1,2}$ | 0.932 | 0.925 | 0.892 | 0.870 | 0.782 | 0.716 | 0.965 | 0.961 | 0.919 | 0.885 | 0.802 | 0.668 |
| | 0.024 | 0.025 | 0.023 | 0.025 | 0.032 | 0.033 | 0.018 | 0.018 | 0.019 | 0.019 | 0.024 | 0.032 |
| $\rho_{1,3}$ | 0.947 | 0.956 | 0.911 | 0.898 | 0.817 | 0.761 | 0.969 | 0.969 | 0.929 | 0.906 | 0.835 | 0.744 |
| | 0.016 | 0.015 | 0.016 | 0.029 | 0.026 | 0.048 | 0.012 | 0.010 | 0.014 | 0.020 | 0.021 | 0.042 |
| $\rho_1$ | -0.021 | -0.021 | -0.021 | -0.021 | -0.021 | -0.021 | -0.021 | -0.021 | -0.021 | -0.021 | -0.021 | -0.021 |
| | 0.097 | 0.097 | 0.097 | 0.097 | 0.097 | 0.097 | 0.097 | 0.097 | 0.097 | 0.097 | 0.097 | 0.097 |
| $\rho_2$ | 0.019 | 0.037 | 0.210 | 0.230 | 0.403 | 0.439 | 0.028 | 0.040 | 0.237 | 0.311 | 0.451 | 0.597 |
| | 0.097 | 0.095 | 0.101 | 0.101 | 0.091 | 0.100 | 0.096 | 0.095 | 0.100 | 0.101 | 0.104 | 0.082 |
| $\rho_3$ | 0.019 | 0.038 | 0.199 | 0.253 | 0.380 | 0.499 | 0.024 | 0.037 | 0.216 | 0.277 | 0.409 | 0.546 |
| | 0.097 | 0.101 | 0.100 | 0.112 | 0.091 | 0.102 | 0.095 | 0.100 | 0.099 | 0.107 | 0.104 | 0.091 |

# From the simplest price formation models to paradigm of agent-based computational finance: a first step

Takashi Yamada and Takao Terano

**Abstract** This paper studies similarities and differences between models based on Monte-Carlo method by focusing on so-called "stylized facts." In this study, we propose a model based on evolutionary algorithm and take the other model based on statistical physics. For this purpose, first we present a genetic learning model of investor sentiment and then several ordinary time series analyses are conducted after generating sample paths. Finally, the price properties are compared to those in the Ising spin model. Our results show that both the Monte-Carlo simulations seem to lead to similar dynamics reported in real markets in that the agents are boundedly rational or have some biases towards the market. However, other time series properties are apparently different since the algorithm of price formation is different.

## 1 Introduction

Agent-based computational finance (hereafter ACF) has been an expected and feasible way to explain micro-macro relations in economic dynamics [6, 25]. The contributions of this area cover from tests of neo-classical economy to searching the possibilities to make more money. If we focus on some of the earlier studies which have shown time series properties, one can fall these studies into three groups in accordance with experimental setups and the methods of time series analyses; First,

---

Takashi Yamada
Department of Computational Intelligence and System Science, Tokyo Institute of Technology, 4259 Nagatsuta-cho, Midori-ku, Yokohama, Kanagawa 226-8502, Japan e-mail: tyamada@trn.dis.titech.ac.jp

Takao Terano
Department of Computational Intelligence and System Science, Tokyo Institute of Technology, 4259 Nagatsuta-cho, Midori-ku, Yokohama, Kanagawa 226-8502, Japan e-mail: terano@dis.titech.ac.jp

the artificial markets in which the external information is available for the agents can generate time series data at daily or longer time scale [2, 12, 14]. The algorithm in these models is usually evolutionary one. Second, in the models based on statistical physics, the agents can perceive mainly the information with respect to price/rate and therefore the data generated are usually considered as intraday ones [9]. We usually call these models "deterministic model". Third, fundamentalist-chartist models, namely a fundamentalist thinks that the asset price converges to a fundamental price and a chartist follows patterns in past series, can generate both the daily and the high frequency data [19, 20, 21], but this does not imply that the time series properties hinge on the number of fundamentalists (or chartists). Those previous studies, in addition, have not taken into account the empirical evidences from the questionnaire studies and the attitudes of speculators to market conditions or to their own unrealized profit/loss, as will be mentioned in the sequel.

On the other hand, the recent development of behavioural economics has enabled us to propose descriptive models with respect to the behaviours of speculators. Besides, some models are proposed from the experimental economic points of view [13, 26], or by applying some evidences of behavioural finance for agent-based computational economic studies [4, 18]. Accordingly, ACF needs to incorporate these findings and contributions into each of the models.

Therefore, the aims of ACF research are to offer the framework/paradigm of how the models are built, and then to contribute to the economic literature. In this study, before presenting such a framework, we explore the similarities and differences between the time series properties of a model model of investor sentiment by Barberis et al. [3] with genetic algorithm, which is often used for learning of agents in ACF models [1, 15, 23], and results in a previous study [24]. In other words, this paper tries to present a multiagent model based on evolutionary algorithm and to report what kind of models is proper to lead to dynamics observed in real markets using Monte-Carlo method and comparing time series properties.

The rest of this paper is organized as follows. The next section introduces a summary of MIS, and shows some conditions requisite for GA to describe MIS. Then, we explored the time series properties of generated sample paths using deducted variables of MIS in section 3, and compared time series properties to those of the Ising spin model in section 4. And finally, section 5 concludes this paper.

## 2 Experimental setup

### 2.1 A model of investor sentiment

Barberis et al. have developed their model of investor sentiment in order to describe representativeness and conservatism of people. The main assumptions are as follows:

- Return process follows a random walk.

**Table 1** Transition probability of MIS

a. Unstable condition

| | $\Delta S_{t+1}=1$ | $\Delta S_{t+1}=-1$ |
|---|---|---|
| $\Delta S_t=1$ | $\pi_L\ (<0.5)$ | $1-\pi_L$ |
| $\Delta S_t=-1$ | $1-\pi_L$ | $\pi_L$ |

$\Delta S_t$: Price movement at time $t$

b. Stable cond ition

| | $\Delta S_{t+1}=1$ | $\Delta S_{t+1}=-1$ |
|---|---|---|
| $\Delta S_t=1$ | $\pi_H\ (>0.5)$ | $1-\pi_H$ |
| $\Delta S_t=-1$ | $1-\pi_H$ | $\pi_H$ |

**Table 2** An abstract of investor's recognition

| $t\backslash t+1$ | Stable condition | Unstable condition |
|---|---|---|
| Stable condition | $1-\lambda_2$ | $\lambda_2$ |
| Unstable condition | $\lambda_1$ | $1-\lambda_1$ |

- The investor in this economy does not realize that the process follows a random walk.
- He/She thinks that the economy is either in a stable state or in an unstable state. While the return process is supposed to be mean-reverting in the unstable economy, the process is believed to have a trend in the stable economy.
- The investor believes "a underlying regime-switching process" but such a process rarely takes place.

More formally, the constitution of MIS is twofold; First, a market is either in a stable state or in an unstable one. If the market is in a stable condition, the probability $\pi_H$ that the price movement will be the same as the previous one is over 0.5. While if the market is in an unstable state, the probability $\pi_L$ is under 0.5 (Table 1). The parameter $\lambda_1$ is the probability of transition from unstable condition to stable one, while $\lambda_2$ is the one from stable condition to unstable one (Table 2). Moreover, Barberis et al. postulate that the sum of $\lambda_1$ and $\lambda_2$ is less than unity and that $\lambda_1$ is smaller than $\lambda_2$. Second, the price movement in the economy is either $+1$ or $-1$. Therefore $q_t$, the probability that the market is unstable, is renewed by equation (1) in case that the price movement is the different from the previous one, or by (2) otherwise.

$$q_{t+1}=\frac{((1-\lambda_1)q_t+\lambda_2(1-q_t))(1-\pi_L)}{((1-\lambda_1)q_t+\lambda_2(1-q_t))(1-\pi_L)+(\lambda_1 q_t+(1-\lambda_2)(1-q_t))(1-\pi_H)} \tag{1}$$

$$q_{t+1}=\frac{((1-\lambda_1)q_t+\lambda_2(1-q_t))\pi_L}{((1-\lambda_1)q_t+\lambda_2(1-q_t))\pi_L+(\lambda_1 q_t+(1-\lambda_2)(1-q_t))\pi_H} \tag{2}$$

where $0<\lambda_1<\lambda_2<1$ and $\lambda_1+\lambda_2<1$.

**Table 3** Opinion updating rule in Sznajd financial market model

| Time | $t$ | | | | $\rightarrow$ | $t+1$ | | | |
|---|---|---|---|---|---|---|---|---|---|
| Label | $i-1$ | $i$ | $i+1$ | $i+2$ | | $i-1$ | $i$ | $i+1$ | $i+2$ |
| (1) | # | + | + | # | | + | + | + | + |
| (2) | # | − | − | # | | − | − | − | − |
| (3) | # | + | − | # | | # | + | − | # |
| (4) | # | − | + | # | | # | − | + | # |

## *2.2 Sznajd model*

Sznajd model is originally one-dimensional and horizontal chain of Ising spin model where each spin stands for one of the two states, up-spin or down-spin. Usually, there are two spin evolution rules proposed. First, if a randomly selected pair has the same opinion, then the two neighbours share the same orientation. Second, if the pair has an opposite opinion, then each of the two neighbours "adopts their opinions from the second nearest neighbours." Therefore, a closed community consists of two contradicting/opposite opinions. The main applications of this model are voter model, small-world networks and so on (for more details, see [5]).

In applying Sznajd model for financial markets, Sznajd-Weron and Weron have modified the second rule as follows: if the selected pair has the different opinion, the opinion of two neighbours takes one of the two directions randomly (Table 6). In this setup, the up-spin is considered as bullish and supposed to do buy-order. On the other hand, down-spin represents bearish and then do sell-order. If the market is more bullish (bearish), then the price goes up (down). Moreover, there is the third market participant in the economy, one fundamentalist. Though all the spins behave as trend-chaser, a fundamentalist knows the system and trade such that the price goes to a fundamental value.

# 3 Implementation

On the one hand, we estimated four parameters of a model of investor sentiment in the following way before generating sample paths:

1. An agent has a binary bit, *judge* (1: stable condition, 0: unstable condition), to judge the market condition and four variables, $stable_{+}$, $stable_{-}$, $unstable_{+}$, and $unstable_{-}$ (1: price-up, 0: price-down), to forecast the next price movement.
   At first, an agent judges the market condition by her *judge*, then she makes a prediction for the next movement, based on that value and on the previous return. Therefore, the roles of the agents in our model is to tell us the possibilities to represent a model of investor sentiment through their learnings.
2. In order to define conditions requisite for GAL to describe MIS, we fed two kinds of price movements ($-+---+++$ and $+-$) to agents. Since the former

**Table 4** Estimated models of investor sentiment

| *pcross* | *pmutate* | fitness | $\lambda_1$ | $\lambda_2$ | $\pi_L$ | $\pi_H$ |
|---|---|---|---|---|---|---|
| .6 | .01 | *fa* | $6.25\times10^{-3}$ | $8.91\times10^{-3}$ | 0.454 | 0.551 |
| | | *fb* | $4.39\times10^{-3}$ | $4.88\times10^{-3}$ | 0.456 | 0.504 |
| | | *fc* | $6.06\times10^{-2}$ | $8.72\times10^{-3}$ | 0.454 | 0.549 |
| | .05 | *fa* | $5.03\times10^{-2}$ | $5.12\times10^{-2}$ | 0.465 | 0.572 |
| | | *fb* | $3.76\times10^{-2}$ | $3.82\times10^{-2}$ | 0.494 | 0.551 |
| | | *fc* | $3.87\times10^{-2}$ | $5.60\times10^{-2}$ | 0.413 | 0.528 |
| .8 | .01 | *fa* | $7.17\times10^{-3}$ | $7.81\times10^{-3}$ | 0.497 | 0.556 |
| | | *fb* | $3.69\times10^{-3}$ | $4.65\times10^{-3}$ | 0.457 | 0.520 |
| | | *fc* | $4.97\times10^{-3}$ | $7.80\times10^{-3}$ | 0.432 | 0.538 |
| | .05 | *fa* | $3.24\times10^{-2}$ | $5.84\times10^{-2}$ | 0.421 | 0.583 |
| | | *fb* | $3.07\times10^{-2}$ | $3.71\times10^{-2}$ | 0.429 | 0.500 |
| | | *fc* | $3.17\times10^{-2}$ | $5.95\times10^{-2}$ | 0.418 | 0.582 |

series has at least the minimum information with respect to diversity, this is used to calculate the variables of MIS. While the latter series is to know how many agents share a common view to the market.

3. The pre-simulation was implemented with the parameters crossover (0.6 or 0.8), mutation (0.01 or 0.05), learning frequency: every step, selection of parents :corresponding fitness value, and three kinds of itness calculation[1].
4. As a result, the following parameter sets were estimated (Table 4).

Then we generated sample paths using the estimated variables of a model of investor sentiment, which means no agent existed or no parameter of genetic algorithm except learning frequencies was used.

- When the previous price movment is up:

$$\Delta S_t = \{ \begin{matrix} \alpha(2p_u - 1), & rnd() < p_u \\ -\alpha(2p_u - 1)/2, & otherwise \end{matrix} \quad (3)$$

  where $p_u = q_t \pi_L + (1 - q_t)\pi_H$ is the probability of price-up.
- When the previous price movement is down:

$$\Delta S_t = \{ \begin{matrix} \alpha(2p_d - 1)/2, & rnd() < p_d \\ -\alpha(2p_d - 1), & otherwise \end{matrix} \quad (4)$$

  where $p_d = q_t(1 - \pi_L) + (1 - q_t)(1 - \pi_H)$ and $rnd() \in (0, 1)$ are the probability of price-up and uniform random number respectively.

Since the above equations are used when the majority forecasts dominate the market, the inequations will be reversed, say $rnd() > p_u$ for (3) and $rnd() > p_d$ for (4).

[1] (fa) She receives +1 if she predicts the asset return and, at the same time, judges the market condition properly. (fb) She recieves +1 if she knows the market condision properly. (fc) She receives +1 if she judges only the market condition properly. While she receives +3 if her expectation is also right about the price movement.

On the other hand, the price in Sznajd model is formed as follows: First, calculate the difference between up-spins and down-spins, say a market-clearing price is defined as the difference between demand and supply. In their literature, the price $x_t$ is written as , where $N$ is the population size and $S_i(t)$ is the opinion of spin $i$. Second, the model determines whether the fundamentalist takes part in the economy, i.e. he/she will buy (take $+1$) with probability $|x_t|$ if $x_t < 0$, and sell (take $-1$) with probability $|x_t|$ if $x_t > 0$.

According to the presented paper, the dynamics generated by the above procedure have successfully replicated the characteristics of actual financial series, fat-tailed distribution of returns, long memory property, and unpredictability of normal returns.

Moreover, if the majority forecast is mighter than a fundamentalist then the price formation rule is opposite to the above one, namely he/she will buy (take $+1$) with probability $1-|x_t|$ if $x_t < 0$, and sell (take $-1$) with probability $1-|x_t|$ if $x_t > 0$.

In short, there are two kinds of price formation rules employed in this study. The one is that the market is dominated by a majority forecast, and the other is that the price movement tends to obey fundamentalist effects or a minority forecast. In this setup, we generated 100 sample paths each of which consists of 20,000 steps.

## *3.1 Stylized facts*

Financial market data contain many statistical properties called "stylized facts" for which traditional economics is difficult to explain. Some of them are about price movements per se and others are the relations between trading volumes, and price movements or volatility. Since the dynamics in this literature is only about price fluctuations, we will focus on the following four properties which seem to be the most popular and significant facts and have been reproduced by some agent-based simulation models:

1. Exchange rates and stock prices have almost unit roots.
2. Returns have fat-tailed distributions.
   Fat-tailed distribution is whose density function decreases in a power-law fashion. But according to Lux and Marchesi [21], the fact is seen for returns at weekly or shorter time scale.
3. Returns per se cannot be predicted, namely they have almost zero autocorrelations.
4. Return distributions shows long memory, namely absolute or squared returns are significantly positive and decrease slowly as a function of the lags.
   There is other methods to investigate the long memory characteristics, Hurst exponents (Hurst 1951), the power law and so on. Hurst exponent is calculated by dividing the sample path into some parts and then by drawing out the data in each part of the mean value. If the Hurst exponents H is over 0.5 then the time series has long memory. Otherwise, the time series is random walk (H= 0.5) or mean-reverting (H< 0.5).

**Table 5** Unit root tests

| | DF | ADF | PP |
|---|---|---|---|
| GA and investor sentiment (majority forecast) | 0.380 | 0.321 | 0.406 |
| GA and investor sentiment (fundamentalist effect) | 0.377 | 0.418 | 0.468 |
| Sznajd model (majority forecast) | 0.443 | 0.448 | 0.585 |
| Sznajd model (fundamentalist effect) | 0.565 | 0.378 | 0.196 |

DF: Dickey and Fuller test
ADF: Advanced Dickey and Fuller test
PP: Phillips and Perron test

**Table 6** Hurst exponents

| | Normal return | Absolute return |
|---|---|---|
| GA and investor sentiment (majority forecast) | 0.517 | 0.636 |
| GA and investor sentiment (fundamentalist effect) | 0.512 | 0.607 |
| Sznajd model (majority forecast) | 0.522 | 0.695 |
| Sznajd model (fundamentalist effect) | 0.524 | 0.692 |

In this paper, we will present the results of normal distribution plots, autocorrelation functions, Hurst exponents, and the BDS statistics respectively, namely we will omit the results of unit root tests because the unit root properties are observed for all the simulation paths.

## 3.2 Computational results

This part of the section reports the results using estimated models of investor sentiment (Table 4). To conduct analyses, we employed 8-term return series, $\log S(t)/S(t-8)$, were employed in accordance with the procedure in Sznajd-Weron and Weron.

First, Table 5 depicts the p-values of three unit root tests, DF test, ADF test, and PP test for generated price series. Those tests prove that no simulation model or setup rejected the null hypothesis of the presence of a unit root.

Second, Figure 1 presents normal probability plots of simulated return series. The curves in those figures are different from the price formation rules.

Especially, that the estimated models with parameter $pcross = 0.8$ appear to replicate real return distribution implies that a kind of trend appears in the market during 8-time price movements. Besides, as equations 3 and 4 show, a positive feed back situation is likely to emerge because the majority opinion tends to dominate the market.

Third, Figure 2 and Table 6 report the results of long memory tests. Both the tests prove that the absolute return series have significantly long memory, but some normal return series show a little bit different dynamics.

Fourth, Table 7 presents the results of the BDS test.

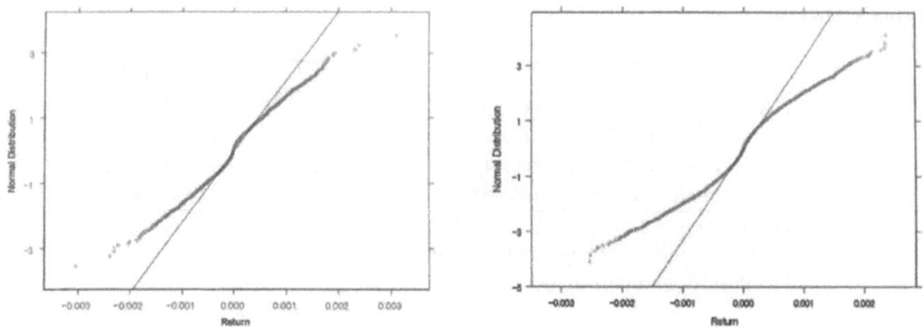


a. GA learning model with investor sentiment

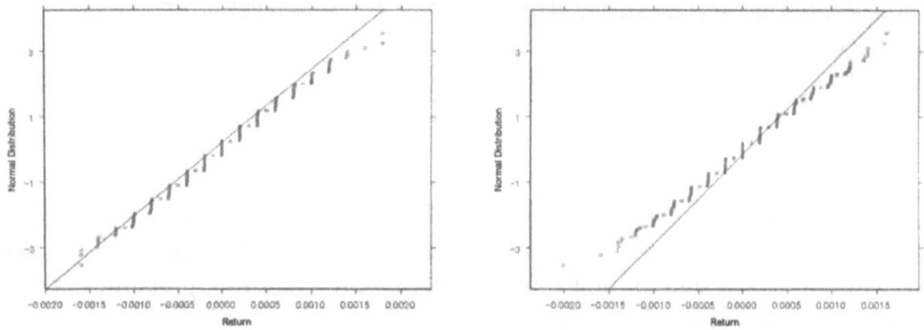


b. Sznajd model

**Fig. 1** Normal probability plots of return series (left panel: majority forecast, right panel: fundamentalist effect)

# 4 Discussion

## *4.1 Similarities and differences*

Table 8 is a brief comparison between the two models. But panel b of this table has been already mentioned. Therefore, in this part of the section, we will review the similarities and differences between two models with respect to models setups and their implications.

On the one hand, similarities or common setups are considered as follows: First, investors are not perfectly rational, namely the agents decide their own behaviour by only the past price movement or the views of their neighbours. Second, the price is formed by the Monte-Carlo method, in particular the models are the simplest ones in that only the expectations determine the price, not a utility function or a market maker. Third, both models do not assume any typical time scale. This enables us to clarify the relations between the frequency of updating/learning of market participants and time series properties.

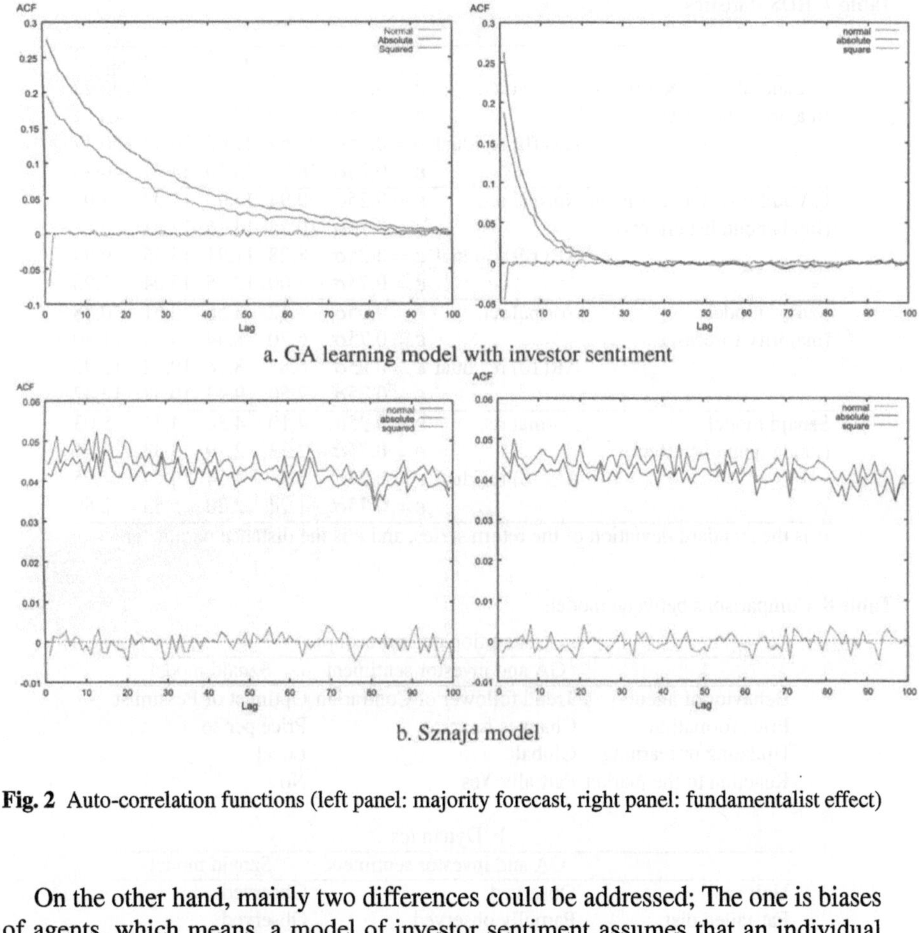


a. GA learning model with investor sentiment

b. Sznajd model

**Fig. 2** Auto-correlation functions (left panel: majority forecast, right panel: fundamentalist effect)

On the other hand, mainly two differences could be addressed; The one is biases of agents, which means, a model of investor sentiment assumes that an individual has one of the two biases; An overreacted investor is considered as a trend-chaser, an underreacted one is likely a contrarian. Whereas an agent in Sznajd model keeps bullish or bearish, say optimist or pessimist, so long as his/her spin shifts from one direction to another. The other difference is timings and meanings of learning. While agents in a genetic learning model of investor sentiment have chances to update their belief in accordance with the past price movements, changes in opinion in Sznajd model are not relevant of changes in prices. Besides, since only a limited number of spins at a time can uptede their opinions, it is that the long memory distinction is hardly created.

## 4.2 Further discussion

It has been over a decade since agent-based computational finance was introduced, and lots of models have been developed. As the time goes by and the number of

**Table 7** BDS statistics

| | | | $m=2$ | $m=3$ | $m=4$ | $m=5$ |
|---|---|---|---|---|---|---|
| GA and investor sentiment | Normal ret. | $\varepsilon=0.25\sigma$ | 21.95 | 41.70 | 78.23 | 152.92 |
| (majority forecast) | | $\varepsilon=0.75\sigma$ | 16.42 | 24.95 | 34.17 | 46.11 |
| | AR(10) residual | $\varepsilon=0.25\sigma$ | 21.67 | 41.05 | 76.97 | 150.17 |
| | | $\varepsilon=0.25\sigma$ | 16.75 | 25.10 | 34.34 | 46.34 |
| GA and investor sentiment | Normal ret. | $\varepsilon=0.25\sigma$ | 9.94 | 14.43 | 18.93 | 24.07 |
| (fundamentalist effects) | | $\varepsilon=0.75\sigma$ | 10.25 | 14.16 | 17.21 | 20.27 |
| | AR(10) residual | $\varepsilon=0.25\sigma$ | 8.28 | 11.91 | 15.56 | 19.94 |
| | | $\varepsilon=0.75\sigma$ | 9.00 | 12.25 | 15.04 | 17.96 |
| Sznajd model | Normal ret. | $\varepsilon=0.25\sigma$ | 7.12 | 8.66 | 9.61 | 10.48 |
| (majority forecast) | | $\varepsilon=0.75\sigma$ | 6.70 | 8.44 | 9.74 | 11.30 |
| | AR(10) residual | $\varepsilon=0.25\sigma$ | 6.82 | 8.98 | 10.82 | 12.96 |
| | | $\varepsilon=0.75\sigma$ | 7.50 | 9.43 | 10.87 | 12.32 |
| Sznajd model | Normal ret. | $\varepsilon=0.25\sigma$ | 4.10 | 4.38 | 4.71 | 5.03 |
| (fundamentalist effect) | | $\varepsilon=0.75\sigma$ | 2.33 | 2.89 | 3.32 | 3.71 |
| | AR(10) residual | $\varepsilon=0.25\sigma$ | 3.51 | 4.04 | 4.54 | 4.95 |
| | | $\varepsilon=0.75\sigma$ | 1.74 | 2.20 | 5.53 | 2.89 |

$\sigma$ is the standard deviation of the return series, and $\varepsilon$ is the distance parameter.

**Table 8** Comparisons between models

a. Assumption and procedure

| | GA and investor sentiment | Sznajd model |
|---|---|---|
| Behavior of agents | Trend follower or Contrarian | Optimist or Pessimist |
| Price formation | Changes in price | Price per se |
| Updating or learning | Global | Local |
| Reaction to the market | Partially Yes | No |

b. Dynamics

| | GA and investor sentiment | Sznajd model |
|---|---|---|
| Unit root | Observed | Observed |
| Fat-tailed dist. | Partially observed | Observed |
| Long memory | Partially observed | (Partially) observed |
| IID process | Rejected | Rejected |

models has increased, cautions, questions, and even criticisms as well as future perspectives have been pointed out. In this part of the section, we will review what is done or not in this study taking some of the comments addressed.

One of the achievements of this study is comparison between models. According to LeBaron [15], "the platforms and structures are not common, and broad comparisons across models are difficult" (LeBaron 2000, p. 696). Similar comments are independently addressed by Duffy [7] and Hommes [11], for example. Or the related conferences are held in 2003, 2004, and 2007 [22]. Through comparison in the preceding section, we observed the relations between market power of fundamentalists or minorities and time series properties. That maybe result from the fact that the two models are the simplest ones, but that implies that there is another aspect done in this study, “models have full of parameters, which get exceedingly complicated” (LeBaron 2006, p. 1222).

“Realism and validation” (LeBaron 2001, p. 259) [16] or attempt to estimate a model on economic and financial data [11] is another important problem to overcome. In a sense, our past researches revealed that a proper setup is consistent with the findings in questionnaire studies, say information requisite for market participants determines the time scale of generated sample paths [27]. But since this paper just reports that the number of fundamentalists or market participants is fixed in the economies, further investigation is required [21].

Unfortunately, there are still problems to deal with in this study. The most substantial one is “the weakness of the behavioral assumptions are particularly apparent in the context of financial markets models” (Durlauf 2005, p. F236) [8], namely that there are stylized facts observed in simulation data means that the arbitrage opportunities are in the market in the long run. We believe that this is because the characters of agents do not change at all in the context of evolution. The other problem is “timing of decisions, information, and trade” (LeBaron 2006, p. 1225) [17]. In this regard, we did not answer this question at all. For instance, it is not a good guess that only a few agents in Sznajd model have chances to update their opinion or all the agents can learn every term in genetic learning model of investor sentiment.

Yet there are still obstacles to be overcome, we hope that in the near future “the research from artificial markets will have a positive spillover into other areas of economic and social science” (LeBaron 2001, p. 260) [16].

## 5 Conclusion

This paper attempts to show whether genetic algorithm can represent a descriptive model of investor sentiment and what distinctions such an agent-based model has compared to the model based on statistical physics. For these purposes, we combined investor sentiment with genetic learning in an agent-based computational economic model, conducted time series analyses using sample paths generated, and compared them to those in the previous studies. Both the time series statistics reveal that a proper setup could lead to dynamics reported in the early studies no matter how the setups are different. Especially, investors are likely to react to a piece of new information and adjust their views to the market. On the other hand, the differences of price formation, even if the agents are boundedly rational, could result in different time series properties in some regards.

## References

1. Arifovic J (2000) Evolutionary algorithms in macroeconomic models. Macroecon. Dyn. 4:373–414

2. Arifovic J, Gençay R (2000) Statistical properties of genetic learning in a model of exchange rate. J. Econ. Dyn. Control 24:981–1005
3. Barberis N, Shleifer A, Vishny R (1998) A model of investor sentiment. J. Finan. Econ. 49:307–343
4. Barberis N, Huang M, Santos T (2001) Prospect theory and asset prices. Q. J. E. 116:1–53
5. Behera L, Schweitzer F (2003) On spatial consensus formation: is the Sznajd model different from a voter model? Int. J. Mod. Phys. C 14:1331–1354
6. Chen S-H (2002) Evolutionary computation in economics and finance. Physica-Verlag, Heidelberg
7. Duffy J (2006) Agent-based models and human subject experiments. in Tesfatsion, L., Judd, K.L. (eds.) Handbook of computational economics: agent-based computational economics volume 2, North-Holland, Netherlands, 949–1012
8. Durlauf S N (2005) Complexity and empirical economics. Econ. J. 115:F225–F243
9. Hirabayashi T, Takayasu H, Miura H, Hamada K (1993) The behavior of a threshold model of market price in stock exchange. Fractals 1:29–40
10. Hommes C H (2001) Financial markets as nonlinear adaptive evolutionary systems. Quant. Finan. 1:149–167
11. Hommes C H (2006) Heterogeneous agent models in economics and finance. in Tesfatsion, L., Judd, K.L. (eds.) Handbook of computational economics: agent-based computational economics volume 2, North-Holland, Netherlands, 1109–1186
12. Izumi K, Ueda K (2001) Phase transition in a foreign exchange market: analysis based on an artificial market approach. IEEE Trans. Evol. Comput. 5:456–470
13. Izumi K, Nakamura S, Ueda K (2005) Development of an artificial market model based on a field study. Info. Sci. 170:35–63
14. LeBaron B, Arthur W B, Palmer R (1999) Time series properties of an artificial stock market. J. Econ. Dyn. Control 23:1487–1516
15. LeBaron B (2000) Agent-based computational finance: suggested readings and early research. J. Econ. Dyn. Control 24:679–702
16. LeBaron B (2001) A builder's guide to agent-based financial markets. Quant. Finan. 1:254–261
17. LeBaron B (2006) Agent-based computational finance. in Tesfatsion, L., Judd, K.L. (eds.) Handbook of computational economics: agent-based computational economics volume 2, North-Holland, Netherlands, 1187–1234
18. Levy M, Levy H, Solomon S: Microscopic simulations of financial markets: from investor behaviour to market phenomena. Academic Press, San Diego.
19. Lux T (1998) The socio-economic dynamics of speculative markets: interacting agents, chaos, and fat tails of return distributions. J. Econ. Behav. Org. 33:143–165
20. Lux T, Marchesi M (1999) Scaling and criticality in a stochastic multi-agent model of a financial market. Nature 397:498–500
21. Lux T, Marchesi M (2000) Volatility clustering in financial markets: a microsimulation of interacting agents. Int. J. Th. Appl. Finan. 3:675–702
22. Third International Model-to-Model Workshop: http://m2m2007.macaulay.ac.uk/index.html
23. Riechmann T (2001) Genetic algorithm learning and evolutionary games. J. Econ. Dyn. Control 25:1019–1037
24. Sznajd-Weron K, Weron R (2002) A simple model of price formation. Int. J. Mod. Phys. C 13:115–123
25. Tesfatsion L, Judd K L (eds.) (2006): Handbook of computational economics: agent-based computational economics volume 2. North-Holland, Netherland
26. Ueda K, Uchida Y, Izumi K, Ito Y (2004) How do expert dealers make profits and reduce the risk of loss in a foreign exchange market? Proc. of the 26th annual conf. of the Cognitive Science Society, Chicago, USA, 1357–1362
27. Yamada T, Terano T (2006) Price formations in genetic learning model of investor sentiment. Proc. of 9th Joint Conf. on Info. Sci., 285–288

# On Emergence of Money in Self-organizing Doubly Structural Network Model

**Masaaki Kunigami, Masato Kobayashi*, Satoru Yamadera*, Takao Terano***

Graduate School of Business Science, University of Tsukuba
Computational Intelligence and Systems Science Tokyo Institute of Technology*

**Abstract** We are conducting the research on the emergence of the money from the barter economy. This paper presents a new model, which consists of a micro-macro doubly structural network reflecting individual recognitions and social connections among agents. This model will show processes in which particular one of goods attains natures of money as a self-organization of the network. We examine this process by mean-field approximation of dynamics and agent-based simulation.

## 1 Introduction

A kind of goods called "money" plays a role dominant as the unique medium for exchange in economy and society, even though that has little practical advantage. Researching on mechanism of 'emergence of money' in the society will bring important knowledge for not only historical study on economics but also construction on E-money or local money.

Like Menger and Polani, many economists have discussed function and emergence of money. In addition, recently, there are literatures approaching on this issue by mathematical models ((Kiyotaki and Wright 1989), (Wright 1995), (Luo 1999), (Starr 2003)) and Agent-based Simulation (ABS) models ((Yasutomi 1995), (Shinohara and Gunji 2001), (Yamadera and Terano 2007). On the other hand, studies on the natures of social connections among agents ((Watts and Strogatz 1998), (Barabási and Albert 1999)), have progressed greatly ((Newman 2003)) in this decade. (Matsuyama et al. 2007) research the p2p economy of network between contents-type goods by heterogeneous double-layered network model. (Matsuyama et al. 2007)

We study on the emergence of money by new model that is composed of a micro-macro doubly structural network and of mutual learning process in the network. In the doubly structural network model, each agent has own micro-level networks each of which represents the corresponding agent's inner recognitions on exchangeability between goods. On the other hand, the macro-level network reflects the social connection among agents. We analyze the contact process in this model by two different approaches i.e. a dynamics with mean-field approximation and a multi-agent simulation.

# 2 The model

## 2.1 Doubly Structural Network

We present a model of contact process represents mutual learning on media of exchange between agents on a social network. However different from other contact process model, state variables of each agent have own network structure. State variables represent agents' inner recognition network on exchangeability of goods. (Fig.1)

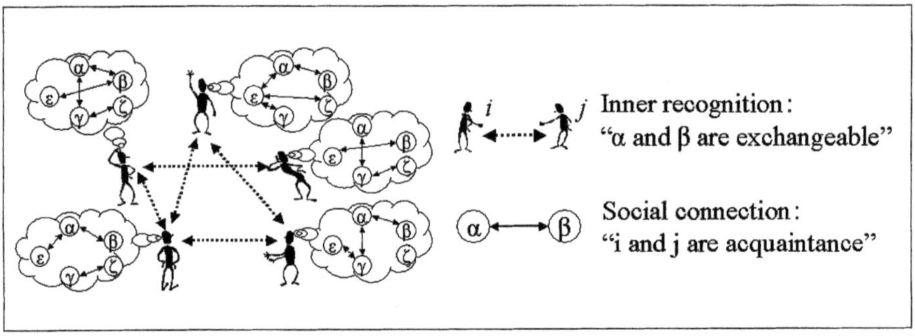


**Fig. 1.** The Doubly Structural Network consists of a social (macro) network and inner (micro) networks.

Our model consists of the doubly structural network as follows. At first, we assume that this inter-agent network is given fixedly and expresses the topology of social connection among agents. Node of the social network represents agent and is identified by suffix *i* or *j* (*i,j*=1~*N*). On the other hand, each agent's inner-network reflects the recognition of the agent, so that each node represents each one of goods that is identified by suffix α, β or γ (1~*M*). In the inner-network of the *i*-th agent, the existence of an edge between α and β represents that the *i*-th agent recognizes the exchangeability of α and β (i.e. agent-*i* is possible to exchange α and β with anyone). We express this state as $e_{\alpha\beta}{}^{(i)}=1$, and otherwise $e_{\alpha\beta}{}^{(i)}=0$. We mention that these inner-networks of agents change with mutual learning on the inter-agents network.

Different from other social agents models by (Epstein and Axtell 1996) (sugar-space) or (Yamadera and Terano 2007), social relations in our model are directly described by inter-agent (macro level) network without using a cellular space structure. Our inter-agent network can represent more general social relations of agents, and allows us to apply methodologies and outcomes on complex networks to the emergence of money.

## *2.2 Emergence of Money*

Among possible definitions for emergence of money, we focus on the "general acceptability" that is usually defined as exchangeability to any kind of goods of any agent. In the time evolution of our model, if a kind of goods becomes the unique one that has "general acceptability", we can mention that this emerges as money.
In our doubly structural network model, the process in which goods α attains general acceptability is represented by the self-organization in which almost inner-networks become similar star-shaped graph with common hub (The node which connects with almost other nodes) α. (Fig.2)
(Although (Starr 2003) and (Yasutomi 1995) pointed out that star-shaped network can represent general acceptability, our doubly structural network model can explicitly handle evolutional processes that star-shaped inner networks spread out in the inter-agent network depending upon its graph topology.)

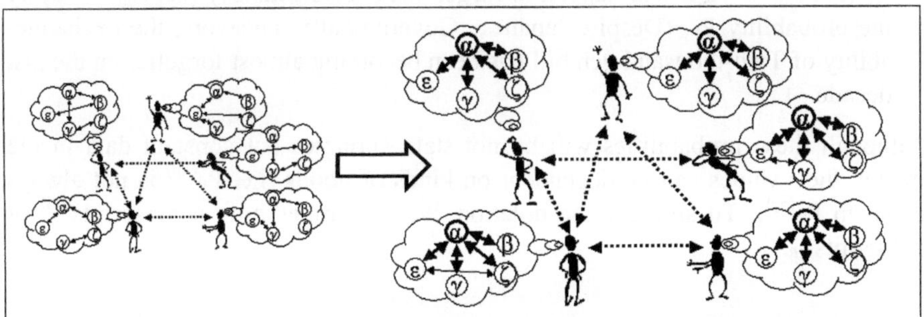

Particular one of goods becomes to common hub among agents.

**Fig. 2.** Emergence of money: the inner networks self-organize into similar star-shaped networks with the common hub.

## *2.3 Interaction of Agents*

Agents in our model interact each other by the following manners in each time step.

1. Exchange: In the social (inter-agent) network, neighboring agents *i* and *j* exchange commodities α and β, if both of them recognize that α and β are exchangeable i.e. $e_{\alpha\beta}^{(i)} = e_{\alpha\beta}^{(j)} = 1$. Then this α-β exchange brings reward to *i* and *j* with probability $P_E$.
2. Learning: Learning process of agents consists of the following four ways.

- Imitation: If an agent *i* has no recognition of α-β exchangeability (i.e. $e_{\alpha\beta}^{(i)} = 0$) and if *i*'s neighboring agents *j* and *j'* get reward in the α-β exchange, then the agent *i* will imitate its neighbor's recognition (i.e. $e_{\alpha\beta}^{(i)}=0 \rightarrow =1$) by the probability $P_I$.
- Trimming: If an agent '*i*' has cycle recognition of exchangeability (ex. $e_{\alpha\beta}^{(i)} = e_{\alpha\gamma}^{(i)} = e_{\gamma\alpha}^{(i)} = 1$), then the agent i will trim its inner-network by cutting randomly one of these cycle edges (i.e. changing one of these three values to 0) by the probability $P_T$. Such elimination of circulation means thrift of deliberation and is also consistent with (Kiyotaki and Wright 1989).

We consider that these two processes are essential for 'emergence of money'. In addition, we introduce two more subsidiary processes to avoid local trapping or over-sensitivity on initial conditions.

- Conceiving: Even if an agent *i* has no recognition of α-β exchangeability (i.e. $e_{\alpha\beta}^{(i)} = 0$), it will happen to conceive this exchangeability (i.e. $e_{\alpha\beta}^{(i)} \rightarrow 1$) by the probability $P_C$.
- Forgetting: Vice versa, even if an agent *i* has recognition of α-β exchangeability (i.e. $e_{\alpha\beta}^{(i)} = 1$), it will happen to forget this exchangeability (i.e. $e_{\alpha\beta}^{(i)} \rightarrow 0$) by the probability $P_F$. (Despite Japanese Government's endeavors, the exchangeability of Two-thousand yen bill has been becoming almost forgotten in the last decade. )

Although these probabilities will be not state variables but constant data in the model, their values can be dependent on kinds of goods (i.e. $P_E^{(\alpha\beta)}$ is not always equal to $P_E^{(\alpha\gamma)}$). To simplify the notation, we sometimes omit superscripts$^{(\alpha\beta)}$ or subscripts $_{\alpha\beta}$.

## 2.4 Micro-Macro Interactions

In the last of this section, we emphasize that this model can represent the micro-macro interactions as follows. Adjacency relation and topology of the macro inter-agent network determines interaction between micro inner networks. Similarity between self-organizing inner networks affects the rates of inter-agent interactions.

In this paper, we simplify that the macro inter-agent network is static, but it is natural to extend the macro inter-agent network to dynamic one. In the next paper, we will discuss a micro-macro co-evolutionary model with the doubly structural network.

# 3 Emergence Scenario in Mean-field Dynamics

## *3.1 Mean-field Dynamics*

To obtain a macro level view of emergence behavior, we derive a dynamics (differential equation system) by a mean-field approximation. In this approximation, states around any agent are approximated by mean of whole agents and nature of social (inter-agent) network is represented by the average degrees of node (to simplify, assuming all nodes have same degree: k), and these variables are identified with probabilities that arbitrary agents have recognition of corresponding exchangeabilities.
The time evolution of these states variables driven by the four interaction processes defined in the section 2.3 is expressed as following formulas.

- $dx_{\alpha\beta}/dt = f_{\mathrm{I}}^{\alpha\beta}(x_{\alpha\beta}) - f_{\mathrm{T}}^{\alpha\beta}(\mathbf{x}) + f_{\mathrm{C}}^{\alpha\beta}(x_{\alpha\beta}) - f_{\mathrm{F}}^{\alpha\beta}(x_{\alpha\beta})$.
  - Imitation: $f_{\mathrm{I}}^{\alpha\beta}(x_{\alpha\beta}) = \mathrm{P_E}^{\alpha\beta}\cdot\mathrm{P_I}^{\alpha\beta}\cdot k(k\text{-}1)(1\text{-}x_{\alpha\beta})x^2_{\alpha\beta}$.
  - Trimming: $f_{\mathrm{T}}^{\alpha\beta}(\mathbf{x}) = \mathrm{P_T}^{\alpha\beta}\cdot x_{\alpha\beta}\cdot\sum_\gamma\{x_{\alpha\gamma}\cdot x_{\beta\gamma}\}$. ( $\mathbf{x}=(x_{\alpha\beta}, x_{\alpha\gamma}, \cdots)$ )
  - Conceiving: $f_{\mathrm{C}}^{\alpha\beta}(x_{\alpha\beta}) = (1\text{-}\mathrm{P_C}^{\alpha\beta})x_{\alpha\beta}$.
  - Forgetting: $f_{\mathrm{F}}^{\alpha\beta}(x_{\alpha\beta}) = \mathrm{P_F}^{\alpha\beta}\cdot x_{\alpha\beta}$.

Here, we obtain the mean-field dynamics.

$$\begin{aligned}\frac{dx_{\alpha\beta}(t)}{dt} &= P_C^{(\alpha\beta)} - \left(P_C^{(\alpha\beta)} + P_F^{(\alpha\beta)} + \sum_{\gamma\neq\alpha,\beta} P_T^{(\alpha\beta\gamma)} x_{\alpha\gamma}x_{\beta\gamma}\right)x_{\alpha\beta} \\ &\quad + P_E^{(\alpha\beta)}P_I k(k-1)x^2_{\alpha\beta} - P_E^{(\alpha\beta)}P_I k(k-1)x^3_{\alpha\beta} \\ &= P_E P_I^{(\alpha\beta)} k(k-1)\Gamma_{\alpha\beta}\left(x_{\alpha\beta}\right).\end{aligned} \tag{3.1}$$

$$\Gamma_{\alpha\beta}\left(x_{\alpha\beta}\right) \equiv b_{\alpha\beta} - a_{\alpha\beta}(\Sigma xx)x_{\alpha\beta} + x^2_{\alpha\beta} - x^3_{\alpha\beta},$$

$$a_{\alpha\beta}(\Sigma xx) \equiv \left(\frac{P_C^{(\alpha\beta)} + P_F^{(\alpha\beta)} + \sum_{\gamma\neq\alpha,\beta} P_T^{(\alpha\beta\gamma)} x_{\alpha\gamma}x_{\beta\gamma}}{P_E P_I^{(\alpha\beta)} k(k-1)}\right),\quad b_{\alpha\beta} \equiv \frac{P_C^{(\alpha\beta)}}{P_E P_I^{(\alpha\beta)} k(k-1)}. \tag{3.2}$$

This mean-field dynamics (3.1) describes time evolution of these state variables (: population ratios) derived from interaction manners mentioned in previous section. Behavior of the dynamics is determined by the sign (positive or negative) and zero points of RHS, so that we introduce a cubic functions $\Gamma_{\alpha\beta}(x)$ that is normalized function of RHS (3.2). Of course sign (positive or negative) and zero points of RHS and $\Gamma(x)$ are same. By some manipulations for $\Gamma(x)$, it is shown that all zero points of $\Gamma(x)$ (at least 1 ~ at most 3 points) exist in the region: [0,1].

## 3.2 Emergence Scenario

Equilibria of our model with mean-field approximation are given by the stable fixed point of the mean-field dynamics, i.e. $\{x \mid \Gamma(x)=0, \Gamma'(x)<0\}$.
The general acceptability of α is represented by the states which separate the zero points of $\Gamma(x)$. In this state, stable zero points of $\Gamma(x)$ related to α are about 1, and stable zero points of $\Gamma(x)$ not related to α are about 0. Except for trivial cases in which initial condition of $x(t)$ or fixed points or shapes of $\Gamma(x)$ have distinct difference, this mean-field dynamics has an interesting emergence scenario described by following steps. (Fig.3)

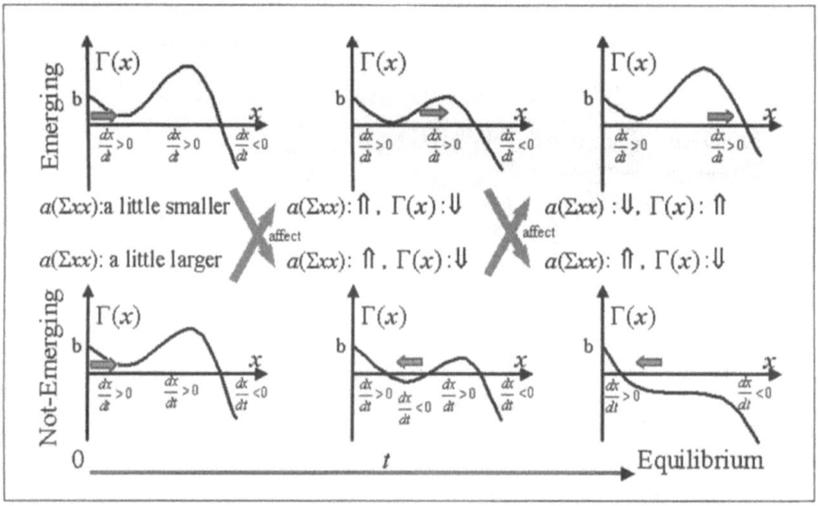


**Fig. 3.** In the emergence scenario, small differences of minima of Γ cause big differences of acceptances of goods.

(i) At first, there are little difference in initial condition of $x(t)$ and in coefficients(shapes and zeros) of $\Gamma(x)$, so that $x(t)$ start the growth from small values.
(ii) According $x(t)$s' growth, values of $a(\Sigma xx)$ also increase, so that corresponding $\Gamma(x)$s shift downward. Then bottom values of some $\Gamma(x)$s go below 0, so that corresponding $x(t)$ are changed to decreasing.
(iii) Decreasing $x$(t) pushes up others' $\Gamma(x)$s through $a(\Sigma xx)$, and these increasing $\Gamma(x)$s shift own zeros and corresponding $x(t)$ to right side. Vice versa, increasing

$x(t)$ shifts other x and zeros of $\Gamma(x)$s to the left. At last, differences of zeros (i.e. some equilibrium points go to 1 and others go to 0.) show accelerative growth.

To work this emergence scenario, it is necessary (not sufficient) that some $\Gamma(x)$ s have bottoms and 3 zeros. (Fig.4)

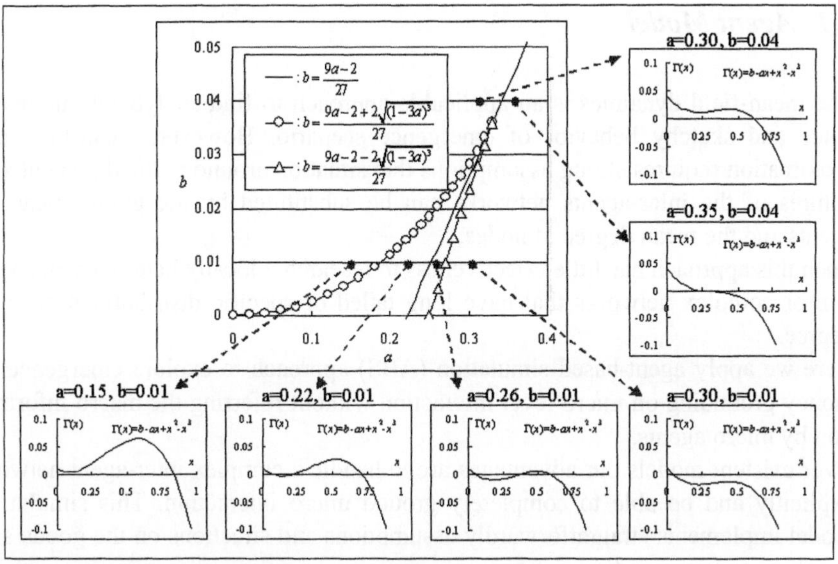


**Fig. 4.** The region in which $\Gamma(x)$ has 3 zeros is between O and Δ.

The numeric example of emergence in the corresponding parameter range is shown in Fig.5

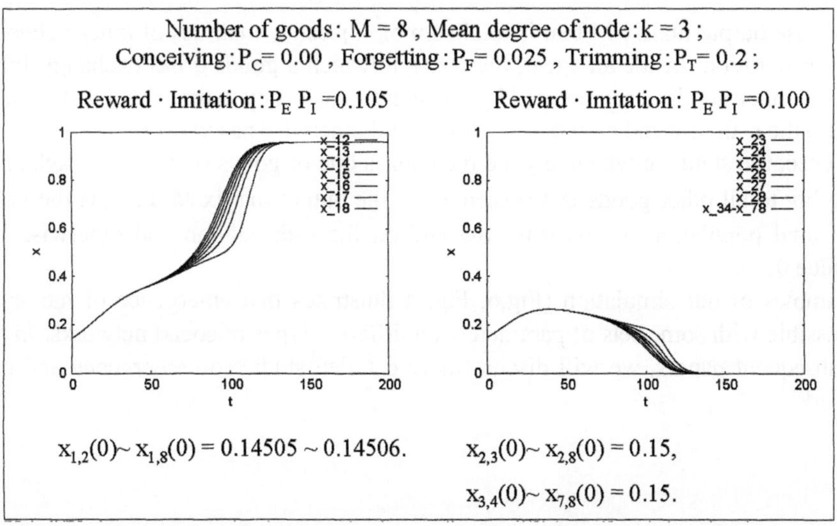


**Fig. 5.** A numeric outcome of the dynamics (3.1) shows the emergence.

# 4 Agent-Based Simulation

## *4.1 Agent Model*

The mean-field dynamics is an applicable approach to find analytically the existence and sketchy behavior of emergence scenario. However, mean-field approximation requires strong assumptions that situation around particular agent and natures of the inter-agents networks can be substituted by the global mean of agents and the mean degree of nodes.
Then this approach has little effectiveness to research a locally heterogeneous system or complex networks that have long tailed or specific distributions of node degree.
Here we apply agent-based simulation (ABS) approach to explore emergence of money grounding on micro level interaction without referring the macro information by micro agents.
Over existent models our advantages are to handle a complex inter-agent network explicitly and be able to completely ground micro interaction. This simulation model implements straightforwardly assumptions and situations on the goods and agents in previous section.

## *4.2 Simulation*

We use output matrix M that is total sum of adjacency matrices of inner-networks of recognition. Inside an agent, the status in which a goods $\alpha$ has exchangeability to all other goods is represented by the agent's adjacency matrix that has the value 1 on the $\alpha$-th row and on the $\alpha$-th column and otherwise has value 0.
Hence, the status in which anyone recognizes one of goods (e.g. $\alpha$) has exchangeability to all other goods is represented by the output matrix **M** that has the value of total population on the $\alpha$-th row and on the $\alpha$-th column and otherwise has value 0.
Samples of our simulation (Fig.6, Fig.7) illustrates that emergence of money is possible with some sets of parameters on different types of social networks. In the subsequent papers, we will discuss more detailed studies on emergence and network.

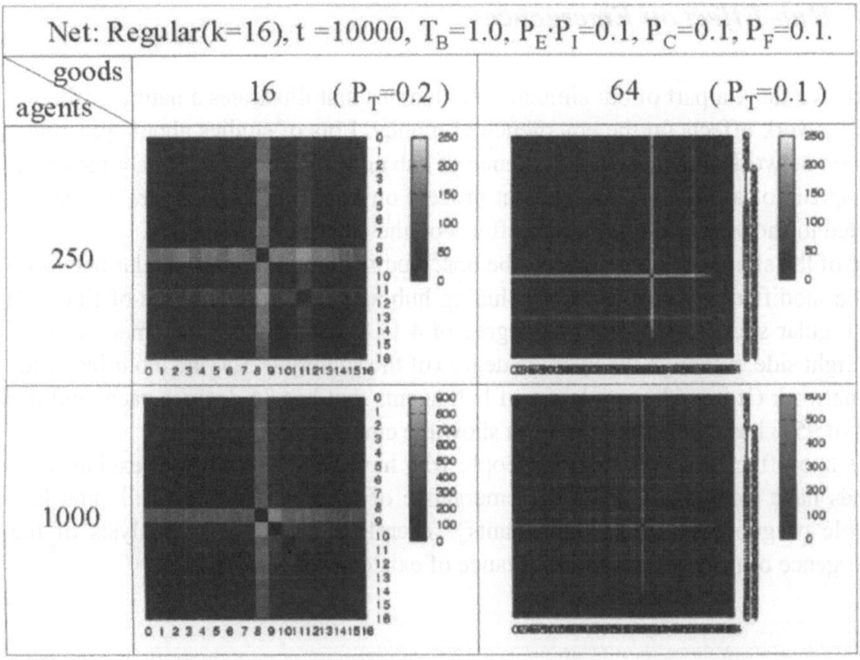


**Fig. 6.** The ABS can show the emergence with wide ranges of the population and the number of goods.: Social network is Regular((Watts and Strogatz 1998)). (The output matrix **M**. White means large population of agents.)

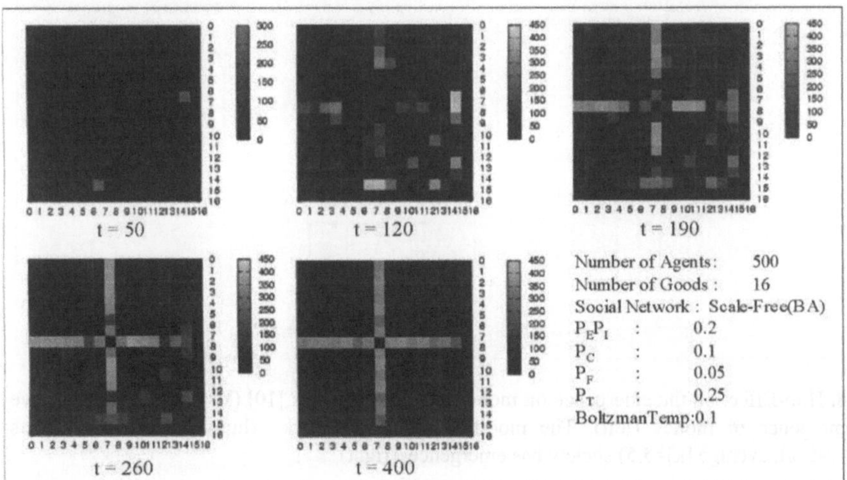


**Fig. 7.** The ABS can show a time sequence of the emergence.: Social network is Scale-free(Barabási and Albert 1999). (Time evolution of the output matrix **M**. White means large population of agents.)

## 4.3 Hub-Effect on Emergence

Here, we show a part of our simulation outcomes that illustrates a nature of the social network affects on the emergence of money. Lots of studies about epidemiology on networks mention that existence of hub nodes is one of important factors in the spread of a disease. As a contact process on a network, our model is also expected to show such a kind of hub-effects on the emergence of money.

One of the simplest hub-effect will be observed in comparison of regular networks and a modified regular network including hub agents. In the left side of figure 8, the regular social networks with degree of 4 to 8 cannot show the emergence. In the right side, although the average degree of the society is 5.5, the modified regular network (It has 5% population of hub agents that has 36 degrees each, and the rest of 95% has 4 degrees each.) can show the emergence.

This hub-effect shows that some people who have high centrality of exchange of goods have important role in the emergence of money. We often call such hub people of goods-exchange "merchants". Therefore the network analysis of the emergence of money implies importance of existence of merchants.

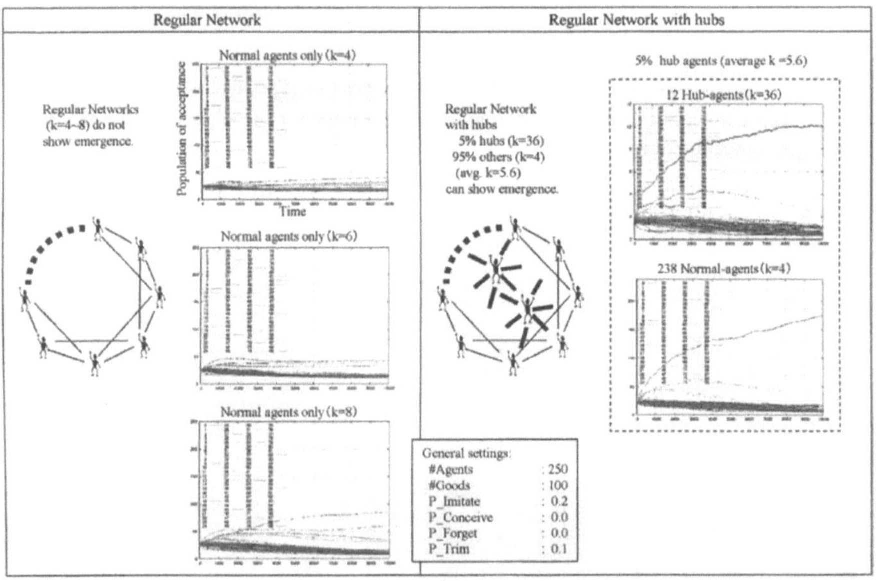


**Fig. 8.** Hub-Effect on the emergence on money: Regular network [10] (k=4, 6, 8) societies have no emergence of money (left). The modified regular network (hubs (k=36): 5%, others (k=4) :95%), average [k]=5.5) society has emergence. (right)

# 5 Summary and Conclusion

In this paper, we present micro-macro doubly structural network model and illustrate the new model is applicable to both analytical and ABS approaches on emergence of money.

The advantages of the presented model are next two points; (i) it can describe a complex network structure of real society or goods, and (ii) it allows dynamical analysis by both differential equations and simulations. Especially by introducing explicitly structured inter-agent network, an important research area has been opened on the relation between the nature of social networks and emergence of money.

In addition, the doubly structural network presents not only a new model of 'emergence of money' but also the new model framework in which each node of the global network is connected to each of interacting local networks that make emergence of some global natures.

# References


Barabási, A. L., Albert, R.: Emergence of Scaling in Random Networks, Science 286, 509-512(1999)

Epstein, J. S., Axtell,R.: Growing Artificial Societies, The Brookings Institution(1996)

Kiyotaki, N., Wright, R.: On Money as a Medium of Exchange, Journal of Political Economy 97, 927-54(1989)

Luo, G.,Y.: The Evolution of Money as a Medium of Exchange, Journal of Economic Dynamics and Control 23, 415-458(1999)

Matsuyama.S, Kunigami,M., Terano,T.: Analysis of the Hub Contents Though Agent-based Simulation, JSAI2007 (2007)

Newman, M. E. J.: The Structure and Function of Complex Networks, SIAM REVIEW 45, 167-256(2003)

Shinohara, S., Gunji, Y., P.: Emergence and Collapse of Money Through Reciprocity, Applied Mathematics and Computation 117, 131-150(2001)

Starr, R.: Why is There Money? Endogenous Derivation of 'Money' as the Most Liquid Asset: a Class of Examples, Economic Theory 21, 455-474(2003)

Watts, D.J., Strogatz, S.H.: Collective Dynamics of "Small-World" Networks, Nature 393, 440-442(1998)

Wright, R.: Search, evolution, and money, Journal of Economic Dynamics and Control 19, 181-206(1995)

Yamadera, S., Terano, T.: Examining The Myth of Money with Agent-Based Modeling, in Edmonds, B. et al.(eds.), Social Simulation - Technologies, Advances, and New Discoveries. Information Science Reference, Hershey, 252-262(2007)

Yasutomi, A.: The Emergence and Collapse of Money, Physica D 82, 180-194(1995)

# Market and Economy II

# Scale-Free Networks Emerged in the Markets: Human Traders versus Zero-Intelligence Traders

Jie-Jun Tseng, Shu-Heng Chen, Sun-Chong Wang and Sai-Ping Li

**Abstract** We design a Web-based prediction market platform to monitor the trading behavior among the human traders in real-time. Two experiments tied to the outcomes of mayoral election in Taiwan are performed in parallel for 30 days. From the accumulated transaction data, we reconstruct the so-called cash-flow networks. We observe that the network structure is hierarchical and scale-free with a power-law exponent of 1.15±0.07. Through carrying out a post-simulation, we also demonstrate that a simple double auction market with "*zero-intelligence*" traders is capable of generating hierarchical and scale-free networks.

## 1 Introduction

Complex networks, exhibit several non-trivial topological features (including a fat-tail in the degree distribution, a high clustering coefficient, community structure at many scales, and a hierarchical structure), emerge in many complex systems such as biological[1], social[2] and technological[3] systems. Although these systems have been modeled as random graphs in the past, more and more empirical evidence suggests that the topology and evolution of these networks are governed by robust organizing principles[4]. We might say that the network topology evolves to fulfill

---

Jie-Jun Tseng
Institute of Physics, Academia Sinica, Taipei 115 Taiwan, e-mail: gen@phys.sinica.edu.tw

Shu-Heng Chen
AI-Econ Research Center and Department of Economics, National Chengchi University, Taipei 116 Taiwan

Sun-Chong Wang
Institute of Systems Biology and Bioinformatics, National Central University, Chungli 320 Taiwan

Sai-Ping Li
Institute of Physics, Academia Sinica, Taipei 115 Taiwan

system requirements. Studying network systems thus help us to have better insight into the complex systems.

In economics, financial markets are complex systems as well. The prices and individual wealth in the market are driven up and down by the so-called "*invisible hand*" coined by *Adam Smith*. Although we know that these fluctuations are resulting from the interactions among traders within the market, it is difficult for us to make any accurate prediction about the markets. In a naive thinking, since the network can be applied to explain the relations or interactions among each element, we shall be able to map the interactions among traders into networks and study them. Therefore, we conduct two experiments for gathering the information about the trading behavior in the market and introduce a new method to map the trading behavior into the networks[5].

In this contribution, we first introduce our platform for market experiments with human traders and show the results of two experiments on this platform. The trading behavior in these two experiments is mapped into a so-called cash-flow network and we then present the observation of these networks. Finally, the results based on a simulation experiment will also be discussed. The model is described by a continuous double-auction (CDA) market with *zero-intelligence* traders.

## 2 Market Experiment with Human Traders

Markets are open systems where intelligent traders interact with each other with some simple trading rules. For an orderly market, the price reflects the underlying value of the market instruments. But when bubbles develop, the orderly behavior breaks down and markets become complex systems. In the bottom-up scheme, if we want to study a complex system, we need to learn how individual elements interact with each other in our system. Following this concept, we build a virtual market for human traders which allows us to monitor their trading behaviors in real-time. We believe that the transactions among traders represent the strength of the interactions (or relations) between them. Therefore, we might have deeper insight into the market with these transaction data.

### *2.1 A Web-based Futures Exchange Platform*

Prediction market[6, 7] is a market designed to run for the primary purpose of mining and aggregating information scattered among traders. The aggregation of the information will therefore be reflected in the form of market prices so as to make predictions about the outcomes of some specific events. From the concept of prediction market, we design a platform which allows the registers to trade the political futures contracts on web and enable us to monitor the transactions among the traders[8, 9, 10]. Although we use the virtual money for the trading, the principles and operation of our platform follow those of major financial exchanges in the real world. Our platform works as a Web-based server which runs for 24 hours a day until we shut it down on the day of liquidation. Any web browser can participate in the trading by on line registration. An account with the user-provided login name is created for the participant after registration. An initial amount of (virtual) money is deposited by the server to the newly created account. The initial wealth for each participant is the same. The demographics about all the registers to date is also updated. The process takes place on the server automatically and a trader can start trading almost immediately after successful registration. The demography, price fluctuation and accumulated volumes plots are open to any Web surfer irrespective of her registration or not. However, only registered users can trade upon login.

Once the user login onto the server, he can buy bundles of contracts from the server for a guaranteed price per bundle or buy the futures contracts from the market directly. In our platform, a given political futures contract is associated with the liquidation price which equals the percentage of votes that a candidate gets on the day of election. A bundle, by design, consists of futures contracts for each candidate in the race as well as for all the invalid casts. After the election, all the futures contracts in the account should be liquidated. The bundle price of 100 is fair since neither the user nor our server loses. Transactions are free in our platform and no further service fees will be charged. Users can place market or limit orders to buy or sell futures contracts. Our platform then stores and sorts the submitted bid (ask) orders in a bid (ask) queue and matches counterpart orders which are compatible with each other's price limits. If no matches are met, limit orders stay in the queue and wait for further matches with new orders. These limit orders would either expire or be canceled by users before the matches. Market orders do not stay if no matches are found. Order matching is via the process of continuous double auction (CDA) which is the price discovery mechanism widely used by exchange markets in the world, including New York Stock Exchange, Tokyo Stock Exchange, SBF-Bourse de Paris, and the Stock Exchange of Hong Kong.

The server records user's trading activities and results. After a transaction, the account assets, including cashes and futures contracts, is balanced immediately. The price of the contract is decided by the current market price. Therefore, even if a user dose not transact anymore, as long as he owns the contracts, his assets would vary depending on current market prices of the contracts, The account in our server earns no interest. When a limit order is placed but the execution is not complete, the platform will block further order submissions by this user. This rule is meant to prevent the server from reckless submissions since there are no transaction fees and the money is virtual.

## 2.2 Experimental Design

In this analysis, we take the data from the experiment on Taipei mayoral election in Taiwan on December 9, 2006. We issued six futures contracts for this experument, which consisted of five candidates ran for Taipei mayor and one for any invalid ballots cast on the election day. Sum of the prices of these six contracts are set to 100 at the beginning. Afterward, the sum should remain 100 if the traders behave rationally or if the market is efficient. The virtual money of amount 30,000 is deposited by the platform for each account to begin with. Two experiments ran in parallel for this event at that time. One is AI-ECON futures exchange (AI-ECON FX [1]) and the other is Taiwan Political Exchange (TAIPEX [2]). AI-ECON FX and TAIPEX are almost identical in design except for the traders of the former one can chose a preliminary software agent for the trading. Both servers started to run 30 days before the day of liquidation. At the end of the experiment, any contracts in the accounts were liquidated using the official result of votes from the government. Money prizes were then awarded to the top ten winners determined by the ultimate wealth in the players' accounts.

By analyzing the change of trading volumes in minutes, we observe that the market was active about 11% of the time in AI-ECON FX and 12% in TAIPEX. The number of registered players increased monotonically with time in both servers. Before the end of the experiments, AI-ECON FX and TAIPEX have accumulated 532 and 628 registrants respectively. The number of successful transactions totaled 7,440 in AI-ECON FX and 8,573 in TAIPEX. We further analyzed the transaction data to distinguish the active players from those who never traded with others throughout the whole experiment. After filtering, we found that there are 366 (427) active players left in AI-ECON FX (TAIPEX), which implies that only about 68% of registrants were active in both servers.

## 2.3 *The Cash-flow Network*

In a previous analysis[10], we showed that such a market, which accumulated typically 400 participants, exhibited power-law distributions of price fluctuation, net wealth and inter-transaction times that are characteristic of real world markets. Furthermore, predictions of the market have so far been consistent with election outcomes. In this work, being inspired by the recent development of complex networks, we introduce a new concept to study the trading behavior in a financial market. Our concept is detailed as follows.

If we treat each trader in the market as a node, and subsequently the transactions among them could be referred to as the edges. Therefore, we can reconstruct a network with traders and transactions. In our experiments, players trade futures contracts. When a transaction was made between traders $i$ and $j$ with volume $v$ and price $p$, an amount of cash $p \times v$ flowed from $i$ to $j$. Because the flow is directional and accompanied with certain amount of cash, the resulting cash-flow networks should be directed, weighted and contains no self-loops. In order to scale down the complex-

[1] http://futures.nccu.edu.tw/exchange/exchange_eng.html

[2] http://socioecono.phys.sinica.edu.tw/exchange/exchange_eng.html

ity of our problem and to extract the essence from the trading behavior, we simplify our cash-flow network into an undirected and unweighted network throughout this analysis. The preliminary results for cash-flow networks with directed and weighted edges are discussed in Ref. [5]. Without the loss of generality, we also assume that non-active players, who never trade during the whole experiment, would scarcely affect our results. We therefore neglect all the isolated nodes in our following analysis. During the experiment, both servers output the accumulated cash flow among traders in every 12 hours, from which we reconstructed 60 networks for each server. To understand the growth rate of the edges in these networks, we plot the value of $\langle k \rangle$, the average number of edges per node, in time series. In Fig. 1, one can see that the value of $\langle k \rangle$ grew with time, topping at 15.94 (18.26) in AI-ECON FX (TAIPEX) on day 30. We observe that the growth rate in our experiment keeps almost a constant (about 0.2 per day) after the first 15 days of the running.

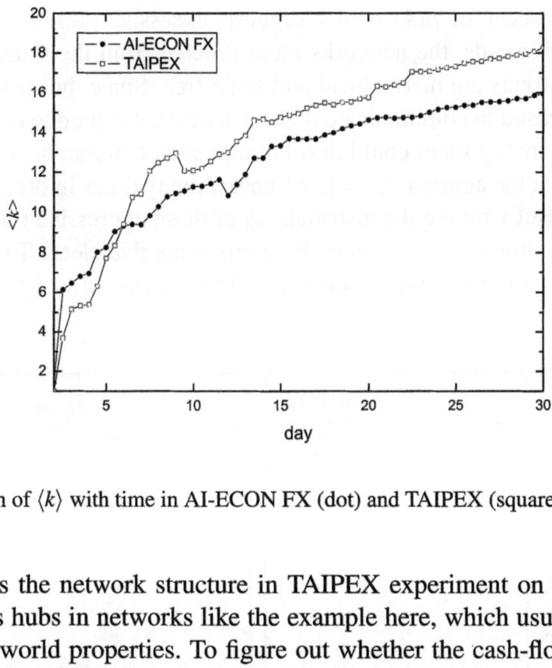


**Fig. 1** The growth of $\langle k \rangle$ with time in AI-ECON FX (dot) and TAIPEX (square).

Fig. 2 shows the network structure in TAIPEX experiment on day 3. One can easily identifies hubs in networks like the example here, which usually accompany with the small world properties. To figure out whether the cash-flow networks are scale-free or not, we calculate the degree distribution of our networks. The degree distribution, $p(k)$, describes the number of nodes to have $k$ edges. In Fig. 3, we show the resulting normalized $p(k)$ of the cash-flow networks on day 15 and day 30 in logarithmic scale with the linear fits. One can found that the degree distributions of these two networks can be well explained by a power-law decay with the form $p(k) \sim k^{-\gamma}$. We have $\gamma = 1.13 \pm 0.08$ and $\gamma = 1.17 \pm 0.06$ for AI-ECON FX and TAIPEX on day 30 respectively. Moreover, we found that these exponents remain almost the same during the last 15 days. This result might be related to the fact that the growth of $\langle k \rangle$ also remains roughly the same rate from day 15 to the end of experiment.

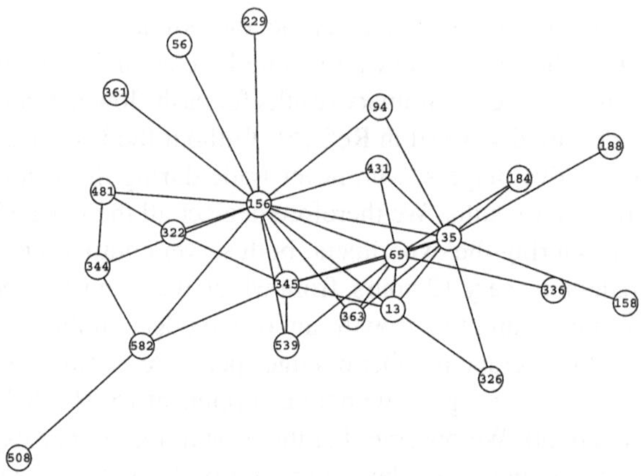


**Fig. 2** The structure of the cash-flow network developed in TAIPEX experiment on day 3. The number on the vertex corresponds to the ID for different traders assigned in the last day while edges denote the transactions among them.

A power-law decay of $p(k)$ with $k$ suggests excessive presence of hubs in our network. In other words, the networks reconstructed from the transactions among traders in our markets are hierarchical and scale-free. Since the traders in our markets are not supposed to communicate with each other, it is hard to imagine that why the transactions among them could develop into such a hierarchical structure. One explanation is that the aggressive traders transact many times in order to make profits from others. But why are the distributions of these aggressive traders (i.e., hubs in our networks) almost the same in both servers is not that clear? To further explain the observed phenomenon, we conduct a simple simulation to figure out whether

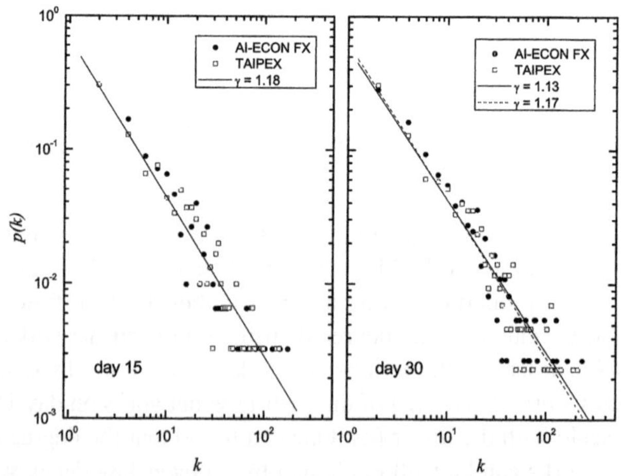


**Fig. 3** The degree distribution of the cash-flow networks on day 15 (left) and day 30 (right) for AI-ECON FX (dot) and TAIPEX (square). On day 15, the $p(k)$ in both servers could be well fitted by a power law decay with $\gamma \sim 1.18$. While on day 30, the best fit is $\gamma = 1.13 \pm 0.08$ and $\gamma = 1.17 \pm 0.06$ for AI-ECON FX and TAIPEX respectively.

the scale-free behavior in our cash-flow networks is due to the interactions among traders or due to the institutional design of our market.

## 3 Market Simulation with Zero-Intelligence Traders

Inspired by the approach of agent-based modeling, in the first attempt, we model our simulation as a continuous double-auction (CDA) market with zero-intelligence traders (ZI traders). The ZI traders, by definition, are agent traders without any intelligence. In the markets, they will submit random bids and offers, therefore the resulting price never converges toward any specific level. We here adopt the definition by *D. K. Gode* and *S. Sunder*[11], which demonstrate that in a symmetrically structured market, by imposing a simple budget constraint (i.e., the ZI-C traders who must profit from the transaction), the allocative efficiency of these transactions could be raised close to 100%. Hence, the trader in our simulation is the ZI trader with a simple budget constraint.

There are many variations of CDA markets, in this toy model, we made two choices to simplify our simulation. Firstly, each bid, offer and transaction is valid for a single item. Secondly, there is no transaction cost and the items are durable. Thirdly, in each duration, every trader could make only one successful transaction (i.e., the buyer could only have one item and the seller only has one item to sell in each duration). The implementation of our simulation is as follows: For the structure of markets, the supply and demand functions are generated from *Smith*'s value mechanism[13] at the beginning for each run and will not change through the end of simulation. The price for the item is ranging from 1 to 100 in units of virtual money. Because the ZI traders could only perform well in a symmetrically structured market[12], we choose the markets of this type in our simulation. In these markets, the intersection of supply and demand curves determines the equilibrium price. For the traders, initially, there is a fixed number of ZI traders in our simulation. Half of them are classified as buyers and the remaining half of the traders are sellers. At each step, one buyer and one seller are chosen for the matching. Due to the budget constraint, the buyer must bid with the price lower than its redemption value given by the demand function and the seller must offer the commodity at the price higher than the cost generated by the supply function. Once the bidding price exceeds the offering price, the transaction between this buyer and seller is made. No transaction will be made otherwise. Whether there exists a successful transaction or not, the platform will move forward to the next step and choose another pair of traders. The simulation lasts for $p$ periods of a specific duration $d$ and terminates after $p \times d$ steps. One transaction represents one edge in the network, but since our network is unweighted, repeated transactions between the same pair of traders will only be counted once. Each simulation runs for 100 times and the resulting degree distribution is an average over these 100 runs.

One should keep in mind that, although the number of traders is fixed at first, not all of them will make a successful transaction with others. The final number of nodes connecting to the whole network (i.e., traders with successful transactions) and $\langle k \rangle$ will also depend on the input value of period and duration. In comparison with the result in the experiments with human traders, we thus require the resulting cash-flow network to have 400 nodes on average with the value of $\langle k \rangle$ around 16. The set of input parameters must satisfied with the above condition. Fig. 4 shows the average degree distributions of the cash-flow networks from the simulations with two different sets of input parameters. We observe that the distribution follows a perfect power law decay with an exponent $\gamma \sim 0.59 \pm 0.04$. The sudden drop of the distribution curve at $k \sim 30$ might be due to the finite size effects.

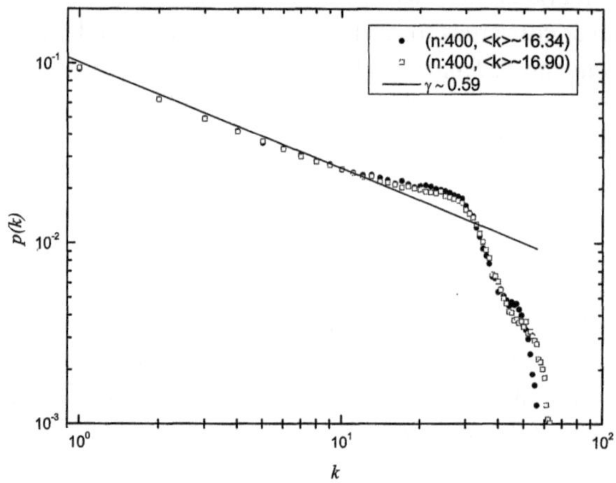


**Fig. 4** The degree distributions of the cash-flow networks resulting from the simulations with the network size $n = 400$ and $\langle k \rangle \sim 16$ . The solid line is the power law fitting with the exponent $\gamma \sim 0.59$.

To further justify this observation, we change the value of the input parameters for obtaining a network with larger network size. One can see that the decay behavior of $p(k)$ remains unchanged even for $n = 3500$ in Fig. 5. The decay exponent for this large network is $\gamma \sim 0.62 \pm 0.02$ which is roughly the same as the exponent in networks with $n = 400$. From the above result, it suggests that the nature of power-law decay of $p(k)$ depends on neither the network size nor the value of input parameters. Although the power-law exponent resulting from the simulation could not explain the observed exponent in the markets with human traders, we might still come up with a conclusion that the scale-free nature of the cash-flow networks dose not rely on the intelligence of the traders. Therefore, we believe that the scale-free nature comes from the institutional design and the structure of markets (i.e., the supply and demand function in the market).

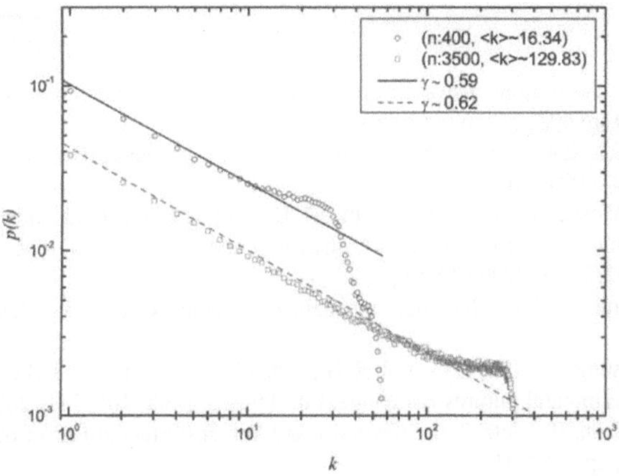


**Fig. 5** The comparison of the degree distributions of the cash-flow networks with different network size. $n = 400$ for the circles and $n = 3500$ for the squares. The exponents for the power-law decay are $\gamma \sim 0.59$ (solid) and $\gamma \sim 0.62$ (dashed) respectively.

# 4 Conclusion

In this work, we introduce a new concept to analysis the trading behavior in a financial market. In order to realize this approach, we design a Web-based futures exchange platform in order to gather enough information about the transactions among traders in a market. Two experiments were conducted with our platform on different servers (AI-ECON FX and TAIPEX) for 30 days. 7,440 (8,573) entries of transactions were accumulated and recorded in AI-ECON FX (TAIPEX). We thus reconstructed the cash-flow networks with these data and found that these networks exhibited hierarchical and scale-free properties with a power-law exponent around 1.15. To further comprehend the underlying mechanism of the observed phenomena, we carry out a simple simulation experiment involving a CDA market with zero-intelligence traders. To our surprise, such a simple market is capable of forming a hierarchical and scale-free network structure. Although the power-law exponent resulting from this toy model, which is only around 0.6, could not explain the observed exponent in the market experiments with human traders. But it reveals that the scale-free nature of the cash-flow networks might rely on the institutional design and the structure of markets rather than on the traders' strategies. In our simulation, all the agents are equiped with the same strategies, therefore, it is the supply and demand function that determines which trader should play as the role of the hub in the network.

**Acknowledgements** The research was supported in part by the National Science Council of Taiwan under grants NSC#95-2415-H-004-002-MY3 and NSC#95-2112-M-001-010.

## References


1. Uetz, P., et al.: A comprehensive analysis of protein-protein interactions in Saccharomyces cerevisiae. Nature **403**, 623–627 (2000)
2. Newman, M.E.J.: The structure of scientific collaboration networks. Proc. Natl. Acad. Sci. U.S.A. **98**, 404–409 (2001)
3. Guimera, R., Mossa, S., Turtschi, A., Amaral, L.A.N.: The worldwide air transportation network: Anomalous centrality, community structure, and cities' global roles. Proc. Natl. Acad. Sci. U.S.A. **102**, 7794–7799 (2005)
4. Albert, R., Barabasi, A.L.: Statistical mechanics of complex networks. Rev. Mod. Phys. **74**, 47–97 (2002)
5. Wang, S.C., Tseng, J.J., Tai, C.C., Lai, K.H., Wu, W.S., Chen, S.H., Li, S.P.: Network Topology of an Experimental Futures Exchange. Eur. Phys. J. B **62**, 105–111 (2008)
6. Berg, J.E., Nelson, F., Rietz, T.A.: Accuracy and Forecast Standard Error of Prediction Markets. working paper, (2003)
7. Berg, J.E., Rietz, T.A.: Prediction Markets as Decision Support Systems. Information Systems Frontiers **V5 N1**, 79–93 (2003)
8. Wang, S.C., Yu, C.Y. , Liu, K.P. , Li, S.P.: A Web-based Political Exchange for Election Outcome Predictions. Proc. IEEE/WIC/ACM International Conference on Web Intelligence (WI' 04), 173–178 (2004)
9. Wang, S.C., Tseng, J.J., Li, S.P., Chen, S.H.: Prediction of Bird Flu A(H5N1) Outbreaks in Taiwan by Online Auction: Experimental Results. New Mathematics and Natural Computation **V2 N3**, 271–279 (2006)
10. Wang, S.C., Li, S.P., Tai, C.C., Chen, S.H.: Statistical Properties of an Experimental Political Futures Market, to appear in Quantitative Finance.
11. Gode, D.K., Sunder, S.: Allocative Efficiency of Markets with Zero-Intelligence Traders: Market as a Partial Substitute for Individual Rationality. J. Polit. Econ. **V101 N1**, 119–137 (1993)
12. Cliff, D., Bruten, J.: Zero is not enough: on the lower limit of agent intelligence for continuous double auction markets. Technical report, HPL-97-141 (1997)
13. Smith, V.L.: Experimental Economics: Induced Value Theory. A.E.R. Papers and Proc. **66**, 274–279 (1976)

# A Model of Market Structure Dynamics with Boundedly Rational Agents

Tatsuo Yanagita and Tamotsu Onozaki

## 1 Introduction

The main purpose of this paper is to investigate the time evolution of the market structure in order to understand how oligopoly and monopoly spontaneously emerge from competition among firms. To this end, the framework of mainstream (i.e., neoclassical) microeconomics is of no use because its paradigm is rigidly static and lacks the dynamical point of view in the true sense of the word. It basically supposes one-shot decision makings by economic agents. Even if intertemporal decision makings are considered, it is always assumed, in order to ensure the rationality of agents in an uncertain world, that agents know all the future information certainly, that agents knows the probability distribution of all the future states, or that agents know the true economic model and form rational expectations consistent with it. In other words, agents know the future states of the economy in advance at least on average. In this sense, it is obvious that the time structure collapses which the mainstream microeconomics premises. It deals with a world essentially without time and not with how a market economy evolves as time goes by. Studying the evolutionary process of a market economy does not make sense unless bounded rationality of agents is assumed. In this study, we investigate the time evolution of a competitive market, transforming from an initial state where there are many, boudedly rational firms into an oligopolistic or monopolistic state.

A decentralized competitive market can be regarded as a typical complex system which consists of a large number of boundedly rational, and therefore, adaptive agents interacting with each other. These micro-level local interactions give rise to

Tatsuo Yanagita
Research Institute for Electronic Science, Hokkaido University, Sapporo 060–0812, JAPAN, e-mail: yanagita@nsc.es.hokudai.ac.jp

Tamotsu Onozaki
Faculty of Management and Economics, Aomori Public College, Aomori 030–0196, JAPAN, e-mail: onozaki@bb.nebuta.ac.jp

a certain macro-level spontaneous order, and then, the macro-order plays the role of binding conditions for micro-behavior. For example, persons who watch and imitate others' apparel beget a fad, and then they get carried away by the fad itself. Complex dynamical behavior emerges as a consequence of recurrent causal chains among individual behavior and the macro-order. This complex two-way feedback between microstructure and macrostructure has been recognized for a very long time, at least since the time of Adam Smith. Nevertheless, until recently, not only economics but other branches of science have lacked the means to model this feedback structure qualitatively in its full dynamical complexity. Researchers are now able to model, with the aid of high-performance computers, a wide variety of complex phenomena in a decentralized market economy, such as business cycles, endogenous trade network formation, market share fluctuations, and the open-ended coevolution of individual behavior and economic institutions. One branch of this research direction has come to be known as agent-based computational economics, i.e., computer-aided studies of economy modeled as an evolving system of autonomous interacting agents (see, e.g., [12, 16]).

In this study, we model a competitive market as a complex adaptive system consisting of locally interacting, boundedly rational firms and consumers[1]. A special attention is paid on the market share dynamics in common with [17], where product differentiation exists and consumer's brand loyalty plays an important role for emerging oligopoly. In this paper, however, it is assumed that a consumer decides from which firm to purchase goods so as to increase his utility, and we employ, as a first step, a statistical physics approach since the number of consumer is large. The aggregate consumer behavior is described by the Bolzmann distribution which is characterized by the 'inverse temperature' indicating how 'greedily' the consumer seeks to increase his utility. A firm, on the other hand, revises production decision and the price so as to raise its profit with the aid of a reinforcement learning algorithm, i.e., by learning through experience. We mainly focus on the dynamical phases which emerge as the 'greediness' of consumers changes, and characterize their statistical properties such as the probability distribution of firms' size and of their growth rates.

## 2 Model

Our model consists of many consumers and many firms that are boundedly rational in the sense that they attempt to increase their utility and profit subject to informational restrictions. Decisions for purchase and production occur at discrete time periods. Goods are homogeneous and perishable within a unit time period.

[1] While the model used in this study is the same as that of [20], we explored the cases of larger size system and found further results.

## 2.1 Consumer's Behavior

We assume the consumers to be identical in the following sense. Each consumer has the same amount of money income at each time step, selects a firm, and is willing to spend all the money to purchase goods from the selected firm[2]. Each consumer's utility $u_i(t)$ at period $t$ is represented by the identical function

$$u_i(t) = U(x_i(t)),$$

where $U(x)$ is a monotonically increasing function of the amount of goods $x_i(t)$ that a consumer obtains from firm $i$ at period $t$. $U$ is specified as $U(x) = x^{\alpha}$, $0 < \alpha < 1$ with $U' > 0$ and $U'' < 0$. The consumer is willing to expense all the money to purchase as much goods as possible because his utility increases as the amount of goods increases. The amount of goods $x_i(t)$ he can obtain from firm $i$, however, depends not only on prices set by firm $i$ but also its output level and the number of customers who select firm $i$.

Each consumer selects a firm so as to raise his utility as high as possible. He compares the utility to be obtained by purchasing goods from firm $i$ or from a randomly selected firm $j$, and selects one according to a transition probability, $\rho(i,j) = \min(1, (u_j/u_i)^{\beta_1})$, from firm $i$ to firm $j$, where $\beta_1$ is a positive parameter that represents how 'greedily' the consumer behaves. This rule is interpreted as follows. If $(u_j/u_i)^{\beta_1} \geq 1$ (this implies $u_j \geq u_i$), the consumer purchases goods from firm $j$ at the next period. Furthermore, even if $(u_j/u_i)^{\beta_1} < 1$, the consumer curiously chooses firm $j$ with a probability $(u_j/u_i)^{\beta_1}$. This rule is the so-called "softmax action selection" in the field of reinforcement learning. It implies that a firm from which consumers can purchase more goods has a higher probability of capturing consumers. The reason why probability is taken into consideration is that exploring other firms may afford consumers the chance to encounter higher utility[3]. Indeed, the softmax rule is used to depict the exploratory decision of human-beings [8].

From a statistical point of view, when a consumer moves from firm $i$ to firm $j$ with a transition probability $\rho(i,j)$, firm $i$'s stationary share distribution of consumers, $w_i^*$, can be written as

$$w_i^*(t+1) = u_i^{\beta_1}(t) / \sum_{j=1}^{M} u_j^{\beta_1}(t), \tag{1}$$

---

[2] For the sake of keeping the model simple, here we assume that money income at a certain period can not be carried over to the next period so as to eliminate the intertemporal allocation problem.

[3] The softmax rule is a modified version of the simplest rule called "greedy action selection", according to which the action with highest estimated value (in our context, the firm with highest utility) is always selected. As described later, the softmax rule indeed corresponds to the greedy rule when $\beta_1 \to \infty$. For this reason, we often characterize the parameter $\beta_1$ by using derived words of 'greedy'. It should be noted, however, that the word 'greedy' is somewhat misleading because it conceals the true meaning of the 'exploring' behavior, and throughout the paper the word is used with single quotation marks so as to call the readers' attention.

where $M$ is the number of firms and the denominator is a normalized constant [6]. Note that we can rewrite this share distribution as a usual form of the Bolzmann distribution: $w_i^*(t) = \exp(-\beta_1 E_i)/Z$, where $E_i = -\log(u_i)$ and $Z = \sum_i \exp(-\beta_1 E_i)$. As stated above, $U$ is specified as $U(x) = x^\alpha, 0 < \alpha < 1$. In our setting, the above market share distribution can be obtained for any value of $\alpha$ as long as $\beta_1$ is rescaled. Thus we fix $\alpha = 0.5$ without loss of generality. We note that the parameter $\beta_1$ corresponds to the inverse temperature in statistical mechanics [6] and expresses how 'greedily' consumer behaves. When $\beta_1 \to 0$, i.e., the temperature goes to infinity the consumers behave in purely random manner irrespective of their utility, whereas all consumers select the same firm that maximizes their utility when $\beta_1 \to \infty$, i.e., the temperature goes to zero.

In our model, consumers' utility varies with time through a change in price and quantity. To take into account this effect, we introduce, for simplicity, a linear relaxation dynamics of firm $i$'s market share, $w_i$, toward the stationary distribution Eq. (1) as follows:

$$w_i(t+1) = w_i(t) - \tau\left(w_i(t) - w_i^*(t+1)\right),$$

where $\tau \in [0,1]$ is a parameter that determines the relaxation time scale.

## 2.2 Firm's Behavior

Owing to bounded rationality, a firm does not know the demand function it faces nor the prices other firms have set, so that it must decide its price $p_i$ and production $q_i$ based only on the restricted local information, i.e., changes in profits. Profit $\Pi_i(t)$ of firm $i$ at period $t$ is defined as

$$\Pi_i(t) = p_i(t)s_i(t) - c\left(q_i(t)\right), \tag{2}$$

where $s_i(t)$ denotes the quantity sold by firm $i$ and $c_i(t) = (q_i(t))^2$ is an identical cost function. The sale $s_i(t)$ is represented as

$$s_i(t) = \min\bigl(q_i(t), w_i(t)T/p_i(t)\bigr), \tag{3}$$

where $T$ is the total money income of all consumers. Eq. (3) comes from the fact that total demand for firm $i$'s products is given by $w_iT/p_i$, and, if the demand differs from the production $q_i$, the sale is determined by the short-side. Thus, the profit of a firm depends upon its decision on price and production, and the demand it faces. Considering Eq. (3), the amount of goods $x_i(t)$ that a consumer can obtain from firm $i$ at period $t$ is written as

$$x_i(t) = s_i(t)/(w_i(t)L) = \min\bigl(q_i(t)/(w_i(t)L), T/(p_i(t)L)\bigr),$$

where $L$ is the number of consumers.

We assume that a firm does not directly control the price and production, but, instead, it determines rates of change of the previous price and production. Firm $i$ chooses a pair of rates of change $(\delta p_i, \delta q_i)$ among all possible options so as to get higher profits. Note that $p_i(t+1) = \delta p_i \cdot p_i(t)$ and similarly for $q_i$. Here the rates of change are given by

$$\begin{cases} \delta p_i = 1 + \Delta p \cos(2\pi n_i/N) \\ \delta q_i = 1 + \Delta q \sin(2\pi n_i/N) \end{cases} \quad \text{for} \quad n_i \in \{0, \ldots, N-1\},$$

where $n_i$ is an integer number from 0 to $N-1$ denoting a strategy of firm $i$, $N$ is the number of possible strategies, and $\Delta p$ and $\Delta q$ are given constants. Therefore, the maximum rates of change in price and production are $1 \pm \Delta p$ and $1 \pm \Delta q$.

A firm selects a strategy $n \in \{0, \ldots, N-1\}$ in pursuit of higher profit, according to a simple reinforcement learning rule which is applied to the "one-armed bandit" problem [18] as follows (here the subscript $i$ of strategy $n_i$ is omitted because there is no possibility of confusion). First, a firm evaluates all of its actions $\{0, \ldots, N-1\}$ by calculating the normalized quantity

$$\tilde{\Pi}_i^n(t) = \left(\hat{\Pi}_i^n(t) - \min_n(\hat{\Pi}_i^n(t))\right) / \left(\max_n(\hat{\Pi}_i^n(t)) - \min_n(\hat{\Pi}_i^n(t))\right),$$

where $\hat{\Pi}_i^n(t)$ is firm $i$'s expectation of its profit at period $t$ when it selects a strategy $n$. Then, firm $i$ selects a strategy $n$, following the "softmax" algorithm, i.e., with a probability proportional to $\exp(\beta_2 \tilde{\Pi}_i^n(t))$, where $\beta_2$ is the inverse temperature determining how 'greedily' the firm behaves. Finally, firm $i$ adaptively revises its profit expectation according to

$$\hat{\Pi}_i^n(t+1) = \hat{\Pi}_i^n(t) - k\left(\hat{\Pi}_i^n(t) - \Pi_i(t)\right), \tag{4}$$

where $k \in [0, 1]$ is the learning rate.

## 3 Simulations

### 3.1 Premises

The absolute level of profit is qualitatively irrelevant to our analysis since the competition among firms drives the dynamics, i.e., only the relative volume of profit matters. Thus we set the total money income of all consumers $T$ to one, and the maximum profit of each firm is rescaled to be one. Similarly, the population $L$ of consumers is qualitatively irrelevant and is set to one.

For the initial condition of profit expectation $\hat{\Pi}_i^n(0)$, we choose an "optimistic" value so that firms revise all their expectations effectively [18]. We set $\hat{\Pi}_i^n(0) = 100$ $\forall i, n$, which is large enough for realizing the maximum profit. The initial prices and productions are $(p_i(0), q_i(0)) = (1+\xi, 1+\xi)$, where $\xi$ is a small random number distributed uniformly in $[-0.01, 0.01]$. The initial market share distribution is the same among firms, i.e., $w_i(0) = 1/M\ \forall i$.

We fix the following parameters throughout simulations: $\alpha = 1.0$, $k = 0.5$, $\tau = 0.1$, $\beta_2 = 3.0$, $N = 10$, $\Delta p = \Delta q = 0.01$. We mainly consider the dependence of consumer's behavior on the inverse temperature $\beta_1$, that is, on how 'greedily' the consumer behaves. As for time scale, we use a non-dimensional time $t/t^*$, where $t^* = N/k$ is a learning time scale estimated from Eq. (4).

## 3.2 Simulation Results

### 3.2.1 Time Series

From a microeconomic point of view, it is easy to imagine that in an "artificial" monopoly case where $M = 1$ the best strategy of the monopolist is to raise the price and to reduce the production so as to increase the profit. As a result, the utility of consumer gradually decreases with time because each consumer has to purchase goods from the monopolist. In a multi-firm system, however, simultaneously raising the price and reducing production is not always the best strategy. Suppose that the market share of one firm is sufficiently larger than that of the others, the market is virtually monopolistic and the dominant firm tends to adopt the optimal strategy for monopolist, namely raising the price and reducing production. When the dominant firm raises the price, its customers' utility decreases. Decrease in utility causes decrease in the number of customers through Eq.(1). As a result, sooner or later, the firm realizes that its strategy is no longer the best through a reinforcement learning process, i.e., by encountering the fact that the profit is actually decreasing. In other words, the monopolist's strategy of raising the price and reducing production is restrained through the feedback from consumers' behavior. With regard to production, the dominant firm slows down the pace of reducing production; otherwise its profit decreases rapidly. Typical time series of these values for the case where $\beta_1 = 1.0$ are shown in Fig. 1, and it is easy to confirm the above explanation from observing the dotted area. The fluctuation range of prices is very wide as compared to other variables, but its absolute scale does not matter because it is the difference between consumers' utility that drives the dynamics.

### 3.2.2 Market Structure Dynamics

To see the temporal behavior of market share, we use the accumulated share $S_k(t) = \sum_{i=1}^{k} w_i(t)$, $k = 1, \ldots, M$, whose typical spatiotemporal patterns are shown in Fig. 2. Dynamics of the market crucially depends on the parameter $\beta_1$, which represents how 'greedily' consumer seeks higher utility. For lower 'greediness' (smaller $\beta_1$), almost all consumers choose firms in a purely random manner, i.e., the choice is

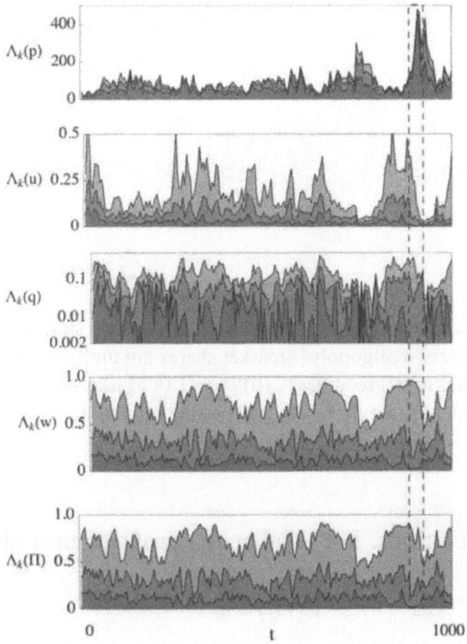


**Fig. 1** Typical time series of price $p_i$, utility $u_i$, production $q_i$, share $w_i$ and profit $\Pi_i$ of top 3 share firms are plotted. Each figure consists of three lines which are $\Lambda_k(x) = \sum_{i=k}^{3} x_{\phi_i}$ for $k = 1, \ldots, 3$, where $\phi_i$ denotes the rank of firms in market share and $x \in \{p, u, q, w, \Pi\}$. The reason why the summation is used is that it makes them easy-to-understand. $\beta_1 = 1.0$ and $M = 100$.

irrelevant to consumer's utility. Thus, the market share distribution is almost uniform among firms and stationary as depicted in Fig. 2(a). In this case, $S_k$ is approximately linear function of $k$. For higher 'greediness' (larger $\beta_1$), almost all consumers choose the "best" firm so as to seek higher utility. Therefore, monopoly emerges as a "quasi-stationary" state, which means that strong monopolists drastically change places with time. One can recognize in Fig. 2(c) some hight ridge lines in the direction of $k$-axis which correspond to the emergence of monopolist. While "quasi-stationary monopoly" is sustained, the time evolutions of the price and production are similar to those of the single-firm system. For intermediate 'greediness', some consumers choose firms in a purely random manner and the other consumers choose the "best" firm. As shown in Fig. 2(b), oligopoly persists owing to balance between these two effects although severe market-share battles among oligopolists are observed[4].

The transition among a uniform, an oligopolistic and a monopolistic market is also statistically characterized by means of the probability distributions of profit

[4] We have calculated the Herfindahl index in order to characterize the market share dynamics further, but have to omit the results due to severe limitations of space. The interested reader should refer to [20].

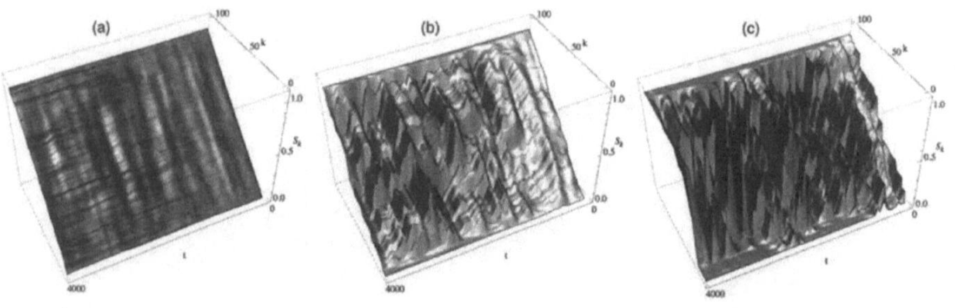

Fig. 2 Spatiotemporal pattern of the accumulated share $S_k(t)$ are shown for different values of $\beta_1$. $M = 100$. (a) $\beta_1 = 0.1$: Pseudo oligopoly – market shares are uniform among firms and stationary because each consumer has low 'greediness'. (b) $\beta_1 = 1.0$: Market-share battle in oligopoly – market shares change dynamically. (c) $\beta_1 = 5.0$: Alternating monopoly – monopolistic firms alternate drastically.

and its growth rate of firms. For smaller $\beta_1$, profit of firm obeys an exponential distribution, i.e.,

$$P(\Pi_i) \propto e^{-\lambda \Pi_i},$$

whose exponent $\lambda$ is approximately 120 as shown in Fig. 3(a). As $\beta_1$ increases, the profit distribution comes to exhibit a long-tail (or fat-tail) and to follow Pareto's law (or Power law), i.e.,

$$P(\Pi_i) \propto \Pi_i^{-\mu}.$$

In fact, as shown in Fig. 3(b) where oligopoly naturally emerges, the profit distribution seems to obey Pareto's law with an exponent $\mu$ being approximately 1. This special case of Pareto's law is known as Zipf's law, which is widely observed in various phenomena including the firm size distribution [5, 9]. As $\beta_1$ increases further, the exponent exceeds 1 (see Fig. 3(c) where $\mu = 2$). It should be noted that these results support the claim that the situation where the exponent of Pareto's law takes approximately 1 is the transition point between the oligopoly phase and the pseudo-equality phase [3, 4, 7, 9].

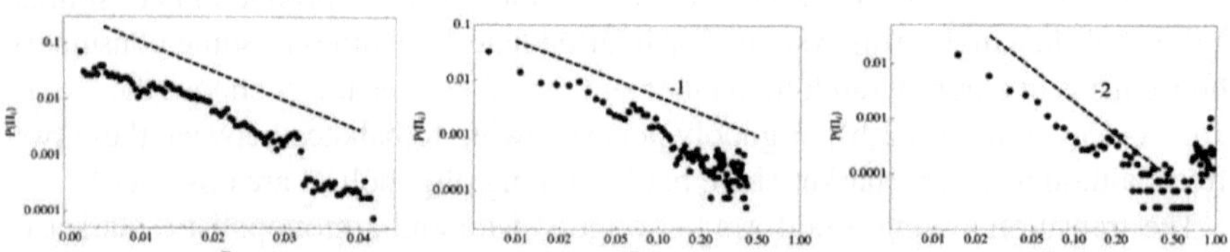

Fig. 3 The probability distributions of $\Pi_i$ for $M = 100$ are shown. Note that (a) is a semi-logarithmic plot while (b) and (c) are double logarithmic plots. The probability is averaged over $4 \times 10^4$ time steps after discarding the $10^4$ initial transient. (a) $\beta_1 = 0.1$. (b) $\beta_1 = 1.0$. (c) $\beta_1 = 5.0$.

The firm size measured by the volume of profit varies with time in the course of the market share dynamics. In order to take a look at the underlying dynamics which governs the firm size fluctuation, let us introduce the notion of logarithmic growth-rate of firm's profit defined as $r_i = \log_{10}(\Pi_i(t+1)/\Pi_i(t))$. The probability distributions of $r_i$ are shown in Fig. 4, which have a high peak at the point of $r_i = 0$ (or the point of no growth). By dividing all the firms into 6 classes $\Lambda_m = \{i \,|\, 10^{-(m+1)} \leq \Pi_i(t) \leq 10^{-m}\}$ $(m = 0, \ldots, 5)$ according to their profits, it turns out that the distribution of $r_i$ is independent of firm size $m$. This fact is often mentioned as Gibrat's law, which states that the conditional probability $Q(r_i \,|\, \Pi_i(t))$ is independent of $\Pi_i(t)$, i.e.,

$$Q(r_i \,|\, \Pi_i(t)) = Q(r_i),$$

and is widely observed in real economies [4, 9, 18].

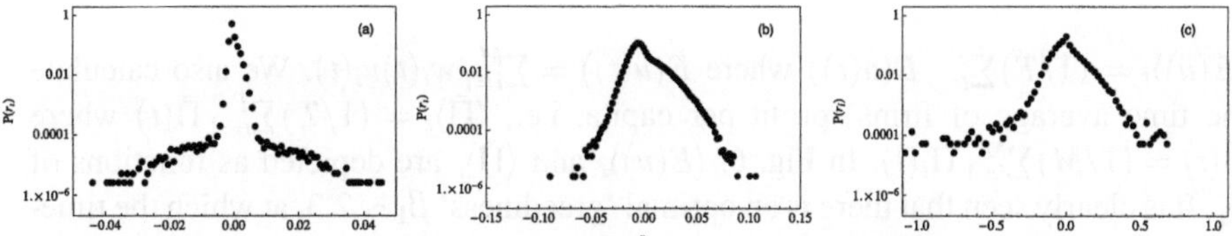

**Fig. 4** The probability distributions of logarithmic growth rate $r_i$ for $M = 100$ are shown. The distribution of growth rate is almost independent of the firm size. (a) $\beta_1 = 0.1$. (b) $\beta_1 = 1.0$. (c) $\beta_1 = 5.0$. Other parameters are the same as in Fig. 3.

The underlying dynamics which governs the firm size fluctuation can also be characterized from a different point of view. To this end, let us decompose the growth-rate of the firm size into $\Pi_i(t)$ and $\Pi_i(t+1)$. The scatter plots of these two variables are shown in Fig. 5, which exhibit strong symmetry between them with respect to the forty-five degree line. Using the joint probability distribution function $P_{t,t+1}(w_i(t), w_i(t+1))$, this symmetry is described as

$$P_{t,t+1}(\Pi_i(t), \Pi_i(t+1)) = P_{t,t+1}(\Pi_i(t+1), \Pi_i(t)),$$

which is called the detailed balance condition in statistical mechanics literature. This implies that a transition probability of a firm from a state where the profit is $\Pi_i(t)$ to a state where it is $\Pi_i(t+1)$ is the same as a transition probability from a state where it is $\Pi_i(t+1)$ to a state where it is $\Pi_i(t)$. If the detailed balance condition holds, the size distributions which satisfy Gibrat's law obey Pareto's law [9].

### 3.2.3 Averaged Utility and Profit

Finally, we investigate the $\beta_1$-dependence of utility and profit. For this purpose, we calculate the time average of the ensemble mean of consumers' utility, i.e.,

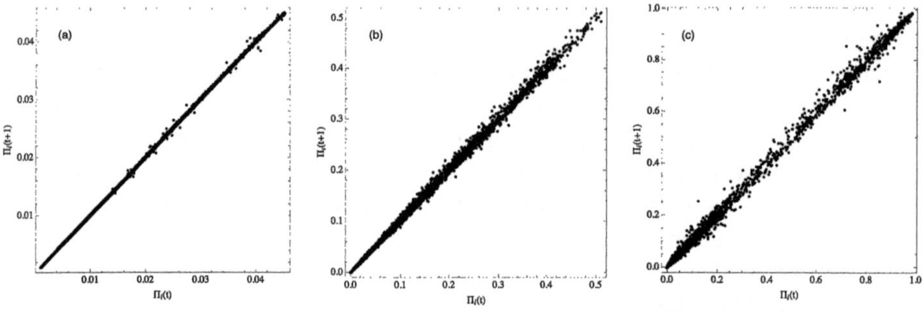


**Fig. 5** The scatter plots of $(\Pi_i(t), \Pi_i(t+1))$ for $M = 100$ are shown. (a) $\beta_1 = 0.1$. (b) $\beta_1 = 1.0$. (c) $\beta_1 = 5.0$. The data show axial symmetry with respect to $y = x$, implying that the detailed balance condition holds. The other parameters are same as in Fig.4.

$\langle E(u)\rangle_t = (1/T)\sum_{t=1}^{T} E(u(t))$ where $E(u(t)) = \sum_{i=1}^{M} w_i(t)u_i(t)$. We also calculate the time average of firms' profit per capita, i.e., $\langle\bar{\Pi}\rangle_t = (1/T)\sum_{t=1}^{T}\bar{\Pi}(t)$ where $\bar{\Pi}(t) = (1/M)\sum_{i=1}^{M}\Pi_i(t)$. In Fig. 6, $\langle E(u)\rangle_t$ and $\langle\bar{\Pi}\rangle_t$ are depicted as functions of $\beta_1$. It is clearly seen that there is an optimal 'greediness' $\beta_1 \sim 2.3$, at which the time-averaged consumers' utility is maximized. For smaller $\beta_1$, the time-averaged utility is very small because each consumer selects a firm in a purely random manner (note that firm's best decision in this case is raising the price and reducing production). With the increase of $\beta_1$, the market-share battle in oligopoly starts to emerge while the time-averaged utility gradually increases and takes an optimal value that gives the maximum utility. Beyond the optimal value, the time-averaged utility gradually decreases because each consumer is too 'greedy' to choose the best firm by seeking higher utility, thereby causing the formation of a monopolistic market. On the other hand, the time-averaged profit per capita $\langle\bar{\Pi}\rangle_t$ gradually decreases with increasing $\beta_1$ and reaches the minimal value at $\beta_1 \sim 2.4$. Beyond the minimum value, the time-averaged profit per capita gradually increases since monopoly starts to emerge. In the vicinity of the optimal 'greediness', oligopoly emerges although its membership changes frequently, and the time-averaged profit per capita of oligopolistic firms reaches the minimal value. Oligopoly is the best state of market in terms of consumers' utility while it brings the minimal profit to participants because of severe competition.

## 4 Concluding Remarks

We have investigated the dynamics of a competitive market consisting of locally interacting, boundedly rational firms and consumers. Instead of relying on demand function, the behavior of consumers is described by the market share distribution, i.e., the stationary distribution of a large number of consumers who employ softmax

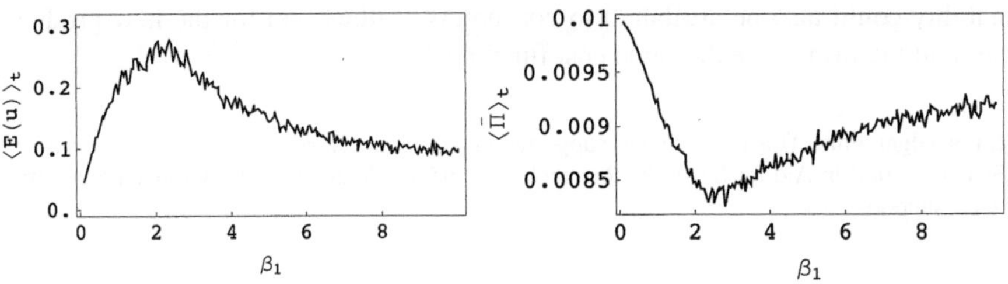


**Fig. 6** The time average of the mean of consumers' utility $\langle E(u)\rangle_t$ and the time average of firms' profit per capita $\langle\bar{\Pi}\rangle_t$ versus $\beta_1$ for M=100. The utility and profit are averaged over $T = 5000$ time steps and are sampled over 20 different initial conditions after initial transients. There is an optimal 'greediness' $\beta_1 \sim 2.3$ which maximizes the time-averaged utility.

strategy. This distribution is characterized by a single parameter $\beta_1$ that represents how 'greedily' the consumers behave. Firms revise their production decisions and prices so as to raise their profit with the aid of a simple reinforcement learning rule. Numerical simulations show the following results:

1. Three phases of market structure, i.e., the uniform-share phase, the oligopolistic phase and the monopolistic phase appear depending upon a key parameter $\beta_1$.
2. In an oligopolistic phase, the market-share distribution of firms follows Zipf's law and the growth-rate distribution of firms follow Gibrat's law.
3. An oligopolistic phase is the best state of market in terms of consumers' utility while oligopoly brings the minimal profit to the firms because of severe competition based on the moderate 'greediness' of consumers.

It would be of interest to compare the first result with the 'industry life-cycle' which has been observed in many industries [1, 2, 10, 11, 13, 14, 15] (especially, see Figure 2.1 of [15], p.28). The early stage is charaqcterized by market share instability and a relatively 'competitive' market structure. In this stage, strong monopolists drastically change places with time or oligopoly persists although severe market-share battles come about. It corresponds to the monopolistic or oligopolistic phase in our results, implying that consumers are 'greedier' for the new product in the early stage. The mature stage is characterized by relatively stable market shares and a more concentrated market structure. It corresponds to the uniform-share phase or relatively less-competitive oligopolistic phase, implying that consumers are less 'greedy' in the mature stage. Such correspondence between the industry life-cycle and $\beta_1$ suggests the possibility of a new explanation about the driving force of market structure dynamics during the course of the industry life-cycle. Previously, high instability in the early phase of the industry used to be attributed to active innovation and cost reduction [2, 13, 14, 15]. Of course this explanation is valid, but it may represent only part of the fact. On the other hand, our model suggests that high instability could also be attributed to consumers' enthusiasm for the new product even if all the firms have the same cost function.

**Acknowledgements** This research was supported in part by the Japan Society for the Promotion of Science, Grant-in-Aid for Scientific Research (C), 19530155, 2007-8. The authors are grateful for their support.

## References

1. Abernathy WJ, Utterback JM (1975): A Dynamic Model of Product and Process Innovation. Omega 3:424–41
2. Acs ZJ, Audretsch DB (1990): Innovation and Small Firms. MIT Press, Cambridge MA
3. Aoyama H (2004): Scaling Laws of Income and Firm-size. Presentation at Workshop on Economics with Heterogeneous Interacting Agents (WEHIA)
4. Aoyama H et al (2007): Pareto Firms. Nippon Keizai Hyoron-sha, Tokyo (in Japanese)
5. Axtell R (2001): Zipf Distribution of U.S. Firm Sizes. Science 293:1818–20
6. Binder K, Heermann DW (1979): Monte Carlo Methods in Statistical Physics. Springer, Berlin
7. Bouchaud J-P, Mèzard M (2000): Wealth Condensation in a Simple Model of Economy. Physica A 282:536–45
8. Daw ND et al (2006): Cortical Substrates for Exploratory Decisions in Humans. Nature 441:876–9
9. Fujiwara Y et al (2004): Do Pareto-Zipf and Gibrat Laws Hold True? Physica A 335:197–216
10. Gort M (1963): Analysis of Stability and Change in Market Shares. Journal of Political Economy 62:51–61
11. Hymer S, Pashigian P (1962): Turnover of Firms as a Measure of Market Behavior. Review of Economics and Statistics 44:82–7
12. Kirman A, Zimmermann J-B (eds) (2001): Economics with Heterogeneous Interacting Agents. Springer, Berlin
13. Klein BH (1977): Dynamic Economics. Harvard University Press, Cambridge MA
14. Klepper S (1996): Exit, Entry, Growth, and Innovation over the Product Life-cycle. American Economic Review 86 (3):153-60
15. Mazzucato M (2000): Firm Size, Innovation and Market Structure. Edward Elgar, Cheltenham UK
16. Namatame A et al (eds) (2006): The Complex Networks of Economic Interactions: Essays in Agent-Based Economics and Econophysics. Springer, Berlin
17. Onozaki T, Yanagita T (2003): Monopoly, Oligopoly and the Invisible Hand. Chaos Solitons and Fractals 18:537–547
18. Sutton J (1997): Gibrat's Legacy. Journal of Economic Literature 35:40–59
19. Sutton RS, Barto AG (1998): Reinforcement Learning: An Introduction. MIT Press, Cambridge MA
20. Yanagita T, Onozaki T (2008): Dynamics of a Market with Heterogeneous Learning Agents. Journal of Economic Interaction and Coordination 3:107–18. doi:10.1007/s11403-008-0038-2

# Agent-based Analysis of Lead User Innovation in Consumer Product Market

**Kotaro Ohori**[1] **and Shingo Takahashi**[2]

**Abstract** In this paper we focus on some phenomena generated by lead user innovation and build a market model based on the lead user's features that are derived from the conventional studies. Then we consider market mechanism and dynamics of the market with the lead user innovation. The innovation changes a conventional market concept that consumers and firms are entities that respectively demand and supply products. It is hard for marketers and managers to comprehend the market dynamics and mechanism generated by the change of market invoked by the innovation. Our simulations with the market model show two propositions: 1) If firms take the conventional strategies based only on conventional marketing strategies and technology management, lead user innovation takes the high market share from firms' products. 2) If firms focus on an innovation community as a new strategy, they can manage lead user innovation.

## 1. Introduction

In recent years innovation researches [3][12] with novel market concepts discuss marketing strategies, which differ from ones noted in conventional marketing theories. The marketing theories mainly focus on attributes of products and consumers. The innovation researches, by contrast, focus on market conditions that are market network or community. In the future firms will have to create novel strategies for managing the rapid innovation and the change of market structure. Then they will need to consider not only marketing theories, but also the innovation researches since their decisions based only on segmentation, targeting, and positioning (STP) in marketing theories especially will have a high risk.

[1] Kotaro Ohori
Department of Industrial and Management Systems Engineering, Waseda University, 3-4-1 Okubo, Shinjuku, Tokyo 169-8555, email: oohori@fuji.waseda.jp
[2] Shingo Takahashi
Department of Industrial and Management Systems Engineering, Waseda University, 3-4-1 Okubo, Shinjuku, Tokyo 169-8555, email: shingo@waseda.jp

This paper tries to approach lead user innovation in consumer product market by using an agent-based approach. The innovation is that a new phenomenon generated by interactions among users including lead users in a consumer population. Hippel[12] has described that the lead users are ahead of the majority of users in their population with respect to an important market trend, and they expect to gain relatively high benefits from a solution to the needs they have encountered there. The lead users, who are both a consumer and supplier, drastically change the market mechanism. Therefore managers and marketers in firms have to create novel strategies to manage lead user innovation.

The purpose of this paper is to consider the possibilities of market changes generated by lead user innovation and to provide two propositions concerning lead user innovation. We apply agent-based approach for archiving the purpose. The approach can analyze some firms' strategies and cases in an artificial market, and consider some tendencies of changes in the future markets that have uncertainty.

## 2. Lead User Innovation for Consumer Products

In this chapter, we explain lead user innovation for consumer products in conventional studies and show key problems of the innovation. Some studies have discussed lead user innovation for industrial products that are OPAC [13], PC-CAD [7] etc. In this case lead users are user firms. Since the innovations developed by a user firm need a long term for the success of them, manufacturing firms can observe the source of the innovations and manage them. On the other hand, the lead user innovation for consumer products on which this paper focuses has a strong possibility that the innovations developed by lead users take market share from manufacturing firms, because the markets with consumer products have three problems as follows: 1) lead user's features, 2) information asymmetry and information stickiness, 3) innovation community.

### *2.1 Lead User's Features*

The studies of lead user innovation in consumer product markets have picked up personal digital assistant (PDA), sports-related products, outdoor-related products etc. Especially some studies with the sports-related and outdoor-related product declared for remarkable lead user's features [5][6]. The lead user will purchase a new product immediately after the product is lunched by manufacturing firms, search a new use of the product, and then develop yet another one for satisfying his/her needs. In particular a third of consumers in out-door related product markets have various ideas for innovations and develop innovations depending on new needs and dissatisfaction at existing products. Moreover consumers who often use the existing products tend to develop a new product. The consumers, who will develop the new product, namely lead users, do not want to make a profit on license of their products and strongly agree to share their products. From these features lead users freely reveal their innovations to the other consumers and lead users.

### *2.2 Information Asymmetry and Information Stickiness*

Hippel [12] took up two key concepts that are relevant to lead user innovation. Firstly, the information asymmetry means that firms and consumers are experts of their own technologies and own needs, respectively. So firms will develop products depending on the technologies, in contrast consumers will prefer and demand the product fitting the needs better than other products that are developed based on the firms' core technologies. Secondly, the problem concerning information stickiness is that firms cannot find the tacit needs and technologies of consumers since they are very sticky and cannot be transferred to firms from consumers by low-cost. Although firms can find the surface needs and technologies, for example, by surveying questionnaires naturally, they are probably useless for firms' innovations.

### *2.3 Innovation Community*

Conventional studies related with innovation communities have dealt with open-source software [14]. A recent study observed innovation communities in markets with sports-related products [4]. Most of members in the sports communities of canyoning, bordercross etc. positively developed innovations. The innovation communities themselves do not have the ability of innovations, but play the key role of assisting among members for innovations and innovation diffusion process.

## 3. Market Model

We build a market model with lead user innovation by agent-based modeling (see Fig.1). The model is supported by CAMCaT (Coevolutionary Agent-based Model for Consumers and Technologies) framework that can be useful for analysis of market dynamics [9]. In the framework consumers and firms have their respective fitness functions, then they achieve coevolution through the product space. The market model in this paper consists of a consumer population, a firm population and a product space.

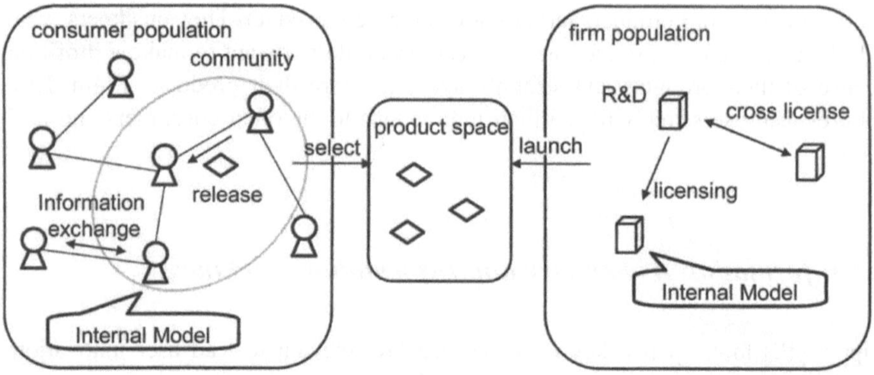


**Fig. 1.** Market model with lead user innovation

### *3.1 Product*

There are some products with 5 attributes in the market. The attributes are defined by $A=(a_{ik})$, where $a_{ik} \in \{1,2,\ldots,100\}$, $i=1,2,\ldots,l$, $k=1,2,\ldots,5$, $i$ identifies an individual product, the maximum number is $l$. $k$ is the attribute number of a product. The products developed by firms and lead users are launched to the product space and consumer population, respectively.

### *3.2 Consumer Population*

In the consumer population there are lead users who will develop or improve a product and choose a product, and users who will only choose a product. The population size of consumer is 100. Each consumer has an internal model consisting of some chromosomes. And each one, based on his/her internal model, decides about choice and development of a product, and revises his/her own internal model in an evolutionary learning process. We call a series of these economic activities "generation," because the revision of internal models is performed by genetic operations. The generation cycle in a simulation is given in 3.2.2 through 3.2.5 described in following sub-subsections.

#### 3.2.1 Internal Model

The internal model of a consumer consists of the potential of development $I$, dependence and sensibility to other consumers $D$, edges with other consumers $E$, link with a community $N$, cutoff value $C$, purchasing weight for product attributes $W$ and possessed technology $T$.

The internal model of consumer $i$ is denoted by $IM_i = (I, D, E, N, C, W, T)$, where $I = (i_i)$ $i_i \in \{0,1\}$, $D = (d_i)$ $0 \le d_i \le 1$, $E = (e_{ij})$ $e_{ij} \in \{0,1\}$, $N = (n_i)$ $n_i \in \{0,1\}$, $C = (c_{ik})$ $c_{ik} \in \{1,2,\ldots,100\}$, $W = (w_{ik})$ $\sum_k w_{ik} = 1$, $T = (t_{ik})$ $t_{ik} \in \{1,2,\ldots,100\}$, $i, j = 1,2,\ldots,100$, $k = 1,2,\ldots,5$; $i, j$ identifies a consumer index, $k$ is an attribute index.

The potential of development $I$ shows that whether the consumer is an innovating consumer (namely lead user) or not innovating consumer. The dependence and sensibility to other consumers $D$ shows that how degree the consumer is affected by a trend product that has maximum number of purchases in the product space. The edges $E$ and the community $N$ represent the information stickiness and information asymmetry that are key concepts of lead user innovation (see Subsection 2.2). The cutoff value $C$ and purchasing weight $W$ for product attributes show the preference of the consumer according to previous our study [9]. The possessed technology $T$ shows the niche technology or tacit knowledge of the lead user that is depending on his/her hobby or occupation.

We build a network model to generate edges and a community. Conventional studies that are related to the innovation community have not discussed the network index. Uchida and Shirayama [10] show that the network index of social networking service (SNS) is similar to the one of connecting nearest neighbor (CNN) model [11]. So the innovation community on web also may be similar to CNN model. Since the network index of many networks in a real world has the structure of scale-free (SF), the innovation community also is possibly similar to BA model [2]. In this paper we adopt BA model, CNN model and Random model.

### 3.2.2 Choice of a Product

Each consumer evaluates products by the evaluation rule of products and chooses one having the maximum utility value bigger than the cutoff values. The evaluation rule is defined by the utility function $u_{ij}$ of consumer $i$ for product $j$ (1).

$$u_{ij} = \Big(\sum_k b_k * a_{ijk} * (d_i) + \sum_k w_{ik} * a_{ijk} * (1 - d_i)\Big) * c_j$$

$$where\ c_j = \begin{cases} 0 & if\ consumer\ i\ cutoff\ product\ j \\ 1 & otherwise \end{cases} \qquad (1)$$

$b_k$ represents the attributes of the trend product. $a_{ijk}$ represents the evaluation value of attribute $k$ of product $j$ that consumer $i$ evaluated.

### 3.2.3 Development of a Product

This phase corresponds to lead user innovation. Each lead user ($i_i = 1$) develops or improves a product and launches the product into the edges and/or community in the consumer population, and not the product space. Since they will develop innovations because of a passing fancy, enjoyment and agency cost, the launch timing of a new product is different among lead users. So we set the launch rate 0.05 as a constant value simply.

The lead user sets the attribute $a_{jk}$ of his/her product $j$ to $t_{ik}$ if $\exists k(c_{ik} < t_{ik})$. This represents that the lead user will develop and improve the attribute if he/she has a higher technology than his/her own cutoff value. Otherwise the lead user does not develop the product. Since this paper does not imply technology values, technology and product attributes are one to one correspondence simply.

### 3.2.4 Evolutionary Learning: Self-evaluation

After choice and/or development of a product, each consumer $i$ evaluates his/her own decision and internal model.

*Evaluation after Choice of a Product*

Each consumer $i$ evaluates his/her own choice and preference by using the fitness function $fp_i$ (2).

$$fp_i = 0.30 * (1 - ncut) + 0.30 * sumcut + 0.10 * (1/\text{maxcut}) + 0.30 * trend \quad (2)$$

*ncut* shows the number of non-cutoff products, *sumcut* shows the sum of cutoff values, *maxcut* shows the maximum cutoff value, *trend* shows the difference between the attributes of the trend product and the consumer $i$ 's preference values. The fitness function represents that a consumer who has a higher evaluation can reduce recognition effort of products attributes and know market trends very well.

*Evaluation after Development of a Product*

Each consumer $i$ evaluates his/her own development and technology by using the fitness function $ft_i$ (3).

$$ft_i = 0.40 * share + 0.25 * overcut + 0.20 * (1/maxtech) + 0.15 * sumtech \quad (3)$$

*share* shows the share of the products that consumer $i$ launched, *overtech* shows the difference between technology values and preference values, *maxtech* shows the maximum technology value, *sumtech* shows the sum of technology values. The fitness function represents that a consumer who has a higher evaluation can gain reputation from other consumers, develop a high spec product and assist the other consumers' innovations.

### 3.2.5 Evolutionary Learning: Selective Crossover and Mutation

Based on the evaluation of a consumer, each consumer revises his/her own internal model in evolutionary learning. The revision of consumers' internal models is not performed by conventional genetic algorithm (selection, crossover and mutation), but are performed by "selective crossover" as a novel genetic operation and "mutation." Evolutionary leaning of a consumer operates preference and technology in the internal model, individually.

*Evolutionary Learning of Preferences*

Consumers' preferences are revised by the selective crossover and the mutation. The selective crossover is our original genetic operation used in a network structure. The operation is a fusion of the selection and uniform crossover (see Fig.2).

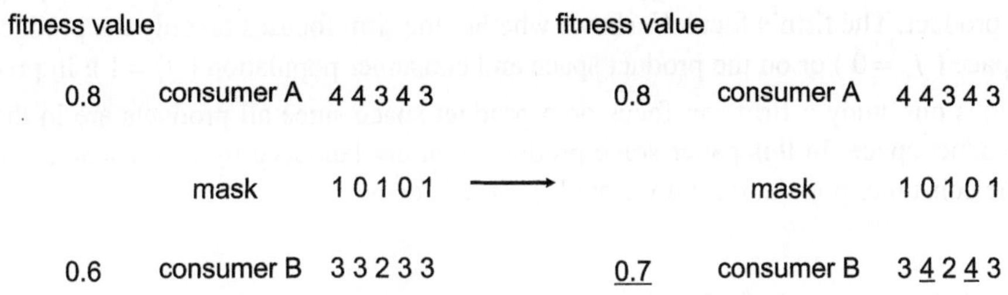


**Fig. 2.** Selective crossover of cutoff values

By one consumer is selected at random in the consumer population. The selected consumer performs selective crossover with the other consumer linked with him/her according to the probability of crossover. The fitness values of the two consumers are compared, then the preference of a consumer who has a higher evaluation copy to the preference of another one.

The operation implies the information exchange among consumers. Lastly each consumer mutates his/her preference, which implies gathering information. The crossover and mutation rates are 0.6 and 0.05, respectively.

*Evolutionary Learning of Technologies*

The evolutionary learning of technologies is performed by the selective crossover and mutation. This implies technological assistance and the notice of technology for innovations. The crossover and mutation rates are 0.6 and 0.05, respectively.

## 3.3 Firm Population

We build the model of firms based on one in previous our study [9]. The population size of firm is 20. Each firm has an internal model as well as consumer, and decides the development and launch of products. The revision of the internal model is performed in evolutionary learning. A generation cycle in a simulation is set by 3.3.2 through 3.3.4 described in following sub-subsections.

### 3.3.1 Internal Model

The internal model of a firm consists of the firm's vision $C$, possessed technology $T$, and firm's focus $F$ .The internal model of firm $i$ is denoted by $IM_i = (C, T, F)$, where $C = (c_{ik})$ $\sum_k c_{ik} = 1$, $T = (t_{ik})$ $t_{ik} \in \{1, 2, \ldots, 100\}$, $F = (f_i)$ $f_i \in \{0, 1\}$, $i = 1, 2, \ldots, 20$, $k = 1, 2, \ldots, 5$, $i$ is a firm index, $k$ is an attribute index.

The firm's vision $C$ shows that what technology the firm thinks core or important. The possessed technology $T$ shows the attributes that are used for developing a product. The firm's focus $F$ shows whether the firm focuses on only the product space ($f_i = 0$) or on the product space and consumer population ($f_i = 1$). In previous our study a firm can focus on a product space since all products are in the product space. In this paper some products that are launched by lead users are in the consumer population. So we modify firms' focus.

### 3.3.2 Development of a Product

Each firm $i$ launches a new product according to the launching rate 0.05. The attribute $A$ of the launched product is set by calculating from the possessed technology $T$ of the firm.

### 3.3.3 Evolutionary Learning: Self-evaluation

After launching the product, each firm $i$ evaluates its own decision by using the fitness function $ff_i$ (4).

$$ff_i = 0.35 * share + 0.25 * (1 - risk) + 0.30 * selfvalue + 0.10 * sumtech \qquad (4)$$

*share* shows the share of the products that firm $i$ launched, *risk* shows that the difference between the trend product and its own technology, *selfvalue* shows the fitness of its own technology with its own vision, and *sumtech* shows the sum of its own technology.

### 3.3.4 Evolutionary Learning: Selection, Crossover and Mutation

The internal models of firms are selected with Baker's linear ranking selection [1] after the evaluation. This implies licensing or M&A. The revision of internal models are performed by crossover and mutation. If $f_i$ is 0, the mutated value is decided by a normal distribution with possessed technology, otherwise, by a normal distribution with the trend product. This implies cross license and R&D. Crossover and mutation rates are 0.6 and 0.05, respectively.

## 4. Simulation Results

This chapter shows some simulation results using two firms' strategies. Firstly every firm, based on the conventional strategy, focuses on only the product space in the market (Subsection 4.1). Secondly every firm, as a new strategy, can focus the product space and the innovation community in the consumer population (Subsection 4.2).

The percentage of lead users in the consumer population is 15 %. Lead users launch their products after 50th generation. This paper shows only some results by using a consumer network based on the CNN model (see Fig.3).

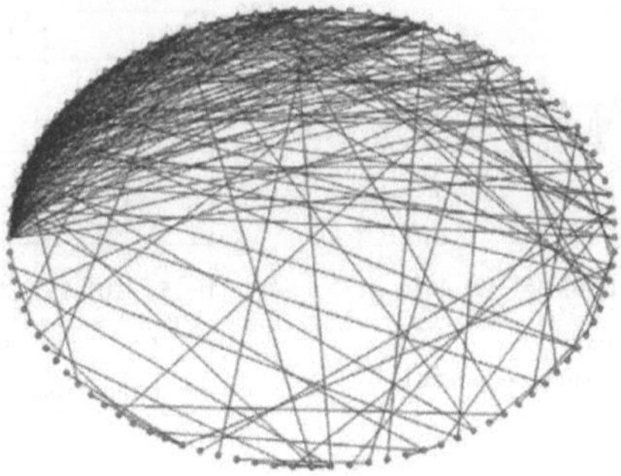

**Fig. 3.** Consumer network with CNN model

### *4.1 Focus on Only the Product Space*

Every firm focuses on only the product space and cannot have access to the consumer population. Lead users develop a new product and introduce it to other consumers. In case 1 lead users introduce their products to only consumers linked with them. Additionally they can introduce to the innovation community in case 2.

*Case 1: Information Propagation among Edges*

Fig.4-(a) shows the transition of the market shares of firms' products and lead users' ones. The lead users' products gradually grab the high market share and spread their products to the consumer population. The main reason for this result is that firms did not aware that lead users' developments and information propagation among consumers because the model can represent the information asymmetry and stickiness by generating the network structure.

*Case 2: Information Propagation in the Innovation Community*

Next we consider the situation that lead users can introduce their products to the innovation community and exchange information in the community. The community consists of lead users and non-lead users. 32 percentages of whole consumers belong to the community. Fig.4-(b) implies that the information exchange can be taken wing in the innovation community and the technologies of lead users exceed firms' one. This corresponds to the phrase "Given enough eyeballs, all bugs are shallow" described as Linus's law [8]. Though each technology of lead users is inferior to firm's one, the sum of lead users' technologies can exceed firms' ones. So the innovation community can promote the evolutionary learning of preferences and technologies.

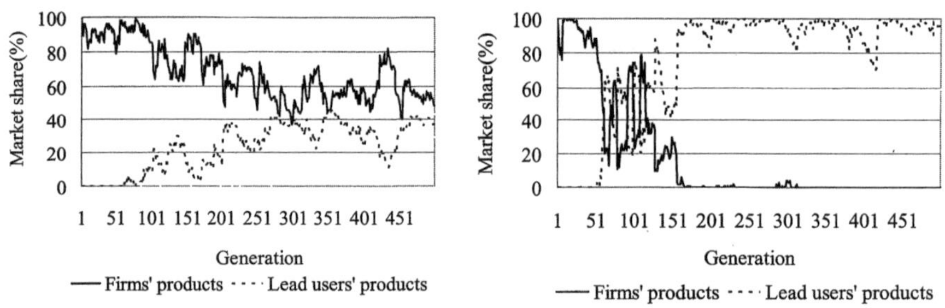


(a) Market share in case 1 (b) Market share in case 2

**Fig. 4.** Transition of market share in case 1 and case 2

## 4.2 Focus on the Product Space and Innovation Community

As we have mentioned above, it is possibly difficult for firms to grab market share in the market with lead user innovation. The firms failed to prevail against lead user innovation mainly because they would consider that the optimum decision is to follows the trend in the product space. Firms should focus on consumers' preferences and technologies to overcome the failure. Since consumers' information is sticky as we noted (see Subsection 2.2), firms cannot collect the information directly. Hence we analyze a new strategy for provision against lead user innovation in this subsection. The strategy is that firms positively focus on a trend product in the innovation community, and not the product space. Since the trend product has consumers' preferences and technologies, firms can gain the consumers' tacit needs and niche technologies that firms cannot develop. Namely it is that firms do not compete with lead users but manage lead user innovation.

Firms can develop and launch a new product by using the trend product in the innovation community. Fig.5-(a) shows the transition of market share of firms' products and lead users' ones. And fig.5-(b) shows the transition of average of firms' fitness values in this strategy and the previous strategy in subsection 4.1. The results imply that firms can grab the market share and manage the lead user innovation by using the community, compared with the previous strategy. Regrettably firms, however, cannot always grab a high market share because the user innovation is presented as the set of niche technologies.

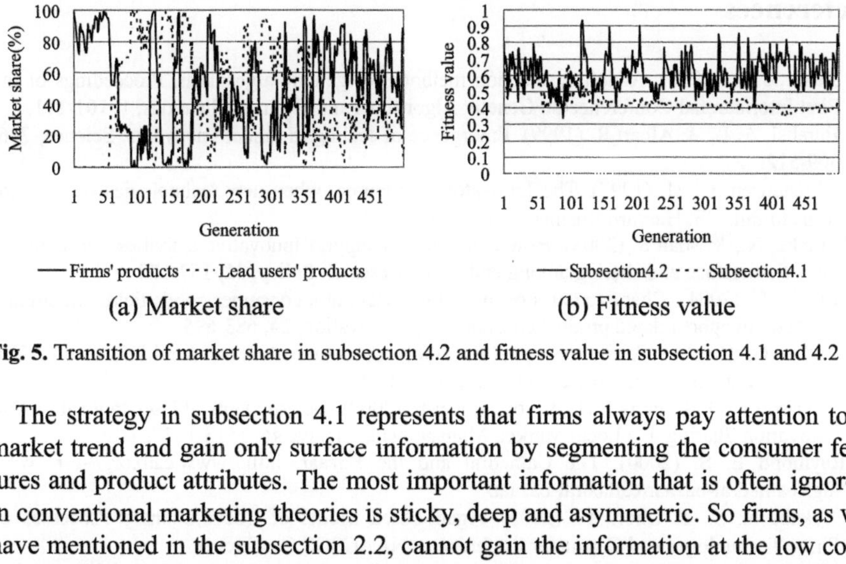


**Fig. 5.** Transition of market share in subsection 4.2 and fitness value in subsection 4.1 and 4.2

The strategy in subsection 4.1 represents that firms always pay attention to a market trend and gain only surface information by segmenting the consumer features and product attributes. The most important information that is often ignored in conventional marketing theories is sticky, deep and asymmetric. So firms, as we have mentioned in the subsection 2.2, cannot gain the information at the low cost. As the result lead user innovation takes the higher market share than firms' one do. On the other hand, firms focus on a trend product in the innovation community in subsection 4.2. From the result firms cannot grab the market a little and manage the lead user innovation.

Then we can infer the two propositions as follows: 1) If firms take the conventional strategies based only on conventional marketing strategies and technology management, lead user innovation takes high market share from firms' products. 2) If firms focus on an innovation community as a new strategy, they can manage lead user innovation.

## 5. Conclusion

In this paper we have proposed the agent-based model for analysis of phenomena of lead user innovation, verified some strategies and cases by computational simulations and provided the two propositions. The results of our simulations indicate that the management of lead user innovation needs to change into the new strategy that is the focus on an innovation community. The phenomena that are similar to the simulation results have been actually observed in some consumer product markets. This could support part of validity of the results to real markets. Agent-based modeling is a role of generators of propositions to unclear phenomena such as lead user innovation. Now we should point out that the results shows only some tendencies of market dynamics, and not real markets directly. Our model is built from important characteristics that are derived from various conventional studies on lead user innovation. So the model is still abstract level. As a future direction of this study, we will analyze specific markets with lead user innovation by specifying the abstract features of the model.

## References

[1] Baker, J. E. (1985). Adaptive selection methods for genetic algorithms. Proceedings of the First International Conference on Genetic Algorithms and Their Applications, 1, 101-111.

[2] Barabsi, A. L., & Albert R. (1999). Emergence of scaling in random networks. Science, 286, 509-512.

[3] Christensen, C. M. (1997). The Innovator's Dilemma, when new technologies cause great firms to fail. US: Harvard Business School Press.

[4] Franke, N., & Shah S. (2003). How communities support innovative activities : an exploration of assistance and sharing among end-users. Research Policy, 32, 157-178.

[5] Luthje, C. (2004). Characteristics of innovating users in a consumer goods field; An empirical study of sport-related product consumers. Technovation, 24, 683-695.

[6] Luthje, C,. Herstatt, C., & von Hippel E. (2005). User-innovators and local information: The case of mountain biking. Research Policy, 34, 951-965.

[7] Morrison P. D., Roberts J. H., & von Hippel E. (2000). Determinants of User Innovation and Innovation Sharing in a Local Market, Management Science, 46, 1513-1527.

[8]Raymond E. S. (2000). The Cathedral and the Bazaar, http://www.catb.org/~esr/ writings/cathedral-bazaar/cathedral-bazaar/

[9] Takahashi, S., & Ohori, K. (2005). Agent-based Model of Coevolutionary Processes of Firms Technologes and Consumer Preferences. NAACSOS Conference.

[10] Uchida, M., & Shirayama, S. (2006). Analysis of Network Structure and Model Estimation for SNS. Transactions of Information Processing Society of Japan, 47, 2840-2849.

[11] Vazquez, A. (2003). Growing network with local rules: Preferential attachment, clustering hierarchy, and degree correlations. Physical Review, 67.

[12] von Hippel, E. (2005). Democratizing Innovation. US:MIT Press.

[13] Urban, G. L., & von Hippel, E. (1998). Lead User Analyses for the Development of New Industrial Product. Management Science, 34, 569 - 582 .

[14] von Krogh, G.S., Spaeth, S., & Haefliger, S. (2005). Knowledge Reuse in Open Source Software: An Exploratory Study of 15 Open Source Projects. Proceedings of the 38th Annual Hawaii International Conference on System Sciences, 198–207.

# Agent-Based Stochastic Model of Barter and Monetary Exchange

Igor Pospelov[1] and Alexandra Zhukova[2]

## 1 Introduction

The main issue addressed in this work is the role of money in the market, in other words is there any use to rational agents to exchange produced goods for intrinsically useless money instead of direct barter.

This question was previously addressed by N. Kiyotaki and R. Wright. In their model [1] the exchange process is described as stochastic. At the same time they consider a continuum set of agents which makes the analysis of dynamics of the stochastic process rather difficult. Instead, they postulate the main properties of stationary states of the system. In this work the modification of the model proposed in [1] with finite number of agents is considered.

Transition from continuum to finite number of agents is not merely a formal procedure, but it is a conceptual modification. The authors of the main model [1] conduct their analysis in the framework of unchangeable and individual preferences of agents. For this reason, when dealing with continuum of goods, a requirement of continuum of consumers of these goods appears. This makes the correct description of stochastic process rather complex and also implies that the nomenclature of commodities in trade is initially infinite.

In the model proposed in this work, consumers may change their preferences. Hence, the connection between the number of commodities and the number of agents is no more required. In fact, the number of commodities continuously grows because it is considered that each producer invents new

Igor Pospelov
Dorodnicyn Computing Center of Russian Academy of Sciences, Vavilov st. 40, 119333 Moscow, Russia

Alexandra Zhukova
Moscow Institute of Physics and Technology, Institutskii per., 9, 141700, Dolgoprudny, Moscow Region, Russiae-mail: alexa10@mail.ru

product not presented in the market. Thus, initially finite number of commodities increases over time through production.

## 2 Agents, their Goods and the Exchange Process

A community of $N + m$ identical agents is considered. Each agent produces, exchanges and consumes one unit of one type from continuum set of goods. After consumption she starts the production of another product. Goods are identified by points on circle $\Omega$ of circumference 2. This assumption is made for convenience of mathematical calculations.

A group of agents (the number of them is $m$) are initially endowed with money. Money is a special type of commodity which has no consumption value. Each agent is able to hold no more than one unit of money.

Goods are being made in the production sector. Potential opportunities to produce a product $\alpha$ from the set of commodities $\Omega$ come to an agent according to Poisson process with fixed arrival rate which we denote $\Lambda$, at random moments of time. The consumption utility $u(\alpha, \omega)$ of commodity $\alpha$ depends decreasingly on arch of circle $\Omega$ between commodity $\alpha$ and some ideal (or most preferred) commodity $\omega$. It is considered, that *the idea of her ideal product $\omega$ comes to an agent randomly during the production process and independently of* $\alpha$ and remains unchangeable till the next moment of production.

In the same way as in the model [1] it is assumed that an agent can not consume her output. She has to proceed to the trade sector and exchange it for one unit of some other commodity and to consume the latter. Agents are randomly matched in pairs according to Poisson process with fixed arrival rate $M/2$. Exchange happens between agents *if and only if they both are agree.* When two agents with money meet nothing happens. When two agents with commodities meet barter occurs (in case of mutual consent). After barter they both consume and proceed to the production sector.

When an agent with money meets agent with a commodity they may exchange (by mutual consent) one unit of product for one unit of money. The agent, who has had money consumes his buying and proceeds to the production sector. The agent who has had a commodity stays in the trade sector with money.

Thus, the number of agents with money is constant and always equal to originally given $m$.

## 3 Rationality of an agent

We assume that each agent tends to maximize her expected discounted consumption utility.

$$K = E\left\{\sum_{\tau_k > t} e^{-r(t-\tau_k)} u(\beta(\tau_k), \omega(\tau_k))\right\}. \tag{1}$$

Here $\tau_k$ are the moments of future changes in their states and $r$ is a discount parameter. Utility of consumption $u(\beta(\tau_k), \omega(\tau_k))$ depends decreasingly on the shortest arch between $\beta(\tau_k)$, and $\omega(\tau_k)$ on the circle, representing the set of commodities. $\beta(\tau_k)$ is a good consumed at moment $\tau_k$, and $\omega(\tau_k)$ is the desirable good which she doesn't have but it brings maximum utility to an agent.

On one hand, it is not profitable to an agent to wait long for meeting to an agent with desirable good, but, on the other hand it is not pays to her to consume the first coming commodity (because it may be far from his favorite product with high probability) . As a compromise, she would better to choose some golden medium between these strategies – the optimal strategy. Since all agents in the model are assumed to be identical, the optimal strategies are the same for everybody.

Optimally acting agents have to come to a particular equilibrium with stationary distribution of a number of money traders, producers and commodity traders.

## 4 Formal Description

We will consider a monetary economy as a general case, since barter economy is a form of monetary with $m = 0$. We build a formal model as a *Markov decision process*, when agents aim at maximizing functionals of expected discounted payments gained from transition from state to another state.

Each individual receives her payment only for turns from state to state, and for this reason we start by describing of individual agent's behavior.

### *4.1 States of an agent and transitions among them*

At any moment of time an agent is in one of the following states: $OP$- state of expectation of a project; $OO(\alpha, \omega)$- state of expectation of exchange, having product $\alpha$ on hand and $\omega$ as ideal; $OM(\omega)$ - state of expectation of exchange, having money on hand and $\omega$ as ideal.

Any possible transition sequence among these states and payments for them are determined in the following way.

**On realization of a project** an agent from the state $OP$ proceeds to one of the states of type $OO(\alpha, \omega)$. This agent receives one unit of a new commodity – her output – and gets a new idea of her ideal product. Payment for this transition is zero.

It is important and unusual in this model that both goods, produced and ideal *come to an agent as a uniformly distributed over the circle $\Omega$ random quantities,* independent one of another and from other agents' states. Thus, the arc between them is uniformly distributed on the interval $[0, 1]$.

**At barter** an agent from state $OO(\alpha, \omega)$ proceeds to the state $OP$. She barters her commodity for another agent's commodity $\beta$ and receives a payment for the transition $u(\beta, \omega)$.

**At selling for money** an agent from state $OO(\alpha, \omega)$ proceeds to the state $OM(\omega)$. Payment for this transition is zero.

**At buying for money** an agent goes from the state $OM(\omega)$ to the state $OP$. At that she buys some commodity $\beta$ and gets a positive payment for this transition, that is consumption utility $u(\beta, \omega)$.

Below, for short, we will call the agents in states $OO(\alpha, \omega)$ *sellers* and the agents in states $OM(\omega)$ - *buyers.* They differ in that sellers are also allowed to exchange their commodities directly through barter.

It is assumed that projects may come to every agent $i = 1, \ldots, N + m$, even trader or seller, but those who are in the trade sector do not react on coming projects. Moments of possible projects form a Poisson process $\xi_{ii}(t)$ individual to each agent $i = 1, \ldots, N + m$ and independent of other processes of the model. Since agents are assumed to be identical, the frequency denoted $\Lambda$ of Poisson processes $\xi_{ii}(t)$ is the same for all agents.

Moments of time when an opportunity of exchange comes to an ordered pair of agents $(i, j)$, where $i, j = 1, \ldots, N + m$, $i \neq j$, form a Poisson Process $\xi_{ij}(t)$, individual for each pair $(i, j)$ and independent of other processes in the model. Since agents are assumed to be identical, the frequency denoted $M/2$ of Poisson processes $\xi_{ij}(t)$ is the same for all pairs $(i, j)$ where $i, j = 1, \ldots, N + m$, $i \neq j$.

## *4.2 Approach to the description of the system*

The dynamics of the system is defined by the set of $(N + m)^2$ independent Poisson processes $\xi_{ij}(t)$. "Diagonal" elements $\xi_{ii}(t)$ describe the production process and have frequency $\Lambda$, "non-diagonal" elements describe exchanges and have frequency $M/2$.

A state of the economy is considered as a distribution of states of agents, that is a point in Cartesian product of spaces of agents' states. Each of these spaces represents a combination of one-point set (state $OP$), a pair of circles (states $OO(\alpha, \omega)$) and one more circle (state $OM(\omega)$).

Here, on account of complexity of a detailed description, we suggest to build the model in the following way.

Firstly, due to symmetry and simplicity we construct a simplified (we call it also short) description of the system. This description aggregates unknown parameters characterizing agent's behavior. Secondly, state the problem of one agent's optimal behavior within the framework of the simplified description. Thirdly, select information restrictions of an agent so that her optimal behavior is characterized by those parameters that are used in the simplified description.

## *4.3 Short description of the system*

We assumed above that production, barter and money trade are random and occurring once, that is that an agent cannot consume her produced output and proceed to generate new goods. Also she cannot stay trader after exchange.

These assumptions allow to conclude that traders in the market have independent and uniformly distributed goods. It is so because what they have is what a random project brought them. And this, in turn, allows to consider the probability of exchange when traders meet not to depend on the state of the system, but only on what they have. Let us denote the probability of barter at the moment of meeting (when a seller meets a seller she exchanges with this probability) as $\theta$, and the probability of selling at the moment when it is possible (when a buyer meets a seller) as $\delta$.

Barter decreases the number of sellers in the market on 2, and sale decreases the number of sellers in the market on 1. Frequencies of meetings are determined only by the number of agents (sellers plus money traders) in the market.

This suggests an idea to consider the number of sellers $n(t)$ as a convenient aggregating variable for simple description of a state of the system. Then one could formulate the dynamics of the system as Markov process for $n(t)$, generated by processes $\xi_{ij}(t)$ and given probabilities $\theta$ and $\delta$.

However, this approach meets an obstacle. It may be advantageous for a trader to wait till a bigger number of traders comes to the market, because then she finds what she wants quicker. If so, the variables $\theta$ and $\delta$ become some (probably very complex) functions of $n(t)$.

In order to avoid this obstacle we put an information restriction on an agent's strategy: ***an agent does not know the number of sellers.***

Equation for change of number of agents

We start from describing a non-stationary version of equation for change of number of agents. Let the probabilities of barter $\theta$ and sale for money $\delta$ (as two traders meet) be constant and the same for every pair of agents. Let $p(t,n)$ denote the probability that there are exactly $n$ sellers in the market at the moment $t$. To write the equation it becomes useful to consider a course-of-value function

$$E\{z^{n(t)}\} = \sum_{k=0}^{N} p\,(t,k)\,z^{k}. \qquad (2)$$

It may be found using the conditional expectation $E\{z^{n(t+dt)}|n(t) = k\}$ that may be calculated as follows. During time period $dt$ only one of two following events may occur with relatively big probability (more than $o(dt)$):

one of the agents i receives a project – it is an event $[i,i]$ and its probability is $\Lambda dt + o(dt)$;
an agent $i$ meets an agent $j$ - it is an event $[i,j]$ and its probability is $(M/2)dt$;

Among the $N+m$ events of type $[i,i]$ under the condition $n(t) = k$, exactly $N-k$ events are attributed to agents in production state $OP$. Therefore, the number of agents in state $OO$ increases on 1 with probability $\Lambda(N-k)dt + o(dt)$ as a result of event of type $[i,i]$.

Among the $(N+m)(N+m-1)$ events of type $[i,j]$ under condition $n(t) = k$ exactly $k(k-1)$ events are attributed to meetings of sellers. Each of these events ends in exchange with probability $\theta$ which decreases the number of agents on 2. Consequently, total probability of two agents entering the trade sector is $\frac{\theta\,\mathrm{M}\,k(k-1)}{2}dt + o(dt)$

Other $km$ events among the $(N+m)(N+m-1)$ meetings are attributed to pairs seller-buyer. Each of the meetings ends in bargain with probability $\delta$, which decreases the number of agents in the market on 1 (buyer receives some commodity, consumes and proceeds to production sector and seller receives money and stays). The total probability that one buyer leaves the market after trade is $\frac{\delta\,M\,k\,m}{2}dt + o(dt)$.

All the remaining meetings (e.g. of two individuals with money), as well as absence of events in the period $[t, t+dt]$ occur with probability

$$1 - \Lambda\;(N-k)\,dt - \frac{\theta\,\mathrm{M}\,k\;(k-1)\,dt}{2} - \frac{\delta\mathrm{M}\,k\,m\,dt}{2} + o(dt). \qquad (3)$$

They do not change the number of individuals in the trade sector. There also exist a probability of superposition of events described above but it is of small order $o(dt)$. Thus we have

$$
\begin{aligned}
&E\{z^{n(t+dt)}|n(t)=k\}= \\
&=\left(1-\Lambda\,(N-k)\,dt-\tfrac{\theta\,\mathrm{M}\,k\,(k-1)dt}{2}-\tfrac{\delta\mathrm{M}\,k\,m\,dt}{2}+o(dt)\right)z^{k}+ \\
&+((N-k)\,\Lambda\,dt+o(dt))\,z^{k+1}+\left(\tfrac{\theta\mathrm{M}\,k\,(k-1)}{2}dt+o(dt)\right)z^{k-2}+ \\
&+\left(\tfrac{\delta\,\mathrm{M}k\,m}{2}dt+o(dt)\right)z^{k-1}+o(dt).
\end{aligned}
\tag{4}
$$

On average over $k$ according to the formulation of mathematical expectation

$$
E\{z^{n(t+dt)}\}=\sum_{k=0}^{N}E\{z^{n(t+dt)}|n(t)=k\}p(t,k). \tag{5}
$$

This gives

$$
\begin{aligned}
\varPhi(t+dt,z)=\sum_{k=0}^{N}\Big[&\left(1-\Lambda\,(N-k)\,dt-\tfrac{\theta\,\mathrm{M}\,k\,(k-1)dt}{2}-\tfrac{\delta\mathrm{M}\,k\,m\,dt}{2}\right)z^{k}+ \\
&+(N-k)\,\Lambda\,dt\,z^{k+1}+\tfrac{\theta\mathrm{M}\,k\,(k-1)}{2}dt\,z^{k-2}+\tfrac{\delta\,\mathrm{M}k\,m}{2}dt\,z^{k-1}\Big]\,p(t,k)+o(dt).
\end{aligned}
\tag{6}
$$

Here is replaced $\varPhi(t+dt,z)=E\{z^{n(t+dt)}\}$.
Setting $dt\to 0$ we get an equation

$$
\begin{aligned}
&\tfrac{\partial\varPhi(t,z)}{\partial z}=-(1-z)\;\Lambda N\varPhi(t,z)+(1-z)\left(\Lambda z+\tfrac{\delta Mm}{2}\right)\tfrac{\partial\varPhi(t,z)}{\partial z}+ \\
&+(1-z)\;(1+z)\;\tfrac{\theta\mathrm{M}}{2}\tfrac{\partial^{2}\varPhi(t,z)}{\partial z^{2}}.
\end{aligned}
\tag{7}
$$

Now it becomes possible to derive the stationary distribution of the number of agents. A non-stationary process described above is a process with a finite number of states, and thus a stationary distribution always exists. Moreover, if $\theta>0$ and $\delta\geq 0$ the probability of transition from any state with $n_1$ agents in the trade sector to any other state with $n_2$ during any finite period of time is not zero. Consequently, the process is ergodic, that is the stationary distribution $p\,(n)$ when $t\to\infty$ is unique and every initial distribution of states converges to $p\,(n)$. Like in [1], we examine the stationary distribution of the process. The stationary distribution is a polynomial

$$
P(z)=\lim_{t\to\infty}\varPhi(t,z)=\sum_{n=0}^{N}p\,(n)\,z^{n}. \tag{8}
$$

It satisfies the equation derived from 7 when $t\to\infty$

$$
0=-\Lambda N\,P(z)+(\Lambda z+\frac{\delta Mm}{2})P'(z)+(1+z)\frac{\theta\mathrm{M}}{2}P''(z), \tag{9}
$$

and the normalization condition $P\,(1)=1$.

## 4.4 Optimal Strategies and Bellman Equation

In order to find optimal strategies and build an equilibrium, it is convenient to use Bellman equation for agents' utilities. Existence of an optimal strategy in Markov decision process with a finite number of states is proven in quite general case [2]. In this model the full formal notation of Kolmogorov equation for every permissible strategy is rather bulky and does not illustrate a lot. For this reason, we derive the Bellman equation straight away proceeding from substantial reasoning.

Distributions of available and ideal products are uniform and independent for every agent and so the maximum of the expected utility must be independent of what an agent has and what she considers as ideal.

Let us denote $K(OP)$, $K(OO(\alpha,\omega))$, $K(OM(\omega))$ the highest possible values of expected utility for a producer, a seller and a buyer correspondingly. We give them the following values ($b$ and $d$ may be positive as well as negative)

$$K(OP) = V, \quad K(OO(\alpha,\omega)) = V + b, \quad K(OM(\omega)) = V + d. \tag{10}$$

Below, these values will be step by step expressed in terms of parameters of the model.

Firstly, find the expression for $K(OP)$.

In the production sector during a time period $dt$ no events happen with probability $(1 - \Lambda dt)$. There are also additional corrections of order $o(dt)$ in the expressions for probability, which we omit for conciseness. With this probability an agent gets the utility $K(OP)$, reduced due to discounting by $\exp^{-r\,dt}$. With probability $\Lambda\,dt$ an agent produces a unit of commodity $\alpha$ and gets an idea of ideal commodity $\omega$. Selling of this commodity brings her utility $K(OO(\alpha,\omega))\,e^{-r\,dt}$.

On average in the state $OP$ she gets

$$K(OP) = (1 - \Lambda dt)\,K(OP)\,e^{-r\,dt} + \Lambda\,dt\,K(OO(\alpha,\omega))\,e^{-r\,dt}. \tag{11}$$

As $dt \to 0$, combined with (10) the latter expression reduces to $V = \frac{\Lambda b}{r}$.

Then we may find the expression for $K(OO(\alpha,\omega))$ by means of new functions of $K(MON(\alpha,\omega))$ and $K(BAR(\alpha,\omega))$, which will be found later.

Consider one seller $S$ in the trade sector. Suppose that at the moment of time t there are $k$ sellers in the market, including $S$. Let us derive Bellman equation for her optimization problem. It is still assumed that there are $m$ buyers (agents with money) in the trade sector.

Only the following events of streams $\xi_{ij}(t)$, occurring during the period $dt$, may effect on a state $OO(\alpha,\omega)$: contacts of $S$ with other $k-1$ sellers and contacts of $k-1$ other sellers with $S$, also contacts of $S$ with m buyers and contacts of $m$ buyers with $S$.

Each of this events has a probability $(M/2)\,dt$. Let $K(BAR(\alpha,\omega))$ denote the maximum expected utility of events of the first type and $K(MON(\alpha,\omega))$ denote the maximum expected utility of events of the second type. Then the expected utility is:

$$\begin{array}{l}(1-(k+m-1)\,\mathrm{M}\,dt)\,K\,(OO(\alpha,\omega))\;\;e^{-r\,dt}+\\+\,(k-1)\,\mathrm{M}\,dt\,K(BAR(\alpha,\omega))+m\,\mathrm{M}\,dt\,K\,(MON(\alpha,\omega))\end{array}. \tag{12}$$

In order to find $K(OO(\alpha,\omega))$ we have to average out this expression. Thus we get

$$\begin{array}{l}K(OO(\alpha,\omega))=\sum\limits_{k=1}^{N}\pi(k)\,[(1-(k+m-1)\mathrm{M})\,K(OO(\alpha,\omega))\;e^{-r\,dt}+\\+\mathrm{M}\,(k-1)\,K(BAR(\alpha,\omega))+\mathrm{M}\,m\;K(MON(\alpha,\omega)))dt]\end{array}. \tag{13}$$

Here $\pi(k)$ are conditional probabilities of that there are $k$ agents in the trade sector under condition that considered agent is in the market

$$\pi(n)=\begin{cases}0, & \text{if } \; n=0;\\ \frac{p(k)}{\sum\limits_{k=1}^{N}p(k)}, & \text{if } \; n\geq 1.\end{cases}$$

Setting $dt\to\infty$ we have

$$K(OO(\alpha,\omega))=\frac{\mathrm{M}\Big(\Big(-1+\sum\limits_{k=1}^{N}\pi(k)k\Big)K(BAR(\alpha,\omega))+mK(MON(\alpha,\omega))\Big)}{r+\mathrm{M}\Big(m-1+\sum\limits_{k=1}^{N}\pi(k)k\Big)} \tag{14}$$

Now find the expression for utility of barter $K(BAR(\alpha,\omega))$.

Consider a seller $A$ with parameters $(\alpha,\omega)$ meeting a seller $B$ with parameters $(\beta,\psi)$. If the deal takes place then the seller $A$ gets utility $u(\beta,\omega)+K(OP)$. If not, then she gets $K(OO(\alpha,\omega))$. Thus, there is no sense for $A$ to consent to a deal if $u(\beta,\omega)+K(OP)<K(OO(\alpha,\omega))$ or, taking into account (10), if $u(\beta,\omega)<b$. So, *a seller barters her commodity $\alpha$ for another commodity $\beta$ only if $u(\beta,\omega)>b$* .

Since the agents' products are uniformly and independently distributed, we have on average the probability that an agent is consent to barter

$$\Theta=\int\limits_{\Omega}\mathrm{I}(u(\beta,\omega)-b)d\beta. \tag{15}$$

$$I(x)>0\;if\;x>0\;and\;I(x)=0\;if\;x<0$$

It would be important to note that $\Theta$ does not depend on $\omega$, since, as it was mentioned in the beginning, the utility $u(\beta,\omega)$ depends only on arc between $\beta$ and $\omega$, so that the integral (15) is the arc of circle where utility values are greater than $b$.

If the deal takes place, then the seller $A$ receives utility

$$\int_{\Omega} \max\left\{u(\beta,\omega)+K(OP),\ K\left(OO(\alpha,\omega)\right)\right\} d\beta = V+U(b). \tag{16}$$

Where we introduce a denotement

$$U(z)=\int_{\Omega} \max(z,u(\beta,\omega))d\beta. \tag{17}$$

Like $\Theta$, this quantity does not depend on $\omega$ and as one may verify that $\Theta = 1-U'(b)$.

Because of symmetry of agents, the probability of $B$ to be consent to barter is the same quantity $\Theta$. By virtue of that distributions of $A$'s and $B$'s parameters are independent, the probability to barter between them when they meet is

$$\theta = \Theta^2 = \left(1-U'(b)\right)^2. \tag{18}$$

Thus the maximum of expected utility of barter deals is

$$K\left(BAR(\alpha,\omega)\right) = \theta\ \left(V+U(b)\right) + (1-\theta)\ \left(V+b\right). \tag{19}$$

Now find the expression for utility of selling $K(MON(\alpha,\omega))$ and buying $K(NOM(\alpha,\omega))$

If a seller exchanges her commodity for money she receives $V+d$ if refuses, receives $V+b$. Thus *a seller is always consent to sell her commodity if* $d>b$ *and never if* $d<b$. One may express the probability of the seller's consent in terms of Heaviside function $\vartheta = \mathrm{I}(d-b)$. If a buyer is consent too, then the seller receives $V+max\{d,b\}$.

Buyer is an agent in state $OM(\psi)$. If she buys a commodity then she receives $V+u(\alpha,\psi)$, if not, she stays with her $V+d$. For this reason *a buyer is ready to buy some product* $\alpha$ *only if* $u(\alpha,\psi)>d$.

On average through all the states of the seller, the buyer is consent to deal with probability

$$\Delta = \int_{\Omega} \mathrm{I}(u(\alpha,\psi)-d)d\alpha = 1-U'(d). \tag{20}$$

Thus the probability that deal between a seller and a buyer takes place is

$$\delta = \vartheta\,\Delta = \mathrm{I}(d-b)\ \left(1-U'(d)\right). \tag{21}$$

On average through seller's goods, the buyer receives

$$K\,(NOM(\alpha,\omega)) = \vartheta \int\limits_{\Omega} \max(u(\alpha,\psi) + V,\, V)d\alpha = \mathrm{I}(d-b)\,(V + U(d))\,. \quad (22)$$

And the seller receives

$$K(MON(\alpha,\omega)) = \delta\ (V + \max\{d,b\}) + (1-\delta)\ (V+b)\,. \quad (23)$$

Now find the $K(OM(\omega))$ by means of utility of buying $K(NOM(\alpha,\omega))$. Suppose that at the moment $t$ there are $k$ sellers in the market sector.

Consider a buyer $B$. Of events of streams $\xi_{ij}$ occurring in period $dt$ only contacts of $B$ with $k$ sellers and $k$ sellers with $B$ effect on value of $OM(\omega)$. Each of events occurs with probability $(M/2)dt$. Average utility of these events is $K(NOM(\alpha,\omega))$. Then if there are $k$ sellers the maximum of expected utility of $B$ is

$$(1 - k\,\mathrm{M}\,dt)\,K(OM(\omega))\,e^{-r\,dt} + k\,\mathrm{M}\,dt\,K(NOM(\alpha,\omega)). \quad (24)$$

In order to find $K(OM(\omega))$ we have to average out this expression over probabilities $\pi(k)$. Then on average a buyer gets

$$\begin{aligned} K(OM(\omega)) = \sum_{k=1}^{N} \pi(k)((1 - k\mathrm{M}\,dt)\,K(OM(\omega))\,e^{-r\,dt} \\ + k\,\mathrm{M}\,K(NOM(\alpha,\omega)))dt \end{aligned} \quad (25)$$

If $dt \to 0$ this leads to

$$K(OM(\omega)) = \frac{\mathrm{M}\,K\,(NOM(\alpha,\omega))\left(\sum\limits_{k=1}^{N} \pi(k)k\right)}{r + \mathrm{M}\left(\sum\limits_{k=1}^{N} \pi(k)k\right)}. \quad (26)$$

The values $d, b$ may be found by solving Bellman equations. These equations derived in the previous section one may reduce to two relatively simple

$$0 = (-\pi\,\theta - \delta\,m - \lambda - \rho)b + \delta\,m\,d + \pi\,\theta\,U(b). \quad (27)$$

$$0 = -\frac{\lambda b}{\rho} - d + \frac{\lambda\,(\pi+1)\,b\mathrm{I}(d-b)}{(\rho+\pi+1)\,\rho} + \frac{(\pi+1)\,\mathrm{I}(d-b)\,U(d)}{\rho+\pi+1}. \quad (28)$$

Here it is denoted $\lambda = \frac{\mathrm{M}}{\Lambda}$, $\rho = \frac{\mathrm{M}}{r}, \pi = \sum\limits_{k=1}^{N} \pi(k)k$.

Quantity $\pi$ may be evaluated by means of the course-of-value function $P(z)$ from the equation (9).

$$P'(1) = (\pi + 1)\ (1 - P(0))\,. \quad (29)$$

## 5 Results

In our work we aimed at tracing of how the difference between $d - b$, that is the difference between average utility of money and of barter depends on the proportion of money traders to other agents in the economy.

For this reason we set $m = \mu N$. One may call the parameter $\mu$ monetization of economy. We solved the equation (9) approximately considering $N$ as a big parameter. Then found $\pi$ from (29) and solved numerically (27), (28). In our computations the utility function was defined as

$$u(\alpha, \omega) = (1 + (\text{arc between } \alpha \text{ and } \omega \text{ of circle } \Omega))^{-\nu}, \quad \nu > 0. \tag{30}$$

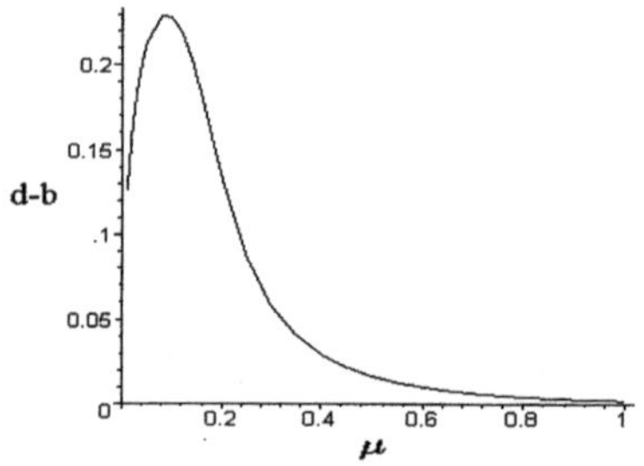


**Fig. 1** The dependence between the difference of $d$ and $b$ from monetization of economy

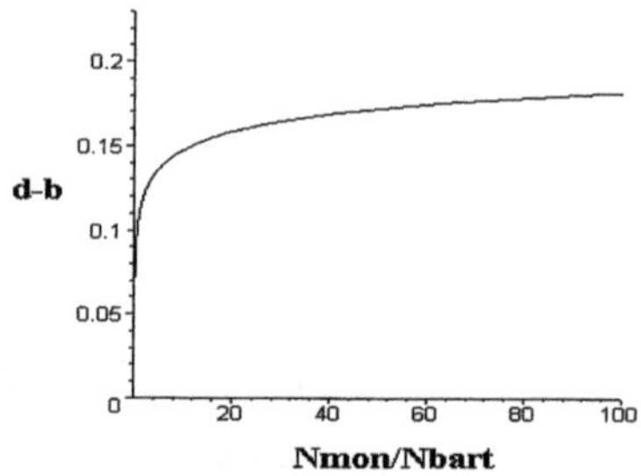


**Fig. 2** The dependence of proportion of monetary exchange to barter $N_{mon}/N_{bart}$ of efficiency $d - b$

One may see from figures 1 that in a wide range of values of monetization money is accepted by agents because it brings a positive payoff and may be even more profitable than ordinary barter. Also, there exists one point at which the preference of money to barter is the highest. Figures 2 show that when $d - b$ is big monetization of the economy is so small that the number of barter is greater than money trades. Although money is profitable, it is in deficit and not easy to get.

## References


1. Kiyotaki, N., Wright, R. On Money as a Medium of Exchange. Journal of Political Economy **97** (1989), 927 – 954 .
2. Prochorov, J. V., Rosanov, J. A. Theory of Probability: Basic Concepts. Limiting Theorems. Stochastic processes. Mathematical Reference Library. (1973) Nauka, Moscow.
3. Kamke E. A Handbook of Ordinary Differential Equations . (1986) Nauka, Moscow.

MIX
Papier aus verantwortungsvollen Quellen
Paper from responsible sources
FSC® C105338

If you have any concerns about our products,
you can contact us on
**ProductSafety@springernature.com**

In case Publisher is established outside the EU,
the EU authorized representative is:
**Springer Nature Customer Service Center GmbH**
**Europaplatz 3, 69115 Heidelberg, Germany**

Printed by Libri Plureos GmbH
in Hamburg, Germany